D1709545

Vision Research

Vision Research

A Practical Guide to Laboratory Methods

Edited by

R. H. S. Carpenter
Physiological Laboratory,
University of Cambridge

and

J. G. Robson
College of Optometry,
University of Houston

Oxford New York Tokyo

OXFORD UNIVERSITY PRESS

1999

Oxford University Press, Great Clarendon Street, Oxford OX2 6DP

Oxford New York
Athens Auckland Bangkok Bogota Bombay Buenos Aires Calcutta
Cape Town Chennai Dar es Salaam Delhi Florence Hong Kong Istanbul
Karachi Kuala Lumpur Madrid Melbourne Mexico City Mumbai
Nairobi Paris São Paolo Singapore Taipei Tokyo Toronto Warsaw

and associated companies in
Berlin Ibadan

Oxford is a trade mark of Oxford University Press

Published in the United States
by Oxford University Press, Inc., New York

A catalogue record for this book is available from the British Library

Library of Congress Cataloging in Publication Data

Vision research : a practical guide to laboratory methods / edited by
R. H. S. Carpenter and J. G. Robson.
Includes bibliographical references and index.
1. Vision—Research—Laboratory manuals. I. Carpenter R. H. S.
(Roger H. S.), 1945– . II. Robson, J. G. (John G.)
QP475.V514 1998 612.8′4′072—dc21 98–19461

ISBN 0 19 852319 X

Typeset by
EXPO Holdings, Malaysia

Printed in Great Britain by
Biddles Ltd, Guildford & King's Lynn

Preface

One of the attractions of working in vision is that it is a field which still demands a certain degree of technical virtuosity. At a time when much of biology—at the molecular end—seems to be tending towards mass-production, with serried ranks of post-docs performing tasks that will no doubt eventually be taken over by industrial robots, experimentation in vision has managed to remain something of a craft as well as a science. An aspiring latter-day Helmholtz needs some familiarity with electronics, skill in programming computers and an understanding of their hardware, knowledge of physical and physiological optics, and a mastery of techniques of experimental design and statistical analysis and inference, and some acquaintance with signal detection theory, information theory, signal processing, Fourier analysis, differential equations, neural networks...; and all this of course in addition to a command of the field of visual physiology and psychology itself, as well as molecular biology. None of this can be learnt overnight; it demands long hours of study and of experimentation at the bench.

There was a time, in that golden age of *Meccano*, *Konstruktor*, of home-made crystal sets, model railways, and astronomical telescopes, of *Amateur Mechanics* and *Practical Electronics*, when the basis of many of these skills could reasonably be expected of students embarking on a career of research. For a younger generation deprived of these pleasures, the prospect of starting out in this individualistic field of study can be extremely daunting: and sometimes almost equally daunting for the seasoned physiologist who suddenly needs to measure a psychophysical threshold, or the psychologist who wants to generate some evoked potentials, or the clinician who would like to record some eye movements. It was all these kinds of people that we had in mind when planning this book; what we did was to approach many experienced practitioners in various aspects of the subject to persuade them to put down on paper some of the practical experience they had accumulated over the years, information of a kind that is seldom written down explicitly in the ordinary course of scientific publishing. We have tried to cover vision research in the widest possible sense, from ERG recording at one end up to the design of psychophysical experiments at the other, from practical hints on setting up optical systems to advice on dealing with children as subjects. There are also chapters on the measurement of eye movements and of accommodation and pupil size; and a recurring theme is of course the use of computers

Now one might well suppose that with the advent of a high-resolution colour display and a fast computer to drive it some of these technical problems might simply have evaporated. Can we simply junk all those filters and beam-splitters, wedges and optical benches, pinholes and iris diaphragms, and leave the computer to get on with displaying stimuli, recording responses, analysing the data, and no doubt writing the paper? There are—unfortunately—labs where precisely this does happen, the computer's owner blissfully unaware of non-uniformities across the screen, of randomly delayed responses to interrupt requests, of display granularity, digitization errors, and line interleaving; and luckily for them, many referees are equally innocent of such matters. Of course, computers can indeed be extremely convenient; but only when set to work by someone who has a clear understanding of their visual and temporal limitations; the chapters in this book by Tom Robson and John Mollon discuss these limitations and show in practical terms how they may be overcome. But for many kinds of experiment—where we need high luminances, accurate modulation, large visual fields, or well-defined spectra—the computer display cannot serve, and we must use specialized light sources and optical systems of the kind described by Walt Makous and John Robson. Visual neurophysiology and visual psychophysics each pose problems of their own, and these are discussed in the chapters by Henk Spekreijse and Frans Riemslag, and Dennis Pelli; and different kinds of subjects each require a particular approach, whether children (Jan Atkinson and Ol Braddick) or animals (Randolph Blake). Finally, Han Collewijn and Stuart Judge provide practical information on how to set about recording eye movements and pupillary responses and accommodation.

We hope that this book may not only serve as a little repository of practical wisdom for those already working in the field, but also encourage newcomers—whether research or project students or visitors from another discipline—to have a go themselves.

Cambridge
November 1997

R. H. S. C.
J. G. R.

Contents

Contributors

Janette Atkinson, Department of Psychology, University College London, Gower St, London WC1E 6BT, UK.

Randolph Blake, Department of Psychology, Vanderbilt University, Nashville, TN 37240, USA.

Oliver Braddick, Department of Psychology, University College London, Gower St, London WC1E 6BT, UK.

Han Collewijn, Faculteit der Geneeskunde en Gezondheidswetenschappen, dr. Molewaterplein 50, Erasmus University, 3000 DR Rotterdam, The Netherlands.

Bart Farrell, Institute for Sensory Research, Syracuse University, New York, NY 13244, USA.

Stuart Judge, University Laboratory of Physiology, University of Oxford, Parks Road, Oxford OX1 3PT, UK.

W. Makous, Centre for Visual Science, University of Rochester, 274 Meliora Hall, Rochester, NY 14627, USA.

J. D. Mollon, Department of Experimental Psychology, University of Cambridge, Cambridge CB2 3EB, UK.

Denis G. Pelli, Psychology Department and Center for Neural Science, New York University, 6 Washington Place, New York, NY 10003, USA.

F. C. C. Riemslag, University of Amsterdam, Ophthalmic Research Institute, PO Box 12141, 1100 AC Amsterdam, The Netherlands.

John G. Robson, College of Optometry, University of Houston, Houston, TX 77204, USA.

Tom Robson, Cambridge Research Systems Ltd, 80 Riverside Estate, Sir Thomas Longley Rd, Rochester, Kent ME2 4BH, UK.

H. Spekreijse, University of Amsterdam, Ophthalmic Research Institute, PO Box 12141, 1100 AC Amsterdam, The Netherlands.

Abbreviations

2afc	two-alternative forced choice
A/D	analog to digital
ACGIH	American Conference of Governmental Industrial Hygienists
CCD	charge-coupled detector
cd	candela
CIE	Commission International de l'Éclairage
cpd	cycles per degree
CRT	cathode-ray tube
CSNB	congenital stationary night blindness
D	dioptre
DAC	digital-to-analog converter
dB	decibel
DLP	digital light processor
DMD	digital micromirror device
E	illuminance
ECG	electrocardiogram
EEG	electroencephalogram
EOG	electro-oculogram
EP	evoked potential
ERG	electroretinogram
eV	electronvolt
f	focal length
ϕ	pitch: vertical angular deviations from the optic axis about a horizontal axis
fcd	foot-candles
fL	foot-lamberts
FPL	forced-choice preferential looking
I/O	input/output
IR	infrared
ISCEV	International Society for Clinical Electrophysiology of Vision
J	joule
L	luminance
L/D	light/dark (ratio)
LDR	light-dependent resistance
LED	light-emitting diode
LGN	lateral geniculate nucleus
lm	lumen
LSB	least significant bit
LUT	look-up table
mks	metre, kilogram, second (system)
mL	millilamberts

MOSFET	metal-oxide-semiconductor field-effect transistor
MPE	maximum permitted exposure
MPEG	Motion Picture Experts' Group
MRI	magnetic resonance imaging
ND	neutral-density log units
OD	*oculus dexter* (right eye)
OKN	optokinetic nystagmus
OP	oscillatory potential
OS	*oculus sinister* (left eye)
PFM	pulse-frequency modulation
PWM	pulse-width modulation
θ	yaw: angular deviations from the optic axis in the horizontal plane about a vertical axis
R	radiance
r	radius
RAM	random access memory
RGB	red–green–blue
rms	root mean square
ROC	receiver-operating characteristic
RPE	retinal pigment epithelium
RT	reaction time
S/N	signal-to-noise (ratio)
SPECT	single-photon emission, computed tomography
sr	steradian
td	troland
TTL	transistor-transistor-logic
UV	ultraviolet
V'_{λ}	luminosity coefficient applied to scotopic (rod) vision (see Chapters 1 and 2)
VEP	visual evoked potential
V_{λ}	luminosity coefficient applied to photopic (cone) vision (see Chapters 1 and 2)
W	watt
ψ	roll: rotation about the optic axis

1

Optics and photometry

W. MAKOUS

1.1 Introductory comments

1.1.1 Purpose

The goal here is to provide the most important and useful information that one is likely to need—especially as a beginner—in order to use optics for vision research. However, sources and some optoelectronic devices are covered in Chapter 2; and colour, in Chapter 4. Some topics are either too specialized (e.g. psychophysical interferometry, Section 1.10), or require more space than can be warranted for detailed treatment here (e.g. fibre optics, Section 1.4.10; and use of polarization as a tool, Section 1.5).

1.1.2 Why optics nowadays?

As the scientific questions that can be answered are limited by the technology available, the trends in science tend to follow advances in technology. Hence, improvements in cathode-ray tubes (CRTs) and the methods of driving them, for example, have led to increasingly complex stimuli that tap mechanisms closer to the realm of perception than vision. However, a survey of the papers on vision—and perception—in two recent years of *Vision Research*, the *Journal of the Optical Society of America A*, *Perception and Psychophysics*, *Visual Neuroscience*, the *Journal of Neuroscience*, and the *Journal of Neurophysiology*, shows that about 20% of the papers use optics involving coaxial systems of lenses. Such optical systems, as opposed to CRTs for example, are still required for high retinal flux densities, colorimetric purity, temporal bandwidth, and spatial homogeneity; and they allow better control over polarization. To study the optical components of the eye, such as the pupillary aperture and iris, the cornea and lens, and also to study movements of the eye—or to subvert their respective influences—requires an optical system. Finally, optical systems allow specialized effects, such as stimulation of the retina by undistorted sine-wave gratings of high contrast and high spatial frequency (LeGrand, 1935; Campbell and Green, 1965; Williams, 1985).

1.1.3 Learning about optics and keeping current

To use optical systems for vision experiments, an understanding of optics at least at the level of an introductory course in physics is probably essential. A one-semester undergraduate course in optics is useful, but familiarizing oneself with the text is more efficient and probably sufficient. A reference book on optics is also of great value. A useful general book is *Optics*, by Hecht (1987); the authoritative source is *Principles of optics*, by Born and Wolf (1970); an intermediate, although somewhat dated, book that I still like is *Light*, by Ditchburn (1963). The *Handbook of optics* (Bass *et al.*, 1995), sponsored by the Optical Society of America, is a mine of information. A glance at many of the chapters, however, shows the gap between the modest needs of most vision experiments and the technical concerns of those in advanced optics. However, the chapter by Burns and Webb (1995), 'Optical generation of the visual stimulus', is a valuable supplement to this chapter and others in this volume, for it differs in approach and coverage.

One can be aware of recent developments by watching such publications as *Optics and Photonics News*, *Laser Focus World*, *Photonics Spectra*, and general sources such as *Science* and *Scientific American*. Perhaps the most useful and focused tutorials appear in the catalogues and brochures for optical equipment (see, for example, Ealing, Melles Griot, Newport, and Oriel).

1.1.4 Specific brands and manufacturers

I have tried to provide specific examples of commercially available devices wherever it seemed helpful, as a place to start one's search for the product best suited for a specific need. These are chance examples, and mention of one product rather than another is not meant to imply that it is better or more desirable. Current addresses and telephone numbers for the sources of the products cited here are listed in Table 1.1. Up-to-date listings of these and other sources not mentioned here can be found in the *Photonics corporate guide* (book 1 of the *Photonics directory*), *The laser focus world buyers' guide*, or *Science*'s *guide to scientific instruments and services* (listed in decreasing order of usefulness). Though the prices reported are the most recent I have, they will be out-of-date by the time this book is published; nevertheless they may provide a more lasting index of *relative* cost.

1.2 Design

1.2.1 Procedure

The method of designing the optical system depends on its complexity. Although a simple Maxwellian apparatus (described in the next section) of three channels or so can be visualized and the details laid out as it is being constructed, a rough sketch is useful to ensure that the space is adequate and that all the necessary components are on hand. Ray-tracing locates the images and establishes their size and orientation.

Novel or complex systems, or those that are subject to many simultaneous constraints, require a more formal design process. Some investigators find scale drawings essential. There are now software applications available, such as Dreams, developed for architects (thanks to Denis Pelli for introducing me to it) and its bigger sibling, MacDraft ($329.95, Innovative Data Design), that significantly assist in any such design process. (The figures for this chapter were made with MacDraft.) Software specifically for optics is available in great variety. Most have more power than is needed to design optics for vision experiments, such as Mathematica's spin-off,

Table 1.1 Current addresses and telephone numbers for suppliers of the products cited in this chapter

Source	USA telephone	UK telephone	Other telephone	Address
C. and H. Sales Co.	800 325 9465			2176 E. Colorado Blvd, Pasadena, CA 91107
Central Scientific Co.	800 262 3626			3300 Cenco Pkwy, Franklin Park, IL 60131
Corion Corp.	508 429 5065			73 Jeffrey Ave, Holliston, MA 01746
Dalsa Inc.	519 886 6000			605 McMurray Rd, Waterloo, Ontario N2V 2E9, Canada
DoALL Ind. Supply	206 623 1191		800 234 0016	PO Box 3683, Seattle, WA 98124
Ealing (Coherent-Ealing)	916 889 5365	01923 242261	800 343 4912	2303 Lindbergh St., Alborn, CA 95602
Eastman Kodak Co.	716 588 2572		800 225 5352	343 State St, Rochester, NY 14652–4115
Edmund Scientific Co.	609 573 6250			101 East Gloucester Pike, Barrington, NJ 08007–1380
E. G. & G. Reticon	408 738 4266			345 Potrero Ave, Sunnyvale, CA 94086–4197
Fisher Scientific	412 490 8300	0171 935 4440	800 766 7000	711 Forbes Ave, Pittsburgh, PA 15219
Gamma Scientific	619 279 8034			8581 Aero Dr, San Diego, CA 92123
Innovative Data Design	510 680 6818			1820 Arnold Industrial Way, Suite L, Concord, CA 94520
Intellimation Library for the Macintosh	800 346 8355		805 968 2291 (Outside USA)	Dept 5SCH, PO Box 1922, 130 Cremona Dr, Santa Barbara, CA 93116
International Light Inc.	508 465 5923			17 Graf Rd, Newburyport, MA 01950–4092
Melles Griot	716 244 7220	01223 420071	800 775 7558	55 Science Pkwy, Rochester, NY 14620
Minolta Corp.	800 724 4075			101 Williams Dr, Ramsey, NJ 07446
Newport Corp.	714 863 3144	01635 521757	800 222 6440	1791 Deere Ave, Irvine, CA 92714
OCLI Inc.	707 545 6440	01383 823631		2789 Northpoint Pkwy, Santa Rosa, CA 95407–7397
Opti-clean	619 758 8250			4093A Oceanside Blvd, Oceanside, CA 92056
Optronics Laboratories Inc.	407 422 3171		800 899 3171	4470 35th St, Orlando, FL 32811
Oriel Corp.	203 377 8282	01372 378822		250 Long Beach Blvd, PO Box 872, Stratford, CT 06497
Parker Hannifin Corp.	800 245 6903			Sandy Hill Rd, Box 500, Harrison City, PA 15636
Photo Research	818 341 5151			9330 DeSoto Ave, PO Box 2192, Chatsworth, CA 91313
Photometrics	520 889 9933			3440 E. Britannia Dr, Tuscon, AZ 85706
Rosco Laboratories	203 708 8900	01633 9220		36 Bush Ave, Port Chester, NY 10573
Rutland Tool & Supply Co. Inc.	818 961 7111		800 727 9787	16700 E. Gale Ave, City of Industry, CA 91745

Table 1.1 *Continued*

Source	USA telephone	UK telephone	Other telephone	Address
Sears Industrial Tools	800 776 8666			
Servo Systems Co.	201 335 1007		800 922 1103	115 Main Rd, PO Box 97, Montville, NJ 07045–0097
Spindler & Hoyer	508 478 6200		800 334 5678	459 Fortune Blvd, Milford, MA 01757–1745
Tektronix, Inc.	503 627 7111		800 835 9433	PO Box 500, Beaverton, Oregon 97077
UDT Sensors Inc.	310 978 0516			12525 Chadron Avenue, Hawthorne, CA 90250
Velmax Inc.	716 657 6151			7550 State Rte. 5 and 20, Bloomfield, NY 14496
Vincent Associates	716 473 2232			1255 University Ave, Rochester, NY 14607
Wolfram Research	217 398 0700			100 Trade Center Dr, Champaign, IL 61820

Optica ($695, Wolfram Research Inc.). At the other extreme is Optics Lab ($39, Intellimation), which may serve as a low-level tutorial but is too limited to be useful for design. Between these extremes is WinLens ($50, Spindler and Hoyer), which computes system parameters and displays cross-sectional drawings of lenses with rays, and has access to the properties of lenses available from Spindler and Hoyer. However, its emphasis is on lenses and associated imaging errors as opposed to reflective components, it is limited to five components, and is available only for Windows.

1.2.2 Basic principles

1.2.2.1 Maxwellian view

A Maxwellian system is one in which an image of the source is focused within the pupil of the eye (Westheimer, 1966). Its principal advantage is that it gathers more light than the pupil of the eye would normally do, and directs it all within the eye. Although no optical system can increase the retinal illuminance above that in the image of a directly viewed source, the added light from a Maxwellian system allows one to expand the area of retina illuminated at that maximum level. In its simplest form, shown at the top of Fig. 1.1, a Maxwellian system consists of a source, such as an incandescent filament, and a single lens. In this case, an image of the lens, filled with light, falls on the retina, perhaps somewhat out of focus depending on accommodation and the distance of the lens.

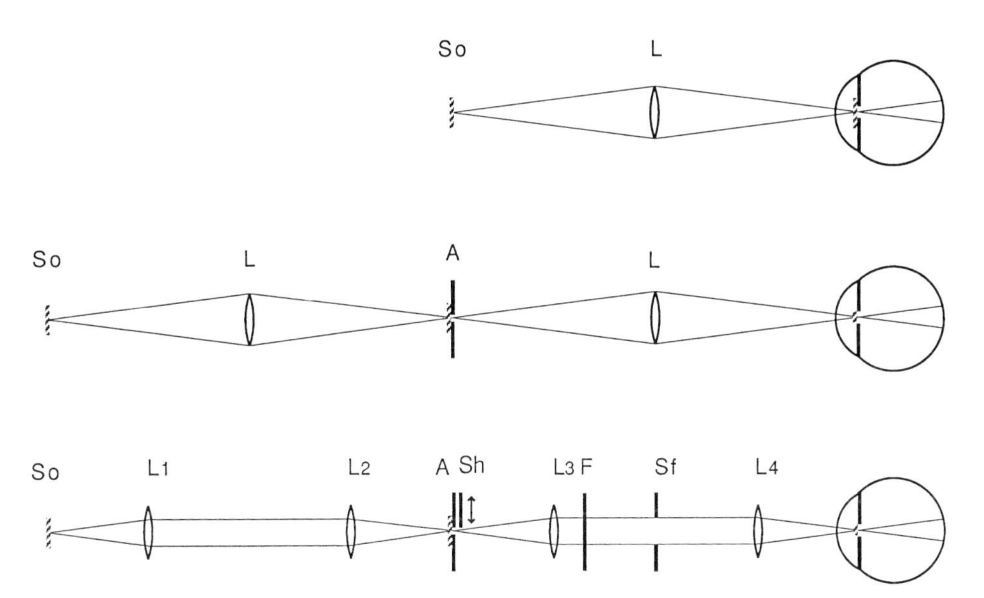

Fig. 1.1 Diagram of Maxwellian systems. The top diagram is the simplest possible example; the middle adds an aperture stop to limit the size of the image within the pupil of the eye; and the bottom shows a single channel of a system such as typically used in practice.The icon at the right represents the observer's eye, the solid, vertical line corresponding to the iris; the cross-hatched lines within the eye and adjacent to the aperture stop represent images of the source. Abbreviations: So, source; L, lens; A, aperture stop; Sh, shutter; F, filter; Sf, field stop.

A second advantage of a Maxwellian system is that variations of pupil size have no effect on retinal illuminance and little effect on image quality as long as the image of the source is smaller than the pupil at its smallest, ordinarily about 2 mm. To keep the image of the source less than 2 mm, it is usually necessary to focus it on an aperture stop that is in turn focused within the pupil of the eye (i.e. is conjugate with the pupil). This requires at least two lenses, such as shown in the middle of Fig. 1.1.

1.2.2.2 Optical relays

It is advantageous to use identical lenses throughout Maxwellian and similar optical systems. A stock of them can be laid in for use anywhere within that system or in any other system in one's laboratory. The resulting systems are simple and seldom yield surprises. Typically, they are used in pairs to provide *optical relays*, in which one lens collimates the light and the other focuses it; hence, the image is relayed from one plane to another with unity magnification. The advantage of using a pair of lenses instead of one is that inexpensive lenses can be used to provide the high-quality images otherwise achievable only by more costly lenses. This, then, increases the number of lenses for the simple system described above to four, as shown at the bottom of Fig. 1.1.

As the cross-section of the optical system is smallest at the aperture stop, locating other components there minimizes the required size. For example, locating a mechanical shutter (Sh in Fig. 1.1) close to the aperture stop (A in Fig. 1.1) minimizes the time it requires to sweep across the beam. However, some components, such as interference filters (F in Fig. 1.1), are designed to be used in collimated parts of the beam (see Chapter 4); and others, such as absorptive filters, may be damaged by the high flux densities where an image of the source is focused. Note also that as reflection depends on incident angle, insertion of a mirror or beam splitter in a converging or diverging part of the optical pathway introduces an intensity gradient across the reflected field and any transmitted field as well.

1.2.2.3 Overdesign and excessive precision

As research is unpredictable, what apparatus will be required for the next experiment is not known. Therefore, design for modification. Assume that the next experiment will require changes in the apparatus. On the other hand, how long the apparatus might be needed is also not known. If it were, one would aim to make the apparatus just robust enough to work until the last observation, but having the apparatus break down before that has obvious drawbacks. So, err on the side of stability and permanence, but not by much; keep it flexible and adjustable; and plan to realign.

1.3 Alignment

1.3.1 Why bother and how good?

Optical systems depend on spatial relationships, and a process of alignment is required to get the components into their proper positions and orientations. Flawed alignment causes flawed results, as described in the next two sections. However, optics is subversive in that multiple errors can be, *in some respects*, mutually compensatory. Hence, adjusting the optics to bring them into alignment according to one criterion may introduce other errors that are not obvious but nevertheless introduce unwanted interactions among variables or degrade image quality (see Section 1.3.5.4.i).

1.3.1.1 Independence of control

A consequence of poor alignment is unintended interactions among variables. For example, lateral movement of an image may change its retinal illuminance or focus, or changing focus may cause lateral movement of the image.

1.3.1.2 Image quality

An almost guaranteed consequence of poor alignment is poor image quality.

Admittedly, the quality of some images, such as those used for backgrounds, may not be important, and images may even be deliberately defocused to remove high spatial frequencies, but incidental image degradation owing to poor alignment is rarely desirable.

If image quality does matter, it may be necessary to evaluate it. An approximation can be obtained by placing a resolution target or other square-wave grating (e.g. Edmund or Spindler and Hoyer) in the plane to be imaged on the retina. A lens is required to construct the image normally erected by the optics of the eye. The highest frequency that can be resolved is the optical cut-off frequency, but this approach tells little about stray light or other losses of contrast. Keep in mind that the quality of the lens used in place of the eye introduces some degradation of the image, but in most cases it is less than that introduced by the eye itself.

A more informative method is to measure the distribution of light in the image of a thin line, i.e. the *line spread function*. This can be measured by placing a thin slit in the plane to be imaged on the retina and erecting an image of it by substituting a lens for the optics of the eye. The slit must be thin enough that decreasing its width affects only the amount of light in the image but not its shape. The distribution of light in the image is measured by covering a photometer with a thin slit (much thinner than the image) and sweeping the slit-covered photometer across the image. The measuring slit is oriented by iteratively rotating and translating it until the reading is at its maximum. Alternatively, the image can be cast on a fine, charge-coupled detector (CCD) array. A serviceable linear array of 256 pixels and 4 mm long that is read out by an oscilloscope (e.g. Fairchild CCD111DB) can be obtained for $135; putting it into a camera and adding the second spatial dimension can easily increase the cost 100-fold, and a computer will be required to read and control it (see Dalsa, Photometrics, E.G. & G. Reticon, Oriel). Once again, the contribution from the imaging or camera lens must be taken into account.

As an alternative to measuring the line spread function, one can measure the contrasts in the images of sinusoidally modulated gratings (see Lamberts, 1963). The ratio of the contrast in the images to that of the object gratings as a function of their spatial frequency is the *modulation transfer function*, which is related to the line spread function. For more on spread functions, transfer functions, and their relationship to one another, see Boreman (1995).

1.3.1.3 Precision

The precision of alignment required depends on such things as the image quality required and the precision of the observations, but it is hard to know the precision necessary for any given experiment, and evaluating it may require time-consuming measurements with special equipment. One can waste vast amounts of time striving to eliminate the last detectable alignment error. In practice, the precision is limited by the mechanical positioners and the misalignment that can be seen by the naked eye. For most apparatus used in vision experiments, errors of 0.1 mm or so for any given adjustment are probably acceptable and can be achieved without undue investment of time. For many experiments of a qualitative nature, errors of 1 mm will probably do little harm, and I have witnessed successful visual scientists using apparatus misaligned by as much as 1 cm.

1.3.2 Stability

Once aligned, a system does not necessarily remain aligned. Although thermal drift is too small[1] to matter for most apparatus used in vision (except laser interferometers), the apparatus must be able to withstand accidental jostling: in the process of adjusting the apparatus, or simply carrying out experiments, one occasionally nudges an aligned optical element. Many apparatus used in vision do not withstand such jostling. Some adjustable devices can even slip under the force of gravity.

The importance of stability is dramatized by a (true) story about a beginning investigator who invited an eminent and more senior investigator into his lab to admire the first apparatus he had built. The senior investigator glanced at the apparatus, then reared back with his foot and delivered a solid kick to the bite-bar. Optics went flying in all directions. 'Too flimsy', he said, and walked out. Designing to withstand the kick test may be overdesign, but the message is clear.

Many adjustments for optical apparatus are held in position by thumbscrews, the implication being that tightening should not exceed the torque one can apply with one's fingers. This is not enough. The thumbscrews can be replaced by Allen screws, or, if one is committed to the quick-and-dirty school, another eighth to quarter turn with a pair of pliers will serve.

1.3.3 Platform

Maintaining the relative positions of the optical elements requires a stable platform. Best, of course, is a steel optical table with tapped holes spaced every 2.5 cm or so, such as is available from Ealing, Newport, and Spindler and Hoyer. They range from smaller tables, called breadboards, 0.15 × 0.3 m and 12.7 mm thick ($260), to tables 1.5 × 3.75 m and 460 mm thick ($12 865). For many purposes, optical benches, or rails, mounted on 6-mm aluminium plate will serve. Triangular rails are inexpensive ($104 for a 1-m aluminium rail, or $250 for cast iron) and satisfactory for most purposes. A theoretical advantage of rails is that well-mounted optics do not require transverse alignment. However, the advantages of rails are undercut if the optics are not well centred on their mounts.

Any sturdy support for the platform will do except for special systems, such as interferometers. In apparatus for psychophysics, the support should be high enough to bring the optical axis to eye level for a comfortably seated observer.

1.3.4 Coordinate system

Discussion of positioning requires a coordinate system with unambiguous terminology, such as that diagrammed in Fig. 1.2.

Here, position along the optic axis is the longitudinal dimension, z; the perpendicular position in the horizontal plane is the transverse dimension, x; and position along the remaining perpendicular is the vertical dimension, y. Owing to ambiguities in the application of most technical reference systems to optics (e.g. whether the reference direction should be straight ahead or upwards), the nautical system of angles is adopted here: angular deviations from the optic axis in the horizontal plane about a vertical axis, θ, are referred to as yaw; vertical angular deviations from the optic axis about a horizontal axis, ϕ, as pitch; and rotation about the optic axis, ψ, as roll.

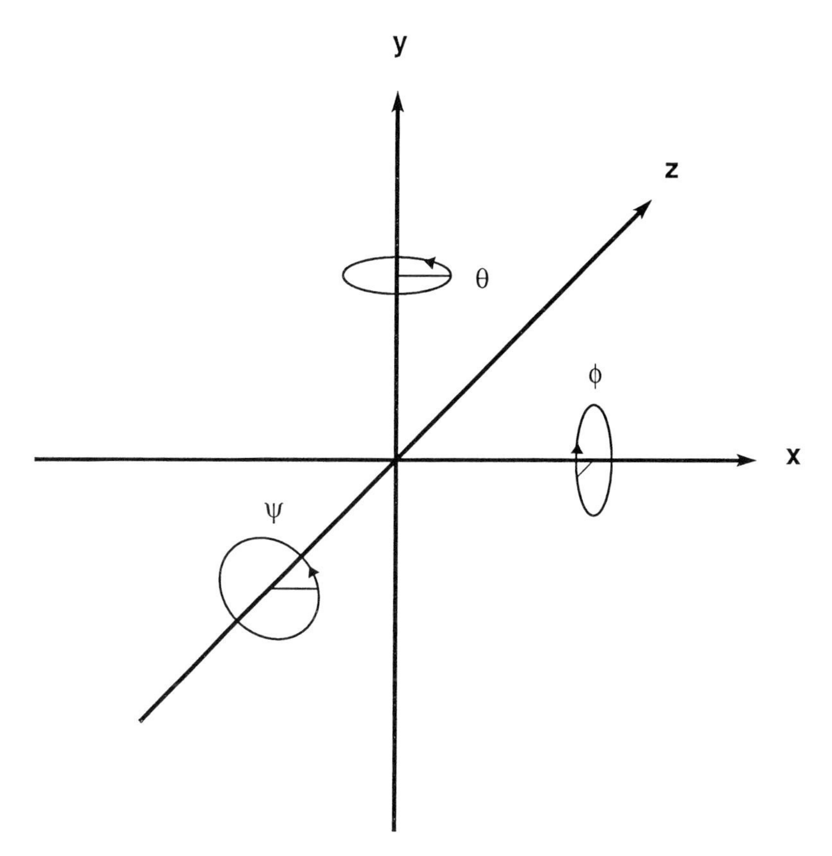

Fig. 1.2 The (nautical) system of coordinates used here. The axes, x and y, are the transverse horizontal and vertical axes, respectively, and z is the dimension parallel to the optic axis. The angles, ϕ, θ, and ψ, are, respectively, pitch, yaw, and roll, as described in the text.

1.3.5 How to align optics

Optical alignment is a skill requiring practice. Students typically cannot produce a satisfactorily aligned, single-channel Maxwellian view even after having it explained and demonstrated. In most cases, some apprenticeship is required in addition to practice.

Alignment begins with layout of the optic axis, followed in turn by positioning of mirrors and prisms, apertures, and, lastly, lenses.

1.3.5.1 Optic axis

The most practical way to lay out the optic axis is by use of a laser. The simplest laser will serve, even a laser pointer, which can be obtained for less than $50. The optical axis generally consists of a series of straight-line segments. The further apart the pair of points that define the segments, the smaller the error of alignment.

The optic axis runs from the source to the retina. In the simplest case, the optic axis is a single segment connecting the two. Aligning the optical system is easiest if the optic axis is parallel to the surface supporting its components. As the location of the eye is typically more constrained than that of the source, it is best to place the laser at the location of the eye and work towards the source.

Locating the fiducial points defining a line requires a reticle that can be manipulated in the dimensions, x and y. Anything that defines a point, such as the prosaic cross-hairs, will serve. Apertures also serve, but they can entirely block a misaligned beam. Apertures within a filter are better (e.g. a photograph of a white disc on a grey background). Best are two, fine patterns of concentric, alternating black and white rings. They allow a misaligned beam to pass, and the moiré pattern formed by superimposition of the image or shadow of one ring pattern on another clearly reveals the presence and axis of any misalignment, even if the patterns are blurred or differentially magnified (Makous, 1974).

As the entire alignment of the system depends on these reticles, align them carefully. Set the beam to pass through one fiducial pattern at the desired height of the optical axis; then replace it with the second reticle, at the same distance from the laser, and adjust its height to that of the beam. As long as the distance from the laser is the same, the beam need not be parallel to the substrate.

The height of the optic axis above the support surface is determined, of course, by the range of adjustment of the hardware used to hold and manipulate the optical components. The closer it is to the surface, the more stable the system.

Alignment of the laser beam that represents the optic axis requires iterative adjustments of y and ϕ, and of x and θ. That is, first translate the beam along the vertical dimension, y, until it passes through the first fiducial (Fig. 1.3, top); then change its pitch, ϕ, until it passes through the second fiducial (Fig. 1.3, bottom). This rotation moves the beam off the first fiducial, requiring readjustment along y. This, in turn, moves the beam off the second fiducial, requiring readjustment of ϕ. Continue translating to align with the first fiducial, and rotating to align with the second until no further adjustment is necessary. The analogous process is applied along the transverse dimension, x, and yaw, θ.

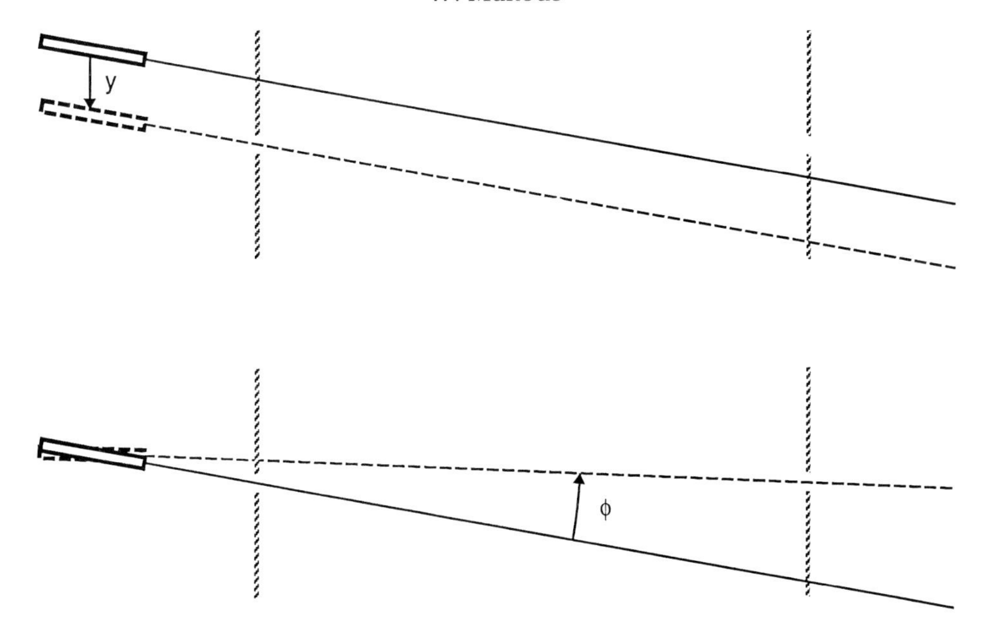

Fig. 1.3 Iterative alignment of a laser beam to two apertures. The laser is translated along y to align with the first aperture, and rotated through ϕ to align with the second. The stops containing the apertures (represented by cross-hatched lines) should not be entirely opaque.

Next check the positions of the fiducial reticles by ensuring that exchanging their locations has no effect on their alignment with respect to the beam (this can be done for the x dimension only if a rail is used).

1.3.5.2 Mirrors and beam splitters

Reflections by mirrors, prisms, and beam splitters break the optic axis into segments. Such reflections need not necessarily form right angles, but alignment and subsequent modifications are easier if they do. This is especially desirable if beam-splitting prisms are used, for the surfaces and reflections of such prisms typically are close to 90°.

A 90° reflection is achieved by placing a pentaprism where the mirror or beam splitter is planned. Such prisms reflect light at 90° no matter what their orientation (θ), as shown in Fig. 1.4.

Set up the pair of reticles on the reflected beam, put the mirror or splitter in place of the pentaprism, and align it to the reticles, as described below.

One can also use the beam-splitting prism to approximate a 90° angle, but the angle is not always exact. The methods described below to orient the surface of a lens normal to the incident laser beam can also be used for a prism. When the incident beam is normal to the surface, the beam reflected by the prism should be at 90° to the incident beam, within the precision of the prism geometry.

Of the six variables defining the location and orientation of any physical object, only three matter for a reflecting surface. Translation in the vertical (y) dimension has no effect on the reflected beam, and translation in the two horizontal dimensions (x and z) have identical effects, as shown in Fig. 1.5. Finally, roll (ψ) has the same

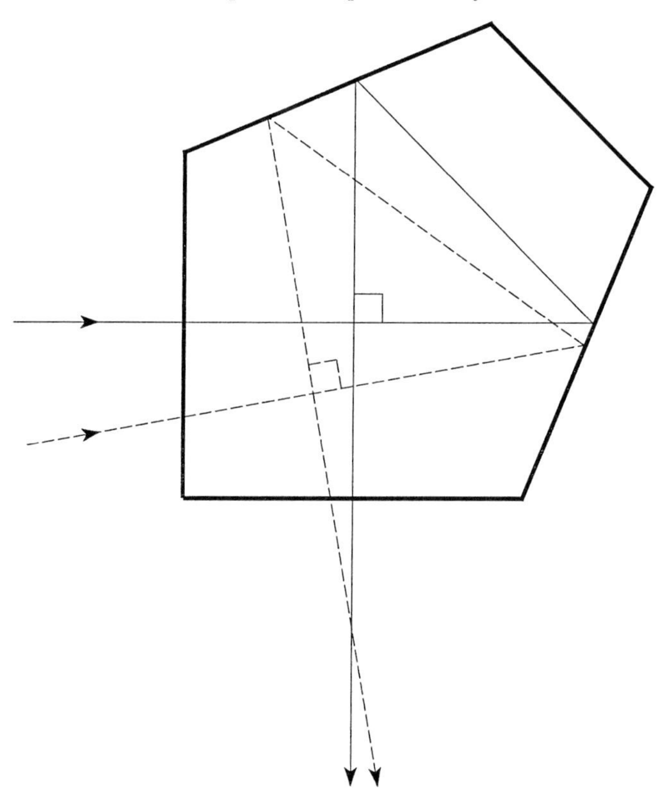

Fig. 1.4 Diagram of a pentaprism. The two pathways of light exemplify the fact that light incident at different angles are all reflected at a 90° angle.

effect as pitch (ϕ). The reflecting surface is aligned by an iterative alternation between translation (x or z) and rotation (θ, and between y and ϕ), analogous to that described for the laser and shown in Fig. 1.3.

Prisms should be used in collimated portions of the beam (e.g. on the source's side of the Maxwellian lens) to avoid the chromatic aberrations they would otherwise introduce.

1.3.5.3 Apertures

Alignment of apertures in the x and y dimensions is so obvious it needs no explanation; pitch and yaw, ϕ and θ, are not critical for apertures, for the errors are in proportion to the cosine of the angular error; and adjustment of roll, ψ, is either obvious or, in the case of circular apertures, irrelevant.

Alignment of apertures along the optic axis (z) is another matter. In Maxwellian systems, apertures should be conjugate either with the pupil of the observer's eye, a case that is addressed below with lens alignment (Section 1.3.5.4.ii), or with the observer's retina. As the optics of the eye focus the image on the retina, placement of objects to be focused there depend on the state of those optics. If the eye is emmetropic and accommodated for infinity, such objects should be placed one focal

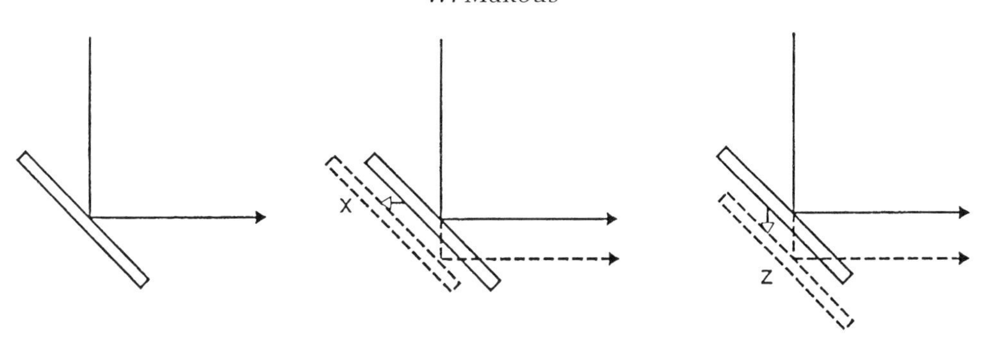

Fig. 1.5 How translation of right-angle mirrors (and beam splitters) along the horizontal dimension, x (middle figure), and the optic axis, z (right figure), are equivalent.

length from the next lens in the optical pathway (see Section 1.3.5.4.ii). Note, however, that the accommodative state of emmetropic college-age observers with eyes at rest (Leibowitz and Owens, 1978) is typically about 1.5 diopters (standard deviation, 0.77 diopters). As the last stage of a Maxwellian system forms a Badal optometer, myopic or hypermetropic correction is achieved by moving the object either towards or away (respectively) from the next lens in the pathway.

1.3.5.4 Lenses

The most critical and most difficult elements to align are the lenses; five dimensions, all except roll (ψ), require alignment.

1.3.5.4.i Transverse (x) *and vertical* (y) *dimensions, and pitch* (ϕ) *and yaw* (θ)
Distributing the refraction as nearly equally as possible over the refractive interfaces minimizes lens aberrations. Hence, the first and simplest rule is to place the more curved surface of the lens on the side of the lens where the beam is more nearly collimated. This rule is not trivial: the change of image quality that can be produced by reversing a lens can usually be easily seen by the naked eye. Figure 1.6 illustrates how placing the curved side of a lens towards the collimated part of the beam distributes the refraction over both surfaces, whereas reversing the lens orientation (or direction of light propagation, as in the figure) places the entire burden of refraction on the curved interface.

The best way to align a lens is by its reflections. Some 4% of the incident light is reflected at each surface. If a laser is used for alignment, the reflected light can be viewed by placing an aperture stop on the optical axis between the laser and the lens to be aligned. The aperture allows passage of the laser beam, but the stop intercepts any reflected light that is not exactly coincident with the beam passing. As the reflected light is at the end of an optical lever, small misalignments are magnified and easily seen. The curved surface of a lens, whether positive or negative, tends to expand the beam. Consequently, if the surface on which the reflection is cast is too distant from the lens, the reflection may be too dim or too large to serve well for alignment; but placing the surface too close to the lens decreases the length of the optical lever and hence decreases the displacement of the reflected spot caused by misalignment. Trial and error quickly establishes a good compromise.

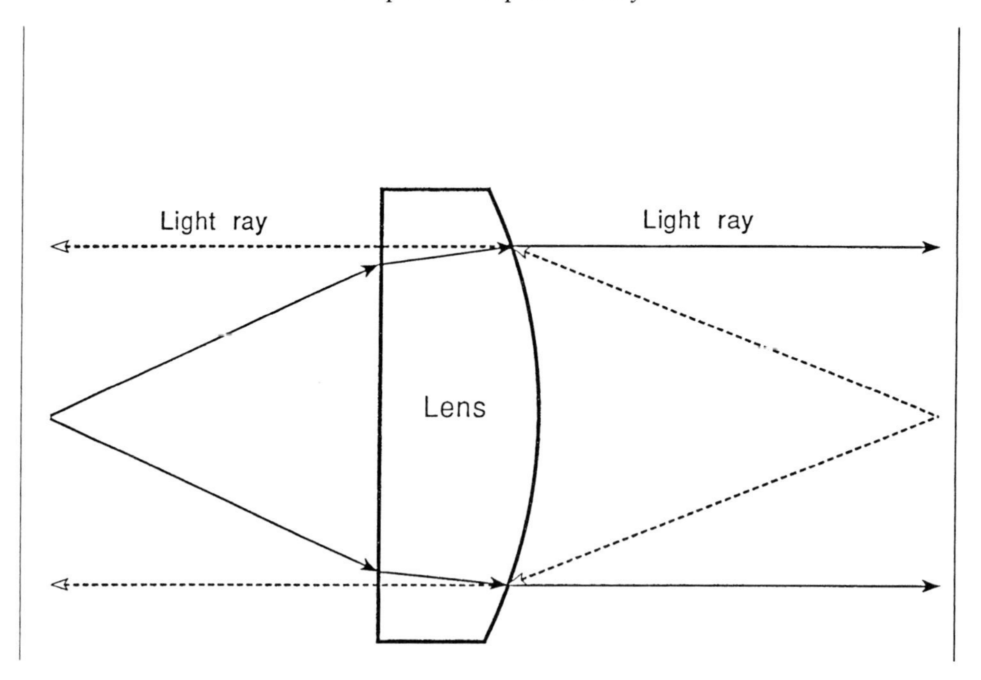

Fig. 1.6 Distribution of refraction between the interfaces of a collimating lens, depending on the direction of light. Diverging light incident on the plane surface is refracted at both surfaces of the lens, but when incident on the curved surface, refraction is restricted to the first surface.

The signature of transverse (x) or vertical (y) misalignment is two reflected spots straddling the aperture through which the laser beam passes, as shown in Fig. 1.7(a). Transverse misalignment turns the image formed by the misaligned lens off-axis in the direction opposite to the lens displacement, or moves the direction of a collimated beam at an angle opposite to the lens displacement. It is not sufficient to align any two lenses, say one to collimate light from a source and another to image it, simply so that the image is centred on the optic axis, for the image can be centred by mutually compensating misalignments of the two lenses, one misaligned to the right and the other to left, for example. Then if a field stop is moved back and forth between two such misaligned lenses, the retinal image moves from side to side, as does the apparent location of the field stop in the visual field. Lateral misalignment of lenses also increases optical aberrations.

The signature of angular misalignment (pitch, ϕ, and yaw, θ) is two reflected spots off the aperture on the same side, as shown in Fig. 1.7(b). Angular misalignment increases lens aberrations, especially coma. The signature of correct alignment is concentric spots centred on the aperture through which the laser beam passes, as shown in Fig. 1.7(c).

The alignment of a series of elements can be checked by viewing them from above or the side, as shown in Fig. 1.8. Scatter at each of the interfaces and virtual images of scatter at other lenses produce a series of bright spots that lie along a straight line when all the lenses are correctly aligned. A translational misalignment of one of the lenses displaces the spots associated with that lens in the direction of misalignment;

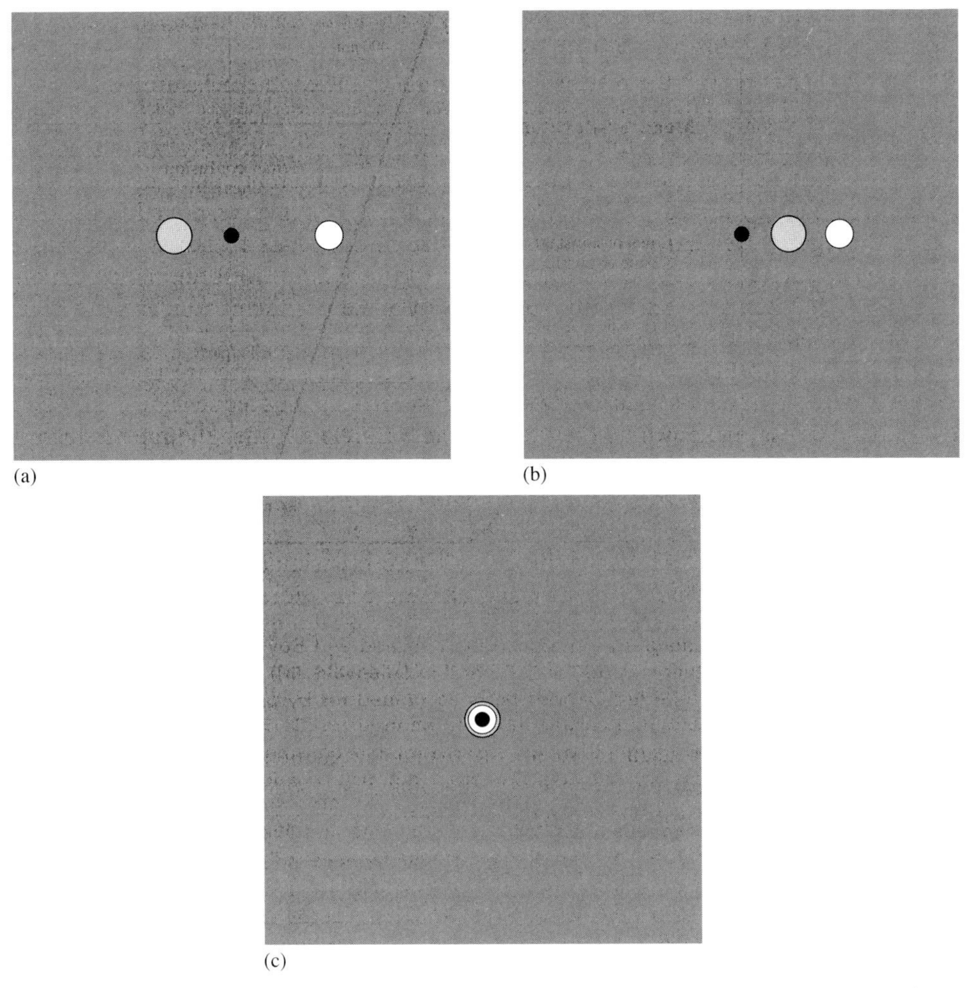

Fig. 1.7 Signatures of misalignment. The grey square represents an aperture stop centred on a laser beam that defines the optic axis. The black circle is the aperture through which the laser beam passes. (As there is no scattering interface in the aperture and as the eye is not in the path of the beam, it looks dark even though a laser beam is passing through.) The white spot shows where the laser beam strikes the aperture stop after reflection from the less curved surface of a lens; the other spot represents the reflection from the more curved surface (the curvature, whether positive or negative, spreads the beam). Part (a) shows that transverse misalignment along the horizontal (x) axis causes the reflections to move off the optic axis in the opposite direction; in part (b), yaw, or angular misalignment of the angle θ, causes both reflections to move off the optic axis in the same direction; and in part (c), correct alignment centres both reflections on the optic axis.

an angular misalignment displaces some spots in one direction and other spots in the opposite direction.

1.3.5.4.ii Longitude, z If the lenses form optical relays (Section 1.2.2.2), then each lens is located one focal length from the source or an image of the source. The lens

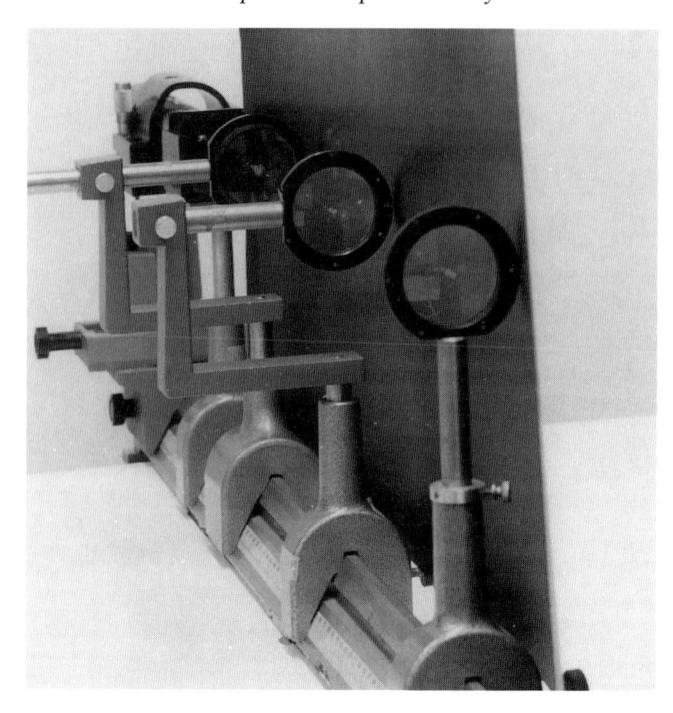

Fig. 1.8 Scatter from the interfaces of aligned lenses. The bright spots within the lenses are formed by light scattered from the laser beam at the surfaces of the lenses. When the lenses are aligned, the spots fall along a straight line. The first lens mount allows adjustment of height (y) and yaw (θ) only. The addition of a right-angle post holder to the second mount allows adjustment of transverse position (x) and pitch (ϕ) as well.

collimates or focuses depending on which side of the lens the source or image lies. If the lens focuses, simply position it along the z dimension where the image, cast on a diffusing surface placed in the focal plane, looks sharpest. To position a collimating lens, place a mirror immediately behind it in the optical path and rotate the mirror slightly so that the reflected light forms an image of the source adjacent to the source. To view this image, place a diffusing surface at the same distance from the lens as the source. Position the lens along the z dimension so that the image of the source is sharpest and of the same size as the source.

1.3.5.5 Filters

There are three aspects of filters that relate to position. The first is that they should be calibrated within the apparatus. This is because multiple reflections from other components of the system can significantly affect the total light transmitted. Second, because the properties of a filter may not be uniform across the filter, only certain parts of the filter are relevant in any given system. A corollary is that a given filter should be inserted into the apparatus in the same orientation whenever used. Filter holders that allow one to slide a set of resident filters in and out of the beam make this easy.

Finally, when glass filters are inserted into an optical path, each filter increases the optical distance of that path by 2 mm or so (and can also introduce a small amount of chromatic aberration). For example, if a filter is placed between the Maxwellian lens and the eye, it displaces the source image that has been so carefully adjusted to be of minimum size as the light passes through the pupil of the eye; this displacement can enlarge the pupillary pathway of light enough to hit the iris. If the filter is placed between the source and the first collimating lens, it somewhat disrupts the collimation of the beam throughout the system. If placed between the field stop and the Maxwellian lens, it can slightly defocus the retinal image. Usually, such effects are negligible if the filters are placed in a collimated part of the pathway between the collimating lens and the field stop focused on the retina. Interference filters should be in a collimated part of the beam, for the pass band for obliquely incident light differs from that of normally incident light, and so the transmitted wavelength for converging or diverging beams varies with distance from the optic axis. To minimize the visual effects of any dust that may happen on to filter surfaces, they should be distant from locations in the optics that are in focus on the retina. Undesirable reflections from filters can be directed outside the optical pathway by giving the filters a slight tilt.

1.3.6 Realigning

When building an optical apparatus, especially for research, assume that it will require realignment. This puts a premium on adjustability. It is a mistake to think that once the system is properly adjusted, it can be set in concrete and left as is. Conditions that affect the alignment change, and the design required for the next experiment, are also likely to change.

Sometimes a single optical element needs replacing, such as a source, and sometimes it is possible to reposition only the replaced component, but often not. Face the fact that the entire system may require realigning. It will save time and a lot of grief in the long run. You may think that you have re-established the original conditions because the end image looks good and is in the right place, but you may have achieved that by introducing mutually compensating errors (see also Section 1.3.5.4.i)

1.3.7 Head stabilization

As the eye is the most important component of the optical system, it must be aligned with the rest of the optics and held securely in place. Stability depends on firm attachment through the most rigid structure attached to the eye. As most of the skull is covered by soft, pliable tissue, the teeth provide the best connection to the structures in which the eyes are embedded. Fixing three points of any rigid structure is sufficient to fix its position in space, and the further apart the points, the greater the stability. This means that the upper incisors and molars, or wisdom teeth if present, should be in firm contact with the optical system. All the teeth can touch the contacting surface if it conforms well with the contour of the teeth. This surface is normally formed by dental wax (Kerr Impression Compound, Type 1, available from any source of dental supplies), which is soft enough to form the impression of the teeth at 56 °C, but is hard and rigid at body temperature. The wax, which comes in irregularly shaped cakes about 82 mm by 57 mm and 5.5 mm thick, can be warmed to the required temperature with a hair-dryer or by holding it in a pan of hot water (the wax sticks to the bottom or sides if you let it touch them when it is soft). When it becomes soft enough to work, wrap it around a U-shaped piece of metal, insert it into the mouth, and have the observer clamp down on it with a fixed bite for about 1 min or until the wax hardens. The observer should bite far enough so that all the teeth make contact with the wax without hitting the underlying metal. One or two cakes provide a good covering for a single bar.

The U-shape follows the contour of the bite, allowing room for the tongue and perhaps a few spoken words, in addition to the prosaic bite-bar grunt. Different observers require bite-bars of different size. Have the observer bite on a piece of paper or cardboard that is cut down to fit into the mouth, to get an idea of that observer's mouth size. This can be used as a rough template for the metal bite-bar or as a basis for choosing one from a premade set of varying sizes. The bar should only be thick enough to be rigid (certainly no more than 3 mm). A series of holes, say, 5 mm or so in diameter, separated by about 12 mm, helps secure the wax to the metal plate.

The wax is rigid enough to hold the head steady, but it chips and cracks easily. This can be avoided by using dental acrylic, but if let to set for too long a dentist may be needed to remove it. Dental work on the observer's teeth may require a change or a new impression, and if the bite-bar is not used for a year or so, the teeth may shift enough to require a new impression.

Some investigators use a forehead rest or other attachment to help stabilize the head. This increases mechanical advantage, but such a support slowly gives and changes due to pressure on the support. This is evident from the depressions visible on the skin after a session using such a support. These depressions presumably are associated with a gradual change of head position during the session as the impressions form and deepen. It is probably better to rely on the more rigid, full-mouth bite-bar, which also benefits from the exquisitely sensitive and relatively unadaptable mechanoreceptors of the teeth that help the observer monitor the distribution of pressure on the bite-bar. No special effort is required of an experienced observer, biting such a bar with good teeth, to keep the pupil within 50 μm of a fixed position.

If more than one observer is used within a week or so, it saves time to attach the bite-bar to an Aloris quick-change tool ($267, Rutland)—made to change tools on such devices as lathes. With these tools the bite-bar can be removed and another attached with two swift movements of a lever, obviating all the screwing and unscrewing otherwise required.

To align the observer's head, graded movement in three dimensions with a precision of 0.1 to 0.025 mm is required. The minimal system (e.g. Velmex) costs about $550, but $150 more buys much in terms of precision, length of travel, and quality. Alternatively, for about $550 (Sears) or $860 (DoALL) one can fashion a larger and sturdier positioner from cross-slides for lathes, but they typically have greater backlash. (Such errors are minimized by making the final adjustment in the same direction.) Be sure that the range of travel in the transverse dimension (x) is at least 70 mm (Hofstetter, 1972), so that either eye can be used. Careful alignment of the eye requires a scale readable to 0.025 mm.

1.3.8 How to align the eye

All six dimensions of eye position are important in vision experiments. Pitch (ϕ) and yaw (θ) are controlled by the fixation point, and roll (ψ), called cyclorotation in reference to the eye, is usually uncontrolled, on the assumption that it rarely changes under the conditions that ordinarily hold with a bite-bar. It is worth noting that, as the pupil is not at the eye's centre of rotation, changing gaze angle by moving the point of fixation also translates the eye's pupil.

1.3.8.1 Longitudinal (z) alignment of the eye

The main objectives of aligning the eye in Maxwellian view are: getting all the light into the eye; maintaining stable stimulus conditions; ensuring independence of stimulus variables; optimizing stimulus variables, such as brightness or acuity; and allowing manipulation of the optical pathway of the light through the eye. All these objectives are served by minimizing the size of the beam as it passes through the pupil of the eye. This is done by focusing an image of the source or aperture stop within the plane of the pupil. Arguments, based on optics, that the focus should be at the first nodal point of the eye, are contradicted by the evidence that the receptors point closer to the centre of the pupil than to the nodal point (Enoch and Laties, 1971; Enoch and Hope. 1972).

The *simplest* way to adjust the distance of the eye from the Maxwellian lens is to move the eye perpendicular to the optic axis until the incident beam strikes the iris, and then adjust the longitudinal (z) position of head, so that the stopped filament image is in best focus on the iris. This differs slightly from the optimal position, for the front surface of the iris is not the plane in which the pupil is smallest, but for most purposes the difference is negligible.

The *best* way to optimize the distance of the eye from the Maxwellian lens is subjective. After getting the beam inside the pupil (this can be done with the help of someone who can see the filament image on the observer's eye), the observer views a lens or field stop while moving his head upwards. The icons on the right of Fig. 1.9 show the appearance of the Maxwellian lens to the observer under the optical conditions diagrammed on the left. In the top row, as the iris cuts into the beam, its shadow falls on the retina, obscuring a part of the lens (Fig. 1.9, top left). As the eye is moved, the darkened region appears to spread across the lens. If the darkened region of the lens is at its top (top right of Fig. 1.9), the eye is too far from the lens; if the darkened region of the lens is at its bottom (middle right of Fig. 1.9), the eye is too close to the lens. If the eye is moved in the appropriate direction, i.e. forwards or backwards, eventually the darkened region appears to flip over to the opposite side of the lens. The point at which it flips over is where the image is in the plane where the pupillary aperture is smallest. When the eye is in this position, the lens appears filled with light, as shown at the bottom right of Fig. 1.9; movement of the iris into the beam causes a gradual, approximately uniform darkening of the lens instead of a movement of a darkened region from one side towards the other. Because the upper eyelid tends to encroach on the top edge of the pupil, it is usually better to use its lower edge for this adjustment.

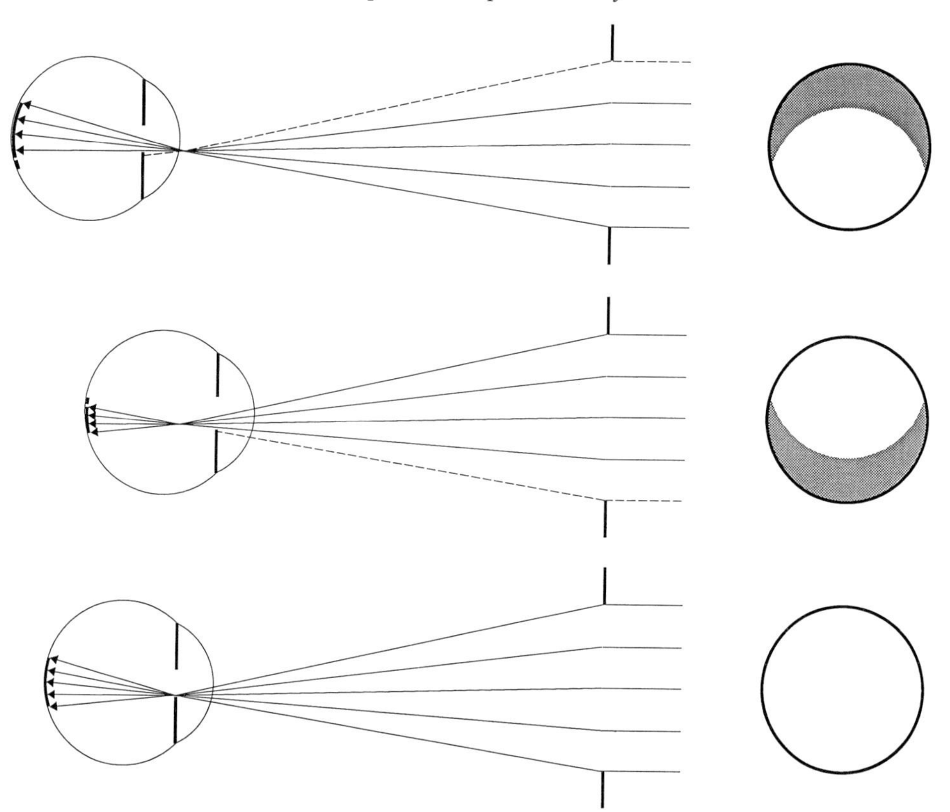

Fig. 1.9 Alignment of the head along the optic axis by the subjective method. The diagrams on the left show the pathway of light incident on the eye, and those on the right, the appearance of the corresponding field of view to the observer. In all three cases, the head and, hence, the eye have been moved upwards until the edge of the iris lies close to the optic axis. If the eye is too far away from the Maxwellian lens (top of the figure), the iris blocks a part of the beam after the rays have crossed and darkens the top part of the Maxwellian lens (the dotted lines represent a blocked ray, and the shaded part of the retinal image, the blocked part of the image). If the eye is too close to the Maxwellian lens (middle of the figure), the iris blocks a part of the beam before the rays have crossed and darkens the bottom part of the Maxwellian lens. When the eye is properly positioned (bottom of the figure), the lens appears entirely filled with light.

1.3.8.2 Vertical (y) and transverse (x) alignment of the eye

The desired location of the eye on the other two translational dimensions (x and y), and how it is aligned, depend on the experiment. If maximal brightness (Stiles and Crawford, 1933) or acuity (Campbell, 1958) is desired, for example, the observer might simply be given an appropriate stimulus to view and asked to find the position that maximizes the variable of interest. It is worth keeping in mind that some hysteresis may occur in both cases; that is, moving the pathway of light through the pupil temporarily increases the light's brightness (i.e. the transient Stiles–Crawford effect; Makous, 1968) and decreases acuity (dubbed transient obtuity; Makous, 1977). Note also that the pupillary pathway that maximizes brightness may not maximize acuity.

In most cases, the main goal is just to get the light through the pupil in a fixed location, just where being unimportant. In such cases, objective techniques are satisfactory. Viewing the pupil remotely through a beam splitter and video camera or telescope works. However, as the pupil is not of fixed shape or size and may constrict asymmetrically, it has drawbacks as a reference for the optical axis; corneal reflection is better. So to monitor eye position, observe the corneal reflection of a fixed source from a fixed position. An infrared source that is invisible to the observer can be viewed, continuously if desired, with a video monitor. Crawford and Granger (1968) describe a system whereby the observer can monitor his own pupillary alignment subjectively.

To establish how close to the edge of the pupil one can focus the image of the source or aperture stop without affecting the retinal image, subjective techniques are best. As the focal plane of a Maxwellian lens lies in or close to the plane of the pupil, the distribution of light there is the convolution of the image of the aperture stop with the Fourier spectrum of the field stop that is imaged on the retina. Light carrying the highest frequencies passes furthest from the optic axis and may not be easily seen by inspection. Consequently, one may not be aware when the iris is filtering these high frequencies on one side of the image. This problem can cause the image of a small spot, for example, to spread towards one side (coma). Having the observer watch for coma in a 1-min target as the head is adjusted may be the best way to avoid it.

The boundaries of the pupil on the horizontal and vertical diameters can be determined subjectively by recording the head (i.e. pupil) positions at which some criterion change in the image occurs. The easiest criterion to see is the point at which the iris just completely occludes the optical pathway, i.e. when the lens goes completely black. This is best for computing the geometric centre of the pupil, but of course overestimates the useful area of the pupil. Alternatively, the observer may look for the first sign of coma or a sudden increase of blurring as the edge of the pupil is approached. Note also, as mentioned above, that the pupil can vary in size during such observations, and that illumination of the contralateral eye also affects the pupil being measured. Some experiments require dilation of the pupil and paralysis of the iris to stabilize the pupil for alignment, but keep in mind that dilation of the pupil is seldom strictly symmetrical.

1.3.9 Fixation targets

The orientation of the eye is normally held in place by the observer, with the aid of a fixation target. If there is a steady field of light, the fixation target usually is a small, opaque dot or cross that can be applied to thin glass, such as a microscope coverslip, which in turn is attached to a field stop conjugate with the retina (Tackiwax, Section 1.4.12, works fine). If foveal stimuli are to be presented, fixation can be guided by two or four small dots spanning the location to be fixated, but note that unless the dots are 1° or more from the foveal stimuli, they can affect the appearance of the stimuli (Jameson and Hurvich, 1967).

If no field is present between test trials, or if one wishes to remove the fixation target during test trials, a luminous target can be used. Usually the observer is given control over its intensity or luminance by means of a potentiometer or transformer and asked to keep it just visible. A light-emitting diode, a small optical fibre, or a small aperture in a piece of shim stock make good fixation targets. It can be introduced into the field by means of a beam splitter, but a microscope slide cover slip serves, for the closely spaced double images associated with the two surfaces seldom matter with a fixation target, or the double image can be eliminated by use of a wedge-shaped mirror (see Crawford and Granger, 1968).

The observer can also use the appearance of the fixation target as a subjective means of *monitoring* accommodation, but it may not be a trustworthy means of *controlling* accommodation (see Westheimer, 1986).

1.3.10 Ergonomics

It need hardly be said that serving as a psychophysical observer can be demanding. One can expect attention to ergonomics to pay rich rewards in quality and quantity of data, and in eagerness of the observers.

The paramount rule is adjustability. In so far as is feasible, the apparatus should be adjustable: first, to accommodate different observers (I have seen a 157-cm observer try to use an apparatus built by a 193-cm investigator: her feet dangled from the chair without support, and she had to stretch to reach the adjustable wedge, not even making contact with the arm support the investigator had provided for himself); and, second, to allow the observer to change position. A bite-bar unavoidably constrains the observer, but, at minimum, the observer requires a comfortable, padded chair of adjustable height with a suitable foot rest. The need to operate a few keys to register a response should leave one hand free and give the hand operating the keys freedom of movement. Although use of the laboratory computer's keyboard may be the easiest way to record the observer's responses, it is not easiest for the observer. The keys of a mouse are much better for the observer; the joysticks used for some computer games are even better (e.g. a Flight Stick Pro, C and H. Sales, $50), some even providing analog input for the control of wedges (see also Sections 1.4.4.3 and 1.7.3).

1.4 Optical components

1.4.1 Positioning and mounting hardware

The catalogues are rich with devices for mounting and positioning optical components, and little need be said here. Nearly all optical components are mounted on a post or pin which fits into a holder that is either part of, or can be attached to, a carrier or base that fits on a rail or table. Unfortunately, the diameter of Ealing's posts is 13.7 mm, and that of Newport posts is 13.0 mm; consequently, Ealing posts do not fit into Newport post holders, and Newport posts fit only loosely in Ealing holders.

For most purposes, a simple post and carrier, such as hold the first lens in Fig. 1.8, are sufficient. A right-angle post holder, such as that holding the second lens in Fig. 1.8, gives all the degrees of freedom necessary for lens alignment. High precision or frequent repositioning may call for more expensive, adjustable carriers and positioners.

1.4.2 Suppliers

Optical components for vision research can be obtained from Corion, Ealing, Edmund Scientific, Melles-Griot, Newport Corporation, Oriel, and Spindler and Hoyer, to mention a few of many. Triangular rails and associated components can be obtained from Ealing as well as Newport; Newport and Spindler and Hoyer also market optical tables and associated components. Newport and Edmund lie at the opposite extremes of cost and quality, with Edmund on the low end, though much that it sells is quite satisfactory for most vision experiments. (See the *Photonics buyers' guide* or *The laser focus world buyers' guide* for other suppliers.)

1.4.3 Lenses

Design and modification of the apparatus is easiest if all the lenses are identical (see Section 1.2.2.2). A focal length of 150–250 mm is about right: too long, and the apparatus gets large and cumbersome; too short, and one has trouble fitting the components into the available spaces, and the field stops are too small to work with easily. Get the largest numerical aperture available at reasonable cost: high numerical aperture increases the visual angle of the usable field, but one is not required to use the entire lens. It is unlikely that anything less than an achromatic doublet lens will serve for most research applications; if the numerical aperture is high, a triplet lens may be required.

Some lenses come mounted in metal rings. This makes it easy to tap the mount and attach a suitable post, but centring the post requires skill. A variety of adjustable lens mounts are available from optical supply vendors.

Although I recommend the use of a single type of lens, it is nevertheless useful to keep a diverse supply of lenses on hand for unforeseen exigencies. A set of trial lenses also finds many uses in the lab; often an inexpensive, used set can be obtained from a retiring optometrist or ophthalmologist.

1.4.4 Intensity

One of the principal advantages of optical systems like a Maxwellian system is the high retinal illuminances that can be achieved. With an incandescent source, 10^6 td is not difficult, and by overdriving the source and filling the pupil with the image of a ribbon filament or other extended surface, 10^7–10^8 td may be possible (see also Section 1.11, on light damage).

These are the maximal troland values (see Section 1.12.1.5, 'Troland value') then, but experiments require a means of controlling flux densities, which is normally done by filters. Filters work either by absorbing or reflecting part of the incident light and can be obtained from a variety of sources, including Corion, Ealing, Edmund, Kodak, and OCLI.

1.4.4.1 Absorbing filters

If control over light flux is the objective, neutral filters are used. They are neutral in the sense that they absorb all wavelengths approximately equally, but this is only approximate, and the filters must be separately calibrated for each separate spectral distribution used with the filter. Absorbing filters tend to be more coloured (less neutral) than reflecting filters, they are less stable, and they are sensitive to heat.

Absorbing filters contain a pigment that is either distributed uniformly through some vehicle such as glass, or as a thin film between protective coatings. Kodak Wratten filters, for example, consist of a 100-μm ($\pm 10\,\mu$m) gelatin film on acetate, with a protective coat of lacquer. They are also available mounted between glass, as a sandwich typically 5 mm thick and 50 mm square. (Note that contact with water may cause the layers to separate.) Many sizes of gelatin filters are available by special order, and larger sizes can be cut to size by placing the filters between fairly stiff paper and cutting them with scissors. Such filters come in optical density steps of 0.1 and 1.0.

1.4.4.2 Reflecting filters

Reflecting filters consist of an evaporated metal film laid on a substrate such as glass. They tend to be more uniform, more neutral, and to have smaller tolerances than absorbing filters, but the reflected light is part of the system until ultimately absorbed, and it can show up at unsuspected times and places. It is exquisitely difficult to avoid spurious interference fringes when reflecting filters are used with the coherent light of lasers.

1.4.4.3 'Wedges'

Filters that vary systematically in optical density along their length are called wedges. Usually the optical density varies approximately linearly with distance. The filter may be laid out in a straight strip, or around the circumference of a disc. The densities go from close to 0, to 1, 2, or 3. They may be absorbing or reflecting.

If the wedge is reflecting, it can be used as a variable beam splitter, which is useful to form two channels with a constant sum and adjustable ratio. However, note that if wedge position changes the optical density (which is a logarithmic measure) linearly in one channel, the other channel must vary in a highly non-linear way. Also, the bearing on which the wedge rotates must hold the wedge in a fixed plane, or else alignment of the reflected channel varies with wedge position.

As optical density varies with location on a wedge, the flux density of transmitted light varies from one side of the beam to the other. If this is significant, as it is likely to be if the wedge is placed in a collimated part of the beam, the non-uniformity must be corrected by use of a compensating wedge. Fortunately, if the wedge density is linear with distance on the wedge, the compensating wedge need only be as wide as the beam, though some wedge systems consist of a pair of matched, opposing wedges that move equally in opposite directions. Problems of compensation can be avoided by locating the wedge at or near an image of the source, but the high flux densities there tend to bleach the pigments. I have seen a white groove burned into a wedge where an incandescent source had been imaged on it during years of use.

1.4.5 Apertures

1.4.5.1 Construction

Apertures of various sizes down to a few micrometres can be bought, but those in the millimetre to centimetre range are easily made from 0.005-gauge shim stock, which can be cut to size with ordinary paper shears. Such apertures make a sharp optical edge and can be easily mounted. Circular apertures can be drilled, with the stock clamped between pieces of wood. If more sturdy material is desired or if a drill of the desired size is unavailable, one can used a thicker metal plate (of aluminium, for example) and adjust the size of the hole by countersinking it, creating the desired sharp edge in the process. Apertures of any size or complexity can be drafted at a convenient scale, photographed with high-contrast film, and the negative mounted on a slide frame.

One can change apertures quickly and easily either with an iris diaphragm or a wheel containing different apertures around its periphery. The wheel can be positioned manually ($535 for an 8-position wheel, Oriel) or under computer control (see Section 1.7.2).

The location of an aperture typically requires frequent, precise adjustment. An inexpensive way to manipulate it is to attach it to a microscope stage (such as is available at present for $69 from Edmund) and mount the stage on a positioning rod. Such stages typically have a Vernier scale with 0.1-mm precision.

1.4.5.2 Aperture stops and artificial pupils

If the cross-section of the pathway of light incident on the eye is too large, then the retinal illuminance fluctuates with changes in the size of the eye's pupil. In a Maxwellian system, the source must be stopped down by focusing it on an aperture that is conjugate with the pupil of the eye. A 2-mm image that is centred and well fixed relative to the pupil of the eye is normally safe from the effects of pupillary constriction, and a 3-mm pupil may often be safe. A 1-mm image practically limits optical degradation to diffraction, so that correction for aberrations of the eye is obviated. A 2.0- to 2.5-mm image yields the best retinal image (see Campbell and Gubisch, 1966).

If viewing is direct instead of Maxwellian, an artificial pupil can be used in place of the imaged aperture. However, as a Maxwellian beam is focused behind the cornea, artificial pupils cannot normally be used in a Maxwellian system. Even in direct view, an artificial pupil vignettes the field of view. (See Troland (1915) on the use of artificial pupils.) The artificial pupil must necessarily be smaller than the natural pupil: Troland recommends 2 mm smaller. The largest field that can be viewed through an artificial pupil is about 60° (assuming a 0.5-mm artificial pupil, 8-mm ocular pupil, and 7 mm between the two). Moving it beyond the tips of the eyelashes (add 3 mm) drops the maximum field size to 42°. If the ocular pupil is decreased from 8 to 2 mm, the maximum field sizes for these two locations dwindle to 12° and 8.5°, respectively; and increasing the size of the artificial pupil from 0.5 to 1 mm shrinks the maximum field sizes (respectively) to 8° and 5.6°. About the same limits hold if the artificial pupil is increased to 2 mm and the ocular pupil to 3 mm. Note that putting an artificial pupil in front of the pupil of the eye increases the effects of diffraction on the retinal image by an amount that depends on the relative sizes and positions of the two pupils.

As pointed out to me by Dave Williams, the purpose of an artificial pupil can be served with none of its disadvantages by viewing the object with a telescope (Galvin and Williams, 1992), putting the pupil of the eye at the exit pupil, and using the entrance pupil of the telescope to control the pathway of light through the pupil of the eye. The entrance pupil can be as small as necessary, and as it is conjugate with the pupil of the eye, no light within the field of view is lost except the 4% reflected at each optical interface. In such a system, an image of the object is formed between the lenses, in the focal plane of the second lens, which puts a virtual image of the object at optical infinity and focuses a real image on the retina. The Maxwellian system at the bottom of Fig. 1.1 can easily be converted into a telescope viewing the source, by removing lenses 1 and 2, and moving lens 3 towards the source until an image of the source is formed in the first focal plane of lens 4. The aperture should be moved the same distance as lens 3 to keep it in the focal plane of that lens. The only disadvantage of this approach is that the usable field of view is restricted by the numerical apertures of the lenses. Although the optics degrade the image somewhat, reasonably good lenses do the image less damage than the eye itself.

1.4.6 Image pattern

In a Maxwellian system, any pattern placed in the first focal plane of the Maxwellian lens is focused on the retina of an emmetropic eye accommodated at infinity. Hence stimuli of any desired spatial complexity are possible, however rarely used in practice.

As discussed above, however, Maxwellian systems severely limit the ease with which one can change the pattern in interesting and useful ways, particularly those representing complex spatiotemporal patterns. Even presentation and manipulation of the properties of a static, undistorted sine-wave grating is a challenge unless one resorts to optical interference. Varying contrast likewise is a challenge, usually requiring one to vary the mixture of two fields while holding their sum constant. This can be done with crossed polarizers (see Section 1.5) or beam-splitting wedges (see also Section 1.4.4.3). Measuring contrast threshold even for such mathematically simple stimuli as moving plaids would be daunting.

Abrupt changes between two patterns, and often between intensities, is usually achieved optically by closing the shutter in one channel while opening one in another. However, attempts to produce a homogeneous field with a Maxwellian system leaves spatial inhomogeneities that cannot be seen during steady viewing, owing to the insensitivity of the visual system to low spatiotemporal frequencies; but alternating between two such fields of any size blatantly reveals the inhomogeneities. If the two channels do not coincide exactly as they pass through the pupil, entoptic differences between optical pathways, local variations in the Stiles–Crawford effect (Stiles and Crawford, 1933), and the transient Stiles–Crawford effect (Makous, 1968) increase the unwanted changes of stimulation caused by exchanging fields. Hence, mutual alignment of the optical axes of separate channels of a Maxwellian system is important.

The most nearly homogeneous field is afforded either by an integrating sphere or by an integrating bar, neither of which is suitable for Maxwellian systems. An integrating sphere contains an aperture for the entry of light, a diffusing glass window (see Section 1.4.9.2) for its exit, a baffle to prevent direct reflections of the entering light to the exit window, and an interior surface with highly reflective and diffusing properties to promote multiple reflection of light within the sphere (see also Section 1.4.9.1). An integrating bar is simply a large light pipe that likewise promotes multiple complete internal reflections.

1.4.7 Shutters

Stimuli are modulated in optical systems such as Maxwellian systems mainly by turning them on and off by means of a shutter. Electromechanical shutters are inexpensive and convenient, but the more mechanical they are, the more subject they are to wear and eventual failure. Mechanical timers also are affected by temperature and humidity.

A leader among the commercially available shutters for vision research is the Uniblitz shutter (Vincent). As they have improved in speed and design, they have also improved in reliability, so that now the manufacturer estimates that they are good for 1 million exposures. At present Uniblitz shutters are available with apertures of 2, 3, 6, 14, 25, and 35 mm, with minimum exposure durations of 1.5–2 ms for the smaller shutters and 10–50 ms for the larger two. Cost ranges from $200 to $570. Vincent also markets an electronic driver ($435) and timing driver ($680) for the shutters. (Be careful: the driver can be damaged by grounding the input.) These shutters make an audible click, which helps the observer know when the stimulus was presented but may subvert a forced-choice task unless masked or disassociated with stimulus presentation, e.g. by a slower, silent vane.

Many investigators fashion their own shutters out of a rotary solenoid, galvanometer, pen recorder, scanner, or audio-speaker, any of which can activate a flag made out of stiff, black paper or other light material attached to the end of a balsa wood or light metal shaft. They can be nearly silent, move across a 2-mm aperture in about 1 ms, and present stimuli as short as 1 or 2 ms. The driving circuits, though usually simple, must be designed and built in-house.

A very fast shutter can be made by tilting a mirror to move the image of a source off an aperture stop. The mirror can be activated by any of the devices mentioned above, but enormous speed can be obtained by tilting it from one side with a piezoelectric crystal. However, the ratio of luminances, on to off, may be limited to 100-fold or so. By varying the degree of overlap between the image and the aperture, one can use such a mirror system to modulate the intensity of the light (e.g. Mandler, 1984).

Shutters and their calibration are discussed extensively by Ruddock (1968), and the general topics of light modulation and ferroelectric shutters are covered in Chapter 2.

1.4.8 Mirrors

To avoid the double reflections from the front and back surfaces that occur with ordinary glass mirrors, those used for optics have the reflective aluminium coating on the front surface. A thin coating, usually silicon oxide, protects it from oxidation and to some extent from abrasion. The inexpensive mirrors from Edmund serve, except for applications like interferometers. Optical adhesives with low shrinkage, to avoid distorting the mirror, are available (e.g. Edmund), but one might not want all bonds to be permanent, and so one does just as well with a silicone adhesive.

Semireflective mirrors, which transmit the light not reflected, can be used to split beams, but only where a double image in the reflected light is acceptable. To avoid the double images and light loss, the mirrors can be made so thin that the double images are practically superposed. Such splitters, or pellicles, transmit 92% and reflect 8% of the light and are sensitive to vibration, even sound. They are also very fragile: I have seen one rupture when suddenly hit by an intense light.

Mirrors that selectively reflect or pass infrared radiation, i.e. hot mirrors and cold mirrors, respectively, can be used to cleanse unnecessary heat from a light source. Analogous mirrors that pass or transmit only certain wavelengths in the visible range are discussed in Chapter 4.

1.4.9 Diffuse surfaces

It frequently is necessary to disperse an incident beam randomly over a broad angle, to view an image, for example. The scattering can be reflective or transmissive.

1.4.9.1 Reflective

With but rare exceptions, reflected light varies with the incident angle, the angle of reflection, and the relationship between the two. The diffusely scattered light typically is mixed with a strong mirror or specular reflection, identified as the component reflected at the angle equal to the incident angle but of reversed sign. The appearance produced by the specular component of reflection from a surface is described as gloss, and it is usually considered to detract from image quality. Surfaces with little or no gloss are described as matt. The ideal case in which the reflected light is equal in all directions, no matter what the incident angle, is called isotropic or Lambertian reflection. This ideal is not realized in practice, but Eastman White Reflectance Standard and Eastman White Reflectance Coating approach it. Matte spray paints, with small particles embedded in the vehicle, form good matt surfaces. Paper is a ubiquitous diffusing surface but always has some gloss.

The proportion of incident light reflected, i.e. the *reflectance factor*, is also important. The whitest white is $BaSO_4$, which forms the basis of paints with reflectance factors of over 0.99 (e.g. the Eastman whites named above). The blackest black is carbon black in oil, which has a reflection factor of 0.003. Other examples are listed in Table 1.2. (See Pompea and Breault (1995) for extensive data on various black surfaces and their use in optical systems.)

1.4.9.2 Transmissive

Most transmissive diffusers used in optics are either opal glasses or ground glass, but a variety of translucent substances can be used, including various kinds of paper. From the viewer's perspective, the distribution of light determines the brightness of the image when viewed from different angles. Smaller angles of distribution produce brighter images but at a cost of a more restricted range of viewing angles.

Opal glasses excel at spreading light widely. Light passing through a ground glass, on the other hand, has a Gaussian distribution with a width at half height of about 15° (Kurtz, 1972). Although the size of the grit used to grind the glass affects image quality, it has little effect on the distribution of transmitted light (however, use of the finest grit does allow a spike of specular, or undeviated, transmission). If scattering the light is all that matters and no image is to be viewed, scatter can be increased by using successive layers of diffusion, and it can be varied by varying the distance between layers.

Table 1.2 Reflectance factors for miscellaneous surfaces

$BaSO_4$	0.99 (WS)
Eastman White Reflectance Coating	0.99 (WS)
Magnesium oxide	0.98 (WS; Hdbk)
Snow	0.93 (Hdbk)
Paint, white	0.90 (WS)
Talcum	0.89 (Hdbk)
Clay (kaolin)	0.82 (Hdbk)
Snow	0.80 (WS)
Paper, white bond	0.75 (Hdbk)
Porcelain enamel, white	0.73 (WS)
'Light grey' ceramic tile	0.60 (WS)
'Light grey' shingle	0.40 (WS)
Newspaper	0.38 (Hdbk)
Dry sand	0.30 (WS)
'Medium grey' paint	0.29 (WS)
'Medium grey' ceramic tile	0.26 (WS)
Lead	0.24 (WS)
'Dark grey' ceramic tile	0.05 (WS)
Paper, black	0.05 (Hdbk)
Paint, black	0.04 (WS)
Black rich soil	0.04 (WS)
Carbon black in oil	0.003 (Hdbk)

WS, Wyszecki and Stiles, 1982; Hdbk, *CRC handbook of physics and chemistry*, 1984.

1.4.10 Fibre optics

Optical fibres, or light pipes, convey light with little loss from one place to another. This is valuable, for instance, when the light source must be distant from the optics or must be isolated from them. For example, it is a good way to control stray light, for the source may require cumbersome cooling equipment not conveniently located close to the optics. Bifurcating fibres work as beam splitters, and fibres can be used to redistribute light spatially, e.g. to change a circular beam to a slit and vice versa. Coherent bundles of fibres preserve the same spatial organization of fibres at both ends, so that the transmitted light carries images, somewhat degraded by the discrete sampling by the fibres.

The advantages of fibre optics lie in their low cost, small size, ruggedness, light weight, and flexibility. The output (or input), whether used as a light source or image, can be easily moved and manipulated. Fibres find application in miniaturization and where such ease of spatial manipulation is important. To give an idea of its value for miniaturization, a dual interferometer occupying an area 1.5 m by 0.9 m on an optical table has been designed (though not built) to fit in a space the size of a flashlight, mainly through the use of fibres and light valves.

The principal disadvantage of fibre optics is loss of information. A lens loses little or no information in transmitting light, for the changes, aside from those caused by imperfections, can be reversed. An optical fibre, on the other hand, selects and sums photons according to location, direction of propagation, and, to some extent, wavelength; and once a photon enters the fibre, all information is lost on its direction of propagation, location, and direction of polarization (unless the fibre is specially designed). Though larger fibres encode some of this information in the structure of mode patterns, the information is lost for all practical purposes, and such larger fibres lose more spatial information. This means that the images carried by fibres are degraded, and the information that makes Maxwellian systems possible, for example, is lost.

1.4.11 Prisms

Prisms are used mainly to reflect or polarize light. For reflection, their advantages over a system of mirrors for the same purpose are compactness, rigidity, protection of the reflecting surface, and low light loss (typically about 8%, but it may be as high as 40%). Their disadvantages include some degradation of image quality, cost, size limitations, and weight. Types of reflecting prisms include: corner cube prisms, which reflect light along the incident pathway no matter what the angle of the prism; right-angle prisms, which do the same thing within a single plane; pentaprisms, which reflect light through 90°, no matter what the orientation of the prism (see Section 1.3.5.2); and roof prisms, which invert images (or erect inverted images). Rotation of a Dove prism rolls an image (through twice the angle of the prism's rotation), but in practice, the field of view is limited, and Dove prisms are difficult to mount and align.

The most frequently used prisms are beam-splitting cubes, which both transmit and reflect light at right angles in varying proportions depending on the prism; such prisms avoid the double reflections of beam-splitting mirrors. Some prisms are designed to polarize the two beams in perpendicular directions (see Section 1.5).

1.4.12 Tackiwax

I admit to no embarrassment in devoting a special section to that wonder of nature and technology, Tackiwax (Central Scientific). Nothing matches its usefulness to one given to the quick-and-dirty mode of research. Essentially beeswax, it can be used to stick anything to anything. It does not harden or lose its grip with age. Objects can be moved or removed with little effort, but they will not move under their own weight once they stick. The only danger is heat: warm it up and it lets go.

I acknowledge that not everyone is so enthusiastic. Profligate users may find it where they did not intend it, and it is not easy to remove from glass surfaces and especially from gelatin filters. This puts a burden on more tidy co-workers.

1.4.13 Cleaning

Some books on practical optics (e.g. Conrady, 1960; Johnson, 1960) go into exquisite detail on how to clean optics. Here I restrict myself to simple concepts.

First, try not to get them dirty. Although a dwindling problem, a cardinal principle is to keep tobacco smoke away from them, for it is especially hard to remove. Also, protect them from dust. This may mean covering them, say, with a plastic sheet (keep the same side down on successive uses) when they are not in use. (Some investigators use a rigid cover, with guy wires and pulleys.) Also, laying components face up allows dust to collect on the surface, and laying them face down makes them vulnerable to scratching; storing them upright, then, is best. Cleaning the floor with a wet mop instead of a broom or vacuum cleaner stirs up less dust.

Some of the dust that does collect can be removed by air forced from an ear syringe or from cans (e.g. Edmund; Fisher) of clean, compressed air or inert gas (be careful about stirring up dust from the substrate), or by a clean brush.

Surface mirrors are covered by a delicate film that requires special care in the cleansing. If a liquid cleaner is used, it is best to rely on surface tension to apply force as the lens tissue is drawn across the surface.

To remove film and stubborn flecks of dust from glass surfaces, standard lens cleaners (e.g. Edmund) with lens tissue will do, but a single application can make things worse, and cleaning optics in place can misalign them. Fingerprints, the residual film left from a previous cleaning, and other minor films and contaminants can be removed with little application of force by using a liquid lens cleaner (Opti-clean polymer) that is peeled off after drying (it also serves as a good protective coat for glass surfaces not in use). Nevertheless, one may have to face the fact that eventually the optics may have to be removed for cleaning, and then realigned. If you ever find yourself between experiments, it may be a good time to clean the optics.

1.5 Polarization

Polarization offers a powerful tool that gives ample range for the play of ingenious minds. For its principles I have space here only to refer the reader to standard texts on optics (e.g. Sears, 1949; Hecht, 1987).

The basic advantage of polarization as a tool is that it makes possible the segregation of light into independent channels, even when they are spatially coextensive and use the same optical elements, and it also provides for precise control over the interaction between channels. For example, if a sine-wave grating is polarized in one direction and a homogeneous field in another, then a variable analyser can change the relative amounts of the two, and hence the contrast of the mixed grating, without changing the total amount of light. Analogous control over other kinds of mixture, such as colour, is also possible, as is the segregation of light destined for opposite eyes, as in stereoscopic or haploscopic presentation of images.

Polarizers are available as plastic Polaroid sheets (e.g. Edmund), crystals embedded in glass (e.g. Newport), or beam-splitting cubes (e.g. Edmund, Newport, and Spindler and Hoyer). Extinction ratios as high as 100,000:1 are available from sheet and crystal polarizers, but those for beam-splitting cubes vary from 500:1 to 10:1. The polarization ratio decreases in the short wavelengths, dropping as low as 10:1 in some cases (Wyszecki and Stiles, 1982, p. 50); this gives light passing through crossed or nearly crossed polaroids a bluish cast. Polarization is lost on diffuse reflection, as on a projection screen, unless an aluminized, polarization-preserving screen is used. It is also lost in fibre optics (Section 1.4.10) that are not specifically designed to preserve it. Hence, it is important to attend to the degree of polarization when designing and ordering for systems based on polarization. Moreover, the state of polarization is often affected by unsuspected optical components, such as mirrors, prisms, and beam splitters, and by the eye itself; and, of course, polarization is the basis of operation for many electro-optic modulators such as liquid crystals and light valves. When working with linear polarization, one can easily overlook components of circular polarization that creep in unsuspected. In practice they are not likely to be detected without specifically looking for them, but they can lead to confusing and surprising effects.

A final consideration in the use of polarization is that the state of polarization is often confounded with the experimental variables—in the example above, varying contrast varies the ratio of light polarized in perpendicular directions—and the assumption is made that this has no effect on the visual system. Every test, excluding the phenomenon of Haidinger's brushes (De Vries *et al.*, 1953), supports this assumption, but it is a conceptual drawback.

1.6 Stray light

1.6.1 Within the apparatus

There are two sources of unwanted light in the retinal image, and they call for opposite remedies. The first source is from within the apparatus itself and derives from unwanted pathways from the experimental light source. Obviously, the first step is to enclose the source, allowing only: an aperture large enough to accommodate the exiting beam; baffled vents for entry and exit of air if the source is hot, e.g. incandescent or arc sources; and access for the electrical power supply, usually along with provision for mounting hardware.

Beam splitters often pass superfluous beams that can be a serious source of stray light. Some 95% of such light can be absorbed by a black matt surface (see Section 1.4.9.1). If that is not enough, the light can be trapped by sending it into an enclosure like an integrating sphere (see Section 1.4.6) that is black inside (like the eye), or a series of black baffles that direct reflected light from one to the other.

Leakage around the optical components can be prevented by enclosing them with baffles perpendicular to the optical pathway. An easy way to do this is to cut holes for the optical components and optical rail (if used) in stiff, black paper and attach it to optics and substrate with black vinyl plastic tape (e.g. Scotch 88) which is completely opaque. (Black masking tape can also be used, but it loses its adhesive power much more rapidly than plastic tape, and it is not opaque.) Paper eventually curls and sags over a period of months to years unless supported. Black velvet or flocked paper (both adhesive or non-adhesive paper are available; Edmund) is an alternative, and if the apparatus is likely to be in use over a long period, more cumbersome but sturdier alternatives, such as black posterboard, plastic, or masonite, are worthwhile (but beware of specular reflections). Use of such baffles, however constructed, reduces stray light and optical cross-talk from within the apparatus more than any other measure.

1.6.2 From within the laboratory

For most experiments, simply turning off the lights is sufficient to reduce stray light from outside the apparatus to acceptable levels. If light is needed, say, for an experimenter to record data, even a dim, shielded light can be a serious source of stray light. There is a great advantage in using light of 600 nm or greater: stray light is mainly a problem for rods, but experimenters can use cone vision, and light of long wavelength is at least 100 times more visible to cones than to rods. Some years ago, Boynton (1966) recommended sequestering both observer and apparatus within a separate enclosure, providing the experimenter with the possibility of comfort and adequate light, but I know of none who followed this excellent advice. Fortunately, the vanquishing of experimenters by computers renders this a diminishing problem.

In its place, however, is glow from the computer monitor, which can be substantial even after darkening by a screen saver or after it has been turned off. Covering it (yes, with black paper or a heavy black cloth, hinged with tape at the top) will usually suffice. Even pilot lights for power supplies and other electronic equipment can be a problem. The pilot lights can be removed altogether or covered by black plastic tape; of course, then it may be hard to tell if the equipment is turned on.

Experiments often require enclosure of the apparatus as well. One can run the optical pathway through tubes, but this increases stray light from internal reflections. Usually it is better simply to build an envelope around the apparatus, but sections of the envelope must be easily removed to allow access to parts of the apparatus for alignment, troubleshooting, and exchange of filters or other changes of experimental conditions.

The last step is tracing the pathway of the last bits of unacceptable stray light. This usually requires complete dark adaptation and a piece of opaque paper to insert here and there while searching for the pathway of the unwanted light, which can be quite surprising.

1.6.3 From outside the laboratory

Some sources of stray light originate outside the laboratory. If one cannot use a room without windows, it may be necessary to seal the windows against light by means of an opaque material such as aluminium foil (sturdy aluminium foil with a black matt surface is available from Rosco Laboratories). Much of the stray light from under and around the door can be baffled with black paper and tape, but if complete darkness is required, a double door is necessary.

1.7 Computer interface

Most vision experiments involving optical systems of the type discussed here require shutter action and the reading of key presses, and perhaps wedge position, on every trial; and changes of wedge position, filters, or apertures between trials. Hardware for such tasks is readily available, as discussed below.

1.7.1 Shutters

Uniblitz shutters can be controlled by manual key, TTL input, or through an RS-232C serial port. Those made in-house obviously require in-house interfacing.

1.7.2 Exchange of optical components

Rotary solenoids can be adapted to insert and remove optical components such as filters, apertures, and lenses. An alternative is a Motorized Filter Wheel System ($1665 from Oriel). It consists of a rod-mounted wheel with five positions that can be aligned to a 25-mm aperture. The circular mounts (also 25 mm) hold filters, apertures, lenses, or anything that can be mounted in a window 25 mm in diameter and 10 mm in thickness. Each of the five wheel positions can be called either by a manual switch or through a parallel digital I/O board.

1.7.3 Wedge positioning

Circular wedges can be controlled by stepping motors or servo systems. Both come in enormous variety. Little torque is required to position a wedge, so both cost and size can be kept in reasonable bounds. For most purposes, stepping motors with 100 steps per revolution are enough, but 200 steps might be desirable for wedges with optical densities covering the range from 0 to 3. The motor and the indexer to control a stepping motor though an RS-232 serial interface can cost from $270 (C. and H. Sales Co.), to $1915 (Parker Hannifin Corp.). A servo system can be controlled through digital to analog converters, and can cost from $10 into the thousands (e.g. Servo Systems).

1.7.4 Reading data

Reading the responses from a mouse switch is trivial, and the commercially available joysticks cited in the section on ergonomics (Section 1.3.10) come with interface cards. The switches on response boxes built in-house are connected through a parallel digital I/O board.

Wedge position can be read, whether positioned by the observer or the computer, through digital shaft encoders (e.g. Parker Hannifin), or by attaching a potentiometer to the shaft, using it as a voltage divider, and reading it with an analog to digital I/O card. An alternative is to deflect some of the light passing through the wedge to a photodetector and read its output, which provides a direct measure of light output but may be troubled by stray light.

1.8 Combining optics with CRTs

One can have the advantages of both CRTs and classical optical systems by combining the two. The simplest way to do so is by reflecting the CRT display with a beam-splitting prism between the Maxwellian or final lens of the optic system.

Note, however, that prisms lose 50–70% of the light in both pathways, and prisms may introduce chromatic aberrations in the Maxwellian field; note also that introduction of a prism increases the length of the optical pathway and may require repositioning (along the z axis) of other components. The disadvantages of a prism can be solved by using a pellicle instead, and a beam-splitting mirror will also solve or mitigate the problems: if the aluminized surface is in front, and the mirror reflects about 90% of the light and transmits 10%, the image of the CRT is reduced only 10%, and the image reflected from the second surface is at least much dimmer than that reflected from the first surface. As the amount of light flux is usually not a problem with Maxwellian systems, the 90% loss usually poses no problem. The double image also may not matter for some applications, as in presentation of homogeneous fields, for example.

A more difficult problem with this approach is that the retinal illuminance contributed by the CRT is influenced by fluctuations in pupil size, and that from the Maxwellian system is not, unless the image of the source is large enough to cover the pupil at its largest. If so, retinal illumination from both fields vary with changes in pupil size, but at least they covary. Note that artificial pupils are not compatible with a Maxwellian system (see Section 1.4.5.2).

A better way (pointed out to me by Dave Williams) to combine a Maxwellian system with a directly viewed object like a CRT is to use a telescope to view the CRT or other object, as done by Williams *et al.* (1994; see also Section 1.4.5.2, on artificial pupils). In this case the lens that serves as the Maxwellian lens for one system, and that also serves as the ocular lens for the telescope, lies between the eye and the beam-splitting cube or pellicle that combines the two systems.

A promising new approach is to place a liquid-crystal display, which is under computer control, at a location in the optics that is conjugate with the retina (David Brainard, personal communication; see also, Bonnardel *et al.*, 1996). Present use may be limited by the speed at which the display can be changed, the relatively large size of the elements of such displays in the retinal image, and problems in controlling graded transmission.

As few light sources used with optics have the colour of a CRT phosphor, the colours of monochrome CRTs must be matched by the use of colour conversion filters (e.g. Kodak filters labelled CC, or Kodak Wratten filters nos. 78–85).

Finally, if brief or flickering stimuli are used, synchronization with the CRT may be a problem, since the phase at which a test flash is presented against a flickering background, like a video raster, affects its visibility even when the flicker itself is not visible (Boynton *et al.*, 1961), and beats can be seen between flickering stimuli that do not appear to flicker (MacLeod and He, 1993). Note that synchronization with raster-driven CRTs also depends on location on the screen. The 67 or 75 Hz typical of most computer-driven displays at present is barely adequate for experiments sensitive to temporal variables. At the time of writing, Apple has released software that can be added to a 7500, 7600, or 8500 Macintosh to run a display at 120 Hz. (See also Chapter 3.)

1.9 Binocular stimulation

Satisfactory control and manipulation of stimuli presented to opposite eyes requires a haploscope (but see Section 1.5). A simple and versatile haploscope consists of a mirror for each eye that can be translated in the transverse direction, x, (to accommodate differences in interpupillary distances) and rotated in pitch, ϕ, (to accommodate differences in phorias) and yaw, θ, (to adjust vergence), as shown in Fig. 1.10.

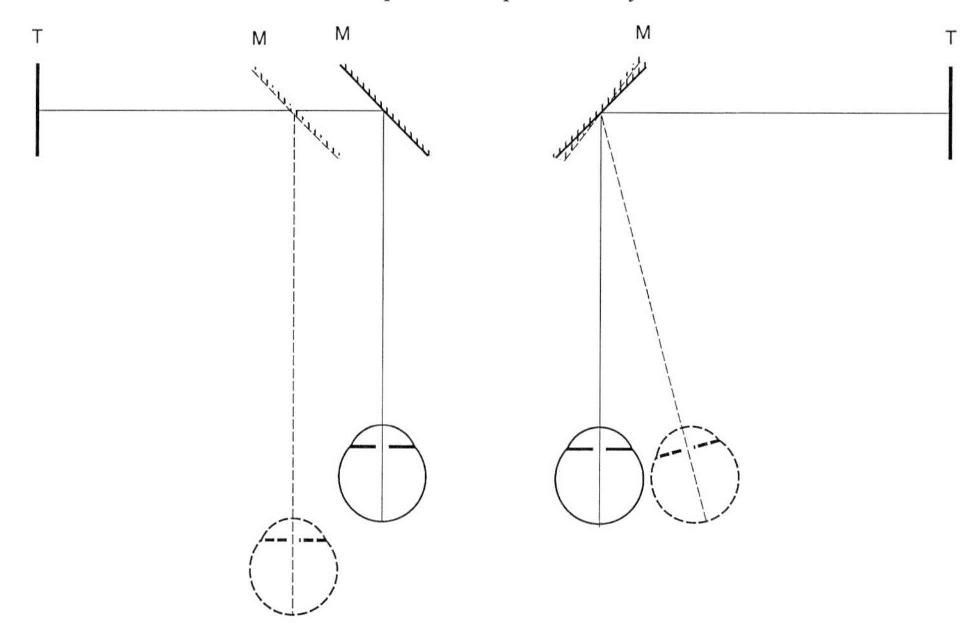

Fig. 1.10 A simple haploscope, with two ways to adjust the interpupillary distane. Separate visual targets (T) are visible to the two eyes in the reflected images of the two mirrors (M). The dashed part of the figure on the right shows how a greater interpupillary distance can be accomodated by rotating the mirror and increasing the vergence angle, and the dashed part of the figure on the left shows how a greater interpupillary distance can be accomodated by translating the mirror and moving the observer back (both mirrors would have to be moved, but the illustration is for the left eye only).

Such a device positions the two stimuli on opposite sides of the observer. If both stimuli must for some reason be in front of the observer, another pair of mirrors is required, but the number of variables requiring adjustment is the same. As long as the distance to the visual targets is long compared to individual differences in interpupillary distances, different observers can be accommodated simply by translating (x) the mirror.

If the two optical pathways are Maxwellian, accommodating differences in interpupillary distances among observers is complicated. The best solution is to mount the optical system for one eye on a moveable platform (e.g. Sturr and Teller, 1973; Makous and Sanders, 1978). An alternative is to direct the beams into the pupils with a pair of right-angle mirrors, placed between the Maxwellian lenses and the eye, that can either be rotated (θ) or translated (x). The disadvantage of rotation is that the vergence angle of a given observer depends on that observer's interocular distance (illustrated in Fig. 1.10, with the right eye), and the disadvantage of translation is that adjusting the interocular distance moves the plane of the image that must be focused within the pupil (illustrated in Fig. 1.10, with the left eye). A range of adjustment from 57 mm to 66 mm is required to accommodate 90% (5th to 95th percentile) of women, and a range of 60 mm to 70 mm is required for men (Hofstetter, 1972).

1.10 Interferometers

Interferometers for visual research (Williams, 1985; MacLeod *et al.*, 1992) are too specialized for detailed treatment here. Principal considerations are the need for superior optical components and for isolation from sources of vibration. Tolerances for alignment are smaller than for other applications, and one must learn a few tricks, for example, in aligning spatial filters. The coherence length of the laser demands attention, and adjustment of acousto-optic modulators requires electronic as well as optical expertise.

1.11 Light damage

Light can damage the retina both by heat and by phototoxicity. To protect the retina from the unnecessary heat of the long wavelengths to which the eye is insensitive, a heat-absorbing filter should normally be placed within the optical pathway. The transmission of heat-absorbing glass typically decreases roughly linearly from 600 to 900 nm, with negligible transmission above 1000 nm. Mainly to protect other optical elements such as filters, but also because the optical quality of such filters is not high, they are best placed close to the source.

The heat in visible light can be damaging as well. The best guidelines on safe limits for broad-band sources are those of the American Conference of Governmental Industrial Hygienists (1995). Any broad-band source of 10 000 cd m^{-2} or less is considered safe. With the standard 7-mm pupil assumed for such guidelines, this is about $10^{5.6}$ td. Applying these guidelines to the spectrum of CIE (Commission International de l'Éclairage) illuminant A (incandescent) yields a threshold limiting value of $10^{8.2}$ td for a source subtending 20°, viewed for 10 s. For other sizes and shorter durations, the threshold is inversely proportional to visual angle and inversely proportional to the 3/4th power of the duration.

The phototoxicity of short wavelengths is greater than that of longer wavelengths, and so separate guidelines apply to short wavelengths; if the source is rich in short wavelengths, the necessary computations (American Conference of Governmental Industrial Hygienists, 1995) should be done. For CIE illuminant A, the risk of thermal damage exceeds that for phototoxicity unless the source is viewed for 20 min or longer. The threshold limiting value of such incandescent light is $10^{7.1}$ td if viewed for 10^4 s, and is inversely proportional to duration for shorter durations.

On the assumption that the viewer would not be fixating the source steadily for the entire duration, the guidelines for short wavelength light are greatly relaxed if the source is smaller than 11 mrad (38 arc min). However, this assumption typically is inapplicable to a vision experiment.

For comparison, I found the luminance of a 48-W, tungsten ribbon-filament source to be about $10^{6.3}$ cd m^{-2}, and the highest luminance from an incandescent source (unless overdriven) is about $10^{7.5}$ cd m^{-2} for a 1200-W projection lamp (*IES lighting handbook*, 1972). If a 2-mm pupil is used to view such a source, these correspond, respectively, to $10^{6.8}$ and $10^{8.0}$ td. (Also, see Table 1.4.)

Standards for safe exposure to laser radiation are somewhat lower than those for broad-band sources and can be found in the American National Standard Z136.1–1993 (American National Standards Institute, 1993). Sources rich in ultraviolet light, of course, should be filtered (e.g. Kodak no.2B or one of the other no.2 filters) to remove it (see also Chapter 4).

1.12 The photometric system

As light is a form of energy, for many purposes it is best described by the terms and units that apply to energy: i.e. by the radiometric system. However, with the advent of artificial light came the problem of deciding how much light was enough. Engineers had to have a way to quantify and specify illumination, and radiometry was not suitable owing to the selective spectral sensitivity of the eye. The practical question, then and now, is how much light is necessary for various visual tasks? The photometric system has evolved to provide the units in which the answer must be couched. It has been successful because it correlates closely with performance of the most demanding visual tasks, and because it is additive in the sense that one can add light of varying wavelength algebraically. It works because the operations by which light is measured for the photometric system excludes the contribution of S cones, which contribute little to performance and enter into non-additive interactions with the M and L cones.

So the photometric system is simply a practical solution to an engineering problem, and it does not necessarily have theoretical or scientific significance. It describes the properties of a statistical fiction called the *standard observer*. Stimuli that are equivalent in the photometric system are equivalent to the standard observer under certain prescribed conditions. Any given observer in an experiment may differ from the standard observer, and even one identical to the standard observer might respond differently to photometrically equivalent stimuli in a given experimental task.

1.12.1 Concepts

1.12.1.1 Luminous flux

Conceptually, the fundamental unit of the photometric system is the *lumen*, a photometric measure of luminous flux analogous to the radiometric measure of radiant flux, the watt. One lumen is defined as the luminous flux of 1/683 W of monochromatic radiation at a frequency of 540 THz (a wavelength of 555 nm in a vacuum).

The ratio of the number of lumens to the number of watts emitted by a monochromatic source is the *luminous efficacy* of the wavelength of the source. The numbers formed by dividing the luminous efficacies at each wavelength by the luminous efficacy of the most efficacious wavelength are the *luminous efficiencies* of those wavelengths (sometimes called luminosity coefficients), V'_λ and V_λ (Table 1.3). V'_λ applies to scotopic (rod) vision, and V_λ applies to photopic (cone) vision. The photometry of mesopic vision has not been worked out.

If one knows the spectral emission curve of a source, P_λ, one can compute the photometric counterpart of any radiometric term (see Wyszecki and Stiles, 1982) by

Table 1.3 The relative sensitivity of the eye to wavelengths at scotopic intensities (V'_λ coefficients) and at photopic intensities (V_λ coefficients) for the CIE standard observer

Wavelength (λ)	Luminosity (V'_λ)	Coefficients (V_λ)
380	0.001	0.000
390	0.002	0.000
400	0.009	0.000
410	0.035	0.001
420	0.097	0.004
430	0.200	0.012
440	0.328	0.023
450	0.455	0.038
460	0.567	0.060
470	0.676	0.091
480	0.793	0.139
490	0.904	0.208
500	0.982	0.323
510	0.997	0.503
520	0.935	0.710
530	0.811	0.862
540	0.650	0.954
550	0.481	0.995
560	0.329	0.995
570	0.208	0.952
580	0.121	0.870
590	0.090	0.757
600	0.033	0.631
610	0.106	0.503
620	0.007	0.381
630	0.005	0.265
640	0.003	0.175
650	0.001	0.107
660	0.001	0.061
670	0.000	0.032
680	0.000	0.017
690	0.000	0.008
700	0.000	0.004
710	0.000	0.002
720	0.000	0.001

summing the products of the emission at each wavelength and the luminous efficacy at that wavelength. Thus, if P_λ is in watts per unit wavelength, *luminous flux*, F, in photopic lumens, is:

$$F = k_m \int_0^\infty P_\lambda V_\lambda \, d\lambda ,$$

where $k_m = 683$ lm W^{-1}; if the conditions are scotopic, V'_λ is used, and k'_m, with a value of 1700 scotopic lm W^{-1}, is used in place of k_m to give F in scotopic lumens.

Measuring the emission spectrum directly is, of course, the only way to be sure of it, but for some purposes one can use the nominal spectral emission curve supplied by the manufacturer of the source, or the approximate luminous output of a particular kind of source can be computed on the basis of general knowledge of its luminous efficiency. For example, the emission spectrum and the overall luminous efficiency of most conventional sources can be obtained from the *IES lighting handbook* (Kaufman and Christensen, 1972), and the data one is most likely to need are also available in Wyszecki and Stiles (1982).

1.12.1.2 Luminous intensity

Luminous intensity refers only to a point source. Any source that subtends less than one minute of arc at the eye can be considered a point source because it casts approximately the same image on the retina as a point source does. For the purpose of computing the illuminance produced on a surface by a particular source, the source can be treated as a point source if its distance from the surface is 10 times the length of the longest perpendicular projection of the source on the surface.

The luminous intensity is the luminous flux radiated by a point source in a particular direction. It is measured in lumens per steradian ($\mathrm{lm\ sr^{-1}}$), which equals the number of candelas, cd; i.e. $1\ \mathrm{lm\ sr^{-1}} = 1\ \mathrm{cd}$. The ratio, lumens per steradian, does not change as the size of the solid angle, ω, decreases. Thus:

$$I = \frac{\mathrm{d}F}{\mathrm{d}\omega}.$$

1.12.1.3 Illuminance

Illuminance, E, is luminous flux density. It specifies the luminous flux passing through a unit area of a plane and is used to specify light falling on a surface; this can be remembered by the similarity of the word *illuminance* to the word *illumination*. The units of illuminance are lumens per square metre ($\mathrm{lm\ m^{-2}}$), which is identical to *lux*. Flux density also does not change as the area considered (A) becomes smaller and smaller. Then:

$$E = \frac{\mathrm{d}F}{\mathrm{d}A}.$$

An extensive list of recommended illuminances for different tasks and environments is provided by the *IES lighting handbook* (Kaufman and Christensen, 1972).

1.12.1.4 Luminance

If the surface illuminated reflects or transmits the incident light, or if it emits light itself, it is *luminous*, and hence it has a certain *luminance*. As mentioned above (Section 1.4.9.1), reflected light varies with the angle of reflection, and so luminance is defined as the flux density emitted in a particular direction. Just as luminous flux per unit angle does not vary as the angle decreases, so the flux density per unit angle does not vary as the angle decreases, and luminance, L, can be given as:

$$L = \frac{dE}{d\omega}.$$

But:

$$E = \frac{dF}{dA}.$$

Then:

$$L = \frac{d(dF/dA)}{d\omega} = \frac{d^2 F}{dA\, d\omega}.$$

Luminance is given in lumens per steradian per square metre (lm sr^{-1} m^{-2}), which is the same as candelas per square metre (cd m^{-2}). To convert to millilamberts (mL), a unit still in use in the United States of America, divide cd m^{-2} by 3.183; i.e.:

$$3.183\ \text{cd m}^{-2} = 1\ \text{mL}.$$

To provide perspective on the scale of luminance, in Table 1.4 I have supplied typical luminances of a variety of surfaces under various conditions.

Table 1.4 Luminances of various surfaces under varying conditions. Those not measured by the author are inferred on the basis of data supplied in the references indicated by the superscripts

Condition	Log cd m^{-2}
Sun's surface	9.5[1]
Projector filament	7.5[2]
Damage: large field, 10 s	6.6[3]
Tungsten-ribbon filament	6.3
Damage: large field, 10 000 s	5.5[3]
Cloud 10° from sun	5.0
White paper in sunlight	4.6
Definitely safe	4.0[3]
Cloud 90° from sun	4.0
White paper in shade	3.4
Blue sky	3.4
Good interior lighting	2.5[4]
Rod saturation	2.5[5]
Typical office desktop (white)	2.1
Computer monitor	1.8
Moderate interior light	1.5[4]
Reading	1.5[1]
Feeble interior lighting	0.5[4]
Lower end of mesopic range	– 1.0[6]
White paper in full moonlight	– 0.7[4]
White paper in starlight	– 3.5[1, 4]
Absolute threshold	– 6.1[7]

1, Riggs, 1965; 2, *IES lighting handbook*, 1972; 3, American Conference of Government Industrial Hygienists, 1995; 4, LeGrand, 1968; 5, Aguilar and Stiles, 1954; 6, Chen and Makous, 1989; 7, Denton and Pirenne, 1954.

1.12.1.5 Troland value

Troland value, T, is a variable used to specify retinal illuminance without assumptions about absorption by the ocular media; the unit of measurement is the *troland*, usually abbreviated *td.* (Troland *value* is to *trolands* as power is to watts.) One troland is equal to the retinal illuminance that would be produced by a surface of 1 cd m^{-2} viewed through a pupil 1 mm^2 in area if there were no losses within the eye. Consequently, to convert from luminance to trolands, td, multiply the luminance, in candelas per square metre, by 10^{-6} times the area of the entrance pupil, A_p, *if the pupil area is measured in square metres* (to be consistent with the mks system):

$$T = 10^{-6} L\, A_p \text{td}.$$

Troland value is related to true retinal illuminance, E_r, as follows. The illuminance at the pupil, E_p, is the luminance of the source times its area, divided by the square of its distance from the surface:

$$E_p = \frac{A_s L}{r^2} \text{ lm m}^{-2},$$

where A_s is the area of the source and r is its distance from the eye; the light flux that enters the eye, F_p, is the illuminance at the pupil times the area of the pupillary aperture (in square *metres*):

$$F_p = E_p A_p = \frac{A_s L A_p}{r^2} \text{ lm};$$

and neglecting losses in the ocular media, the retinal illuminance is equal to the flux divided by the area of the retina, A_r, over which it is distributed:

$$E_r = \frac{F_p}{A_r} = \frac{A_s L A_p}{A_r r^2} \text{ lm m}^{-2}.$$

We assume for present purposes that the solid angle, ω_s, subtended by the source equals the angle subtended by the image on the retina, A_r; i.e. assuming the image distance to the retina is 0.01668 m:

$$\omega = \frac{A_s}{r^2} = \frac{A_r}{0.01668^2}.$$

Substituting equal angles yields:

$$E_r = 0.01668^{-2} L\, A_p \text{ lm m}^{-2}.$$

Then, from the definition of troland value, above, the number of trolands equals 278 times the retinal illuminance in lm m^{-2} (lux):

$$T = 278\, E_r \text{ td}.$$

In Maxwellian systems, the flux passing through the pupil can easily be measured directly. Hence, from above:

$$T = \frac{F_p}{\omega}(10^6)\ \text{td.}$$

The use of troland values entails certain caveats. It represents neither retinal illuminance nor photometric brightness. The properties of the ocular media affect retinal illuminance but not trolands; and, owing to the Stiles–Crawford effect, the part of the pupil through which light enters affects its brightness but not trolands.

LeGrand (1968) and Moon and Spencer (1944) have suggested ways to correct for the Stiles–Crawford effect, but they are rarely used, if ever. A simpler approach that introduces no more than a 5% error, is to describe the relative effectiveness of a troland as a linear decrease with increasing area of pupil illuminated, beginning 0.5 mm from the centre of the Stiles–Crawford effect. That is, if the diameter of the cross-section of the pathway of light through the pupil, d_p, is 1 mm or more, then:

$$T_p = \frac{\pi}{4}L - s(d_p - 1);$$

where s is the slope of the decrease. From the norms established by Applegate and Lakshminarayanan (1993), one can determine that for 95% of a population of healthy young observers (2.5th to 97.5th percentile), s lies between 0.05 and 0.10. A complication is that the centre of the Stiles–Crawford effect varies among observers, with a standard deviation of 0.68 mm (Applegate and Lakshminarayanan, 1993). The subscript in T_p is used to limit the formula to photopic trolands, for there is no Stiles–Crawford effect under scotopic conditions (Crawford, 1937; Alpern *et al.*, 1983), so no such correction is required for scotopic trolands.

Note also that the use of trolands, whether one corrects for the Stiles–Crawford effect or not, assumes that the effects of light passing through different parts of the pupil sum linearly, which may not be true (Enoch, 1958).

1.12.2 Quantitative relationships

1.12.2.1 Luminous intensity and illuminance

To compute illuminance from luminous output of the source, first compute luminous intensity; that is:

$$I = \frac{F}{4\pi};$$

which derives from the fact that 1 candle emits 4π lumens.

Luminous intensity is specified in lumens per unit solid angle. Then, by definition, a source that emits n lumens per steradian, *il*luminates n lumens per unit perpendicular area at one unit distance (as one steradian subtends one square *metre* at a distance of a *metre*, or one square foot at a distance of one foot); hence, the illuminance of 1 cd (1 lm sr^{-1}) at 1 m is 1 lm m^{-2} (1 lux). The area, of course, increases in proportion to the square of the distance, and as a fixed luminous flux is spread over a larger area, the illuminance follows the inverse square law. Therefore, to compute illuminance from luminous intensity, divide the luminous intensity by the square of the distance, r^2.

$$E = \frac{I}{r^2}.$$

If r is in metres, illuminance is in lumens per square metre (lm m^{-2} or lux). If r is in feet, illuminance is in foot-candles (fcd). Since:

$$1\ \mathrm{m}^2 = 10.76\ \mathrm{ft}^2$$

$$1\ \mathrm{fcd} = 10.76\ \mathrm{lm\ m}^{-2}\ \mathrm{(lux)}.$$

To compute illuminance from luminance, multiply the luminance by the area of the source, A_s, and divide by the square of the distance, r, to the illuminated surface. Thus:

$$E = \frac{A_s L}{r^2}.$$

If the distance is less than ten times the diameter (D) of the source, i.e. if $10\,D < r$, then integrate across the area:

$$E = \int_0^{A_s} \frac{L_s}{r^2}\,\mathrm{d}A.$$

In any of these calculations of illuminance, if the surface is obliquely illuminated, the illuminance E_q is

$$E_q = E \cos \theta,$$

where E is the illuminance computed as described above, and θ is the incident angle of the light.

1.12.2.2 Luminance

To compute the luminance of an illuminated surface, merely multiply the illuminance by the reflection coefficient, R:

$$L = RE,$$

where L is in candelas per square metre (cd m^{-2}) or footlamberts (fL), and E is in lumen per square metre (lm m^{-2}) or foot-candles (fcd), respectively. To convert from footlamberts to millilamberts (mL), multiply the number of footlamberts by 1.076 (i.e. 1 fL = 1.076 mL).

In practice, many of the factors described here vary with other variables. For example, the reflection coefficient typically varies with incident angle and wavelength. Thus, it often is specified as $R_{\theta\lambda}$.

If one has measurements or data in units other than those used here, first convert to these before performing computations. Tables of conversion factors are widely available (e.g. Judd, 1951; Wyszecki and Stiles, 1982). Use LeGrand (1968) to check units seen in the older literature that are seldom used today.

1.12.3 Calibrations

1.12.3.1 General considerations

As elsewhere in this chapter, I take the space here to state the obvious for the sake of emphasis, namely: calibrations are the only way to know what stimuli have been or will be presented; manufacturers' specifications cannot usually be taken at face value; the cost (in time and other resources) of recalibration is balanced against the probability and cost of repeating observations made after the last calibration; the accuracy of calibration required is inversely proportional to the experimental error; and calibrations of relative stimulus values require greater accuracy than those of absolute values (perhaps a corollary of the previous point).

On the frequency of recalibrating the calibrating instruments themselves, one usually is advised to recalibrate annually, but the instrument cannot be used when it is away being calibrated (some experiments require daily measurement of the stimulus). Many investigators find drift over 2 to 3 years tolerable, but if so it is valuable to have a colleague with equipment for making the same calibrations, with whom one can stagger calibrations and make cross-comparisons.

Most light meters nowadays are wonderfully linear, but their linearity is nevertheless worth checking, especially at the top of their ranges. This can be done by ensuring that the ratio of measurements of a beam made with and without a given filter is constant as the intensity of the beam is otherwise varied.

1.12.3.2 Luminance

Luminance is measured simply by pointing a telescopic, or 'spot,' meter at the location to be measured and reading the output. A graticule in the field of view indicates the area over which the measurement is averaged. Keep in mind that luminance depends on the direction from which it is measured, and on the direction of any incident light.

Such meters are widely available (e.g. International Light, Gamma, Minolta, Optronics Laboratories, Tektronix). Most meters scale automatically to the level being measured, but they do vary in sensitivity, angular size of the area that can be measured, closest distance at which measurements can be made (which, with angular spot size, determines the smallest stimulus area that can be measured), other options such as the capacity to integrate light flashes, and, of course, cost. Acceptable meters start at around \$1500 (e.g. Photo Research), pass through a middle range from just under \$3000–6000 (e.g. Minolta), up to \$30 000 for cooled detectors that perform spatial and spectral analysis under computer control (e.g. Photo Research).

1.12.3.3 Illuminance and flux

Illuminance can be measured by measuring the luminance of a surface with calibrated reflectance. The reflection factor of a surface freshly coated with a $BaSO_4$

paint (e.g. one of the Eastman Whites discussed in Section 1.4.9.1), is close enough to unity that for most purposes it may not require calibration.

Alternatively, one can use a meter that measures illuminance directly (e.g. Photo Research, Photonics, UDT). If the incident beam is larger than the entrance aperture of the meter (typically a circular area equal to 1 cm^2), the meter measures illuminance; if the incident beam is smaller, it measures light flux (see Section 1.12.3.5). Flux measurements, of course, require uniform sensitivity across the entrance aperture.

1.12.3.4 Light flashes

Photodetectors trade speed for sensitivity, and those in photometers typically cannot follow the time-course of a typical shutter action. To measure the time-course of a rapidly changing stimulus, one needs a fast photodetector (check UDT or almost any catalogue), an oscilloscope, and a lot of light. This means measuring close to the shutter or other modulator, without filters and perhaps with some field stops removed. Extrapolating to the conditions of the experiment usually means depending on filter calibrations.

On the other hand, one may simply want to know the integrated light in a flash. Some meters (e.g. Photo Research, Photonics) that measure either luminance or illuminance are equipped to integrate over time.

1.12.3.5 Maxwellian systems

Before the availability of reliable and precise photometers, elaborate (Westheimer, 1966) and not always simple or accurate (Buck and Makous, 1982) procedures were required to measure retinal illuminance in Maxwellian systems. Nowadays there is a clear and simple way to do it: measure the total light flux in the image that passes through the pupil of the eye, and divide by the solid angle subtended by the source (see also Section 1.12.1.5).

Photodetectors come in a profusion of types and characteristics (Zalewski, 1995), and for most vision experiments nearly any would serve. Attractive features of devices based on silicon photodetectors (e.g. UDT's QED 100 and 200) are their total internal quantum efficiency and capacity for self-calibration (Zalewski and Duda, 1983; Zalewski, 1995). When arranged as a light trap, so that light reflected from one detector is directed to another (and so on), well over 99% of the incoming light is absorbed. These features combine to make the detectors spectrally flat and accurate to 0.1% from about 400 to 900 nm. Converting from radiometric to photometric measurements requires insertion of a filter that mimics either the photopic spectral sensitivity curve (part of any photometer) or the scotopic spectral sensitivity curve (available from International Light Ltd and Photo Research, for example).

Note, however, that if one is measuring narrow band stimuli, substantial errors can occur. For example, most detectors are accurate within 2%. This means that their sensitivity at each wavelength, λ, is within ± 0.02 of V_λ, but at a wavelength of 700 nm, for example, V_λ is 0.004, and an error of ± 0.02 is a 500% error. Note also that the CIE photometric system is inappropriate for nearly all animals.

A final word of advice: be sure to calibrate the source before starting the experiment, lest it die before it is calibrated.

Acknowledgements

I thank John Robson and Roger Carpenter for their invaluable contributions to this chapter; Xiafeng Qi for comments on an earlier draft; Peter Bex, Alan Russell, David Sliney, Robert Steinman, and Stephen Burgart for technical advice; and, especially, Teresa Williams for her careful reading of the manuscript, for her help in ferreting out many specific facts, and for her many contributions to my education on equipment and its sources. This work was supported by US Public Health Service grants EY-4885 and EY-1319.

References

Aguilar, M. and Stiles, W. S. (1954). Saturation of the rod mechanism of the retina at high levels of stimulation. *Optica Acta*, **1**, 59–65.

Alpern, M., Ching, C. C., and Kitahara, K. (1983). The directional sensitivity of retinal rods. *Journal of Physiology* (London), **343**, 577–92.

American Conference of Governmental Industrial Hygienists (1995). *Threshold limit values (TLVs) for chemical substances and physical agents and biological exposure indices (BEIs)*. American Conference of Governmental Industrial Hygienists, Cincinnati, OH.

American National Standards Institute (1993). *American national standard for safe use of lasers*. The Laser Institute of America, Orlando, FL.

Applegate, R. A. and Lakshminarayanan, V. (1993). Parametric representation of Stiles–Crawford functions: normal variation of peak location and directionality. *Journal of the Optical Society of America A*, **10**, 1611–23.

Bass, M., van Stryland, E. W., Williams, D. R., and Wolfe, W. L. (ed.). (1995). *Handbook of optics*. McGraw-Hill, New York.

Bonnardel, V., Bellemare, H., and Mollon, J. D. (1996). Measurements of human sensitivity to comb-filtered spectra. *Vision Research*, **36**, 2713–20.

Boreman, G. D. (1995). Transfer function techniques. In *Handbook of optics*. Vol. II: *Devices, measurements, and properties* (2nd edn) (ed. M. Bass, E. W. van Stryland, D. R. Williams, and W. L. Wolfe), pp. 32.1–32.10. McGraw-Hill, New York.

Born, M. and Wolf, E. (1970). *Principles of optics: electromagnetic theory of propagation interference and diffraction of light* (4th edn). Pergamon, New York.

Boynton, R. M. (1966). Vision. In *Experimental methods and instrumentation in psychology* (ed. J. B. Sidowski), pp. 273–330. McGraw-Hill, New York.

Boynton, R. M., Sturr, J. F., and Ikeda, M. (1961). Study of flicker by increment threshold technique. *Journal of the Optical Society of America*, **51**, 196–201.

Buck, S. L. and Makous, W. (1982). Calibrating Maxwellian-view optical systems. *Journal of the Optical Society of America*, **72**, 960–2.

Burns, S. A. and Webb, R. H. (1995). Optical generation of the visual stimulus. In *Handbook of optics*. Vol. 1: *Fundamental, techniques, and design* (2nd edn) (ed. M. Bass, E. W. van Stryland, D. R. Williams, and W. L. Wolfe), pp. 28.1–28.29. McGraw-Hill, New York.

Campbell, F. W. (1958). A retinal acuity direction effect. *Journal of Physiology* (London), **144**, 25P-26P.

Campbell, F. W. and Green, D. G. (1965). Optical and retinal factors affecting visual resolution. *Journal of Physiology* (London), **181**, 576–93.

Campbell, F. W. and Gubisch, R. W. (1966). Optical quality of the human eye. *Journal of Phisiolology* (London), **186**, 558–78.

Chen, B. and Makous, W. (1989). Light capture by human cones. *Journal of Physiology* (London), **414**, 89–109.

Conrady, A. E. (1960). *Applied optics and optical design*. Dover, New York.

Crawford, B. H. (1937). The luminous efficiency of light entering the eye pupil at different points and its relation to brightness threshold measurements. *Proceedings of the Royal Society* (London), **B124**, 81–96.

Crawford, B. H. and Granger, G. W. (1968). Some worked examples. In *Techniques of photo-stimulation in biology* (ed. B. H. Crawford, G. W. Granger, and R. A. Weale), pp. 235–71. Elsevier, New York.

Denton, E. J. and Pirenne, M. H. (1954). The absolute sensitivity and functional stability of the human eye. *Journal of Physiology* (London), **123**, 417–42.

De Vries, H., Spoor, A., and Jielof, R. (1953). Properties of the eye with respect to polarized light. *Physica*, **19**, 419–32.

Ditchburn, R. W. (1963). *Light* (2nd edn). Wiley, New York.

Enoch, J. M. (1958). Summated response of the retina to light entering different parts of the pupil. *Journal of the Optical Society of America*, **48**, 392–405.

Enoch, J. M. and Hope, G. M. (1972). An analysis of retinal receptor orientation. III. Results of initial psychophysical tests. *Investigative Ophthalmology*, **11**, 765–82.

Enoch, J. M. and Laties, A. M. (1971). An analysis of retinal receptor orientation. II. Predictions for psychophysical tests. *Investigative Ophthalmology*, **10**, 959–70.

Galvin, S. J. and Williams, D. R. (1992). No aliasing at edges in normal viewing. *Vision Research*, **32**, 2251–9.

Hecht, E. (1987). *Optics* (2nd edn). Addison-Wesley, New York.

Hofstetter, H. W. (1972). Interpupillary distances in adult populations. *Journal of the American Optometric Association*, **43**, 1151–5.

Jameson, D. and Hurvich L. M. (1967). Fixation-light bias: an unwanted by-product of fixation control. *Vision Research*, **7**, 805–9.

Johnson, B. K. (1960). *Optics and optical instruments*. Dover, New York.

Judd, D. B. (1951). Basic correlates of the visual stimulus. In *Handbook of experimental psychology* (ed. S. S. Stevens), pp. 811–67. Wiley, New York.

Kaufman, J. E. and Christensen, J. F. (ed.) (1972). *IES lighting handbook* (5th edn). Illuminating Engineering Society, New York.

Kurtz, C. N. (1972). Transmittance characteristics of surface diffusers and the design of nearly band-limited binary diffusers. *Journal of the Optical Society of America*, **62**, 982–9.

Lamberts, R. L. (1963). The production and use of variable-transmittance sinusoidal test objects. *Applied Optics*, **2**, 273–6.

LeGrand, Y. (1935). Sur la mesure de l'acuité visuelle au moyen de franges d'interférence. *Comptes rendus de l'Academie des sciences, Paris*, **200**, 490–1.

LeGrand, Y. (1968). *Light, colour and vision* (2nd edn). Chapman and Hall, London.

Leibowitz, H. W. and Owens, D. A. (1978). New evidence for the intermediate position of relaxed accommodation. *Documenta Ophthalmologica*, **46**, 133–47.

MacLeod, D. I. A. and He, S. (1993). Visible flicker from invisible patterns. *Nature*, **361**, 256–8.

MacLeod, D. I. A., Williams, D. R., and Makous, W. (1992). A visual nonlinearity fed by single cones. *Vision Research*, **32**, 347–63.

Makous, W. (1968). A transient Stiles–Crawford effect. *Vision Research*, **38**, 1271–84.

Makous, W. (1974). Optimal patterns for alignment. *Applied Optics*, **13**, 659–64

Makous, W. (1977) Some functional properties of visual receptors and their optical implications. *Journal of the Optical Society of America*, **67**, 1362.

Makous, W. and Sanders, R. K. (1978). Suppressive interactions between fused patterns. In *Visual psychophysics and physiology* (ed. J. Armington, J. Krauskopf, and B. R. Wooten), pp. 167–79. Academic Press, New York.

Mandler, M. B. (1984). Temporal frequency discrimination above threshold. *Vision Research*, **24**, 1873–80.

Moon, P. and Spencer, D. E. (1944). On the Stiles–Crawford effect. *Journal of the Optical Society of America*, **34**, 319–29.

Pompea, S. M. and Breault, R. P. (1995). Black surfaces for optical systems. In *Handbook of optics*. Vol. II: *Devices, measurements, and properties* (2nd edn) (ed. M. Bass, E. W. van Stryland, D. R. Williams, and W. L. Wolfe), pp. 37.1–37.70. McGraw-Hill, New York.

Riggs, L. A. (1965). Light as a stimulus for vision. In *Vision and visual perception* (ed. C. H. Graham), pp. 1–38. Wiley, New York.

Ruddock, K. H. (1968). Control of the time pattern of the stimulus. In *Techniques of photo-stimulation in biology* (ed. B. H. Crawford, G. W. Granger, and R. A. Weale), pp. 109–43. Elsevier, New York.

Sears, F. W. (1949). *Optics* (3rd edn). Addison Wesley, Menlo Park, CA.

Stiles, W. S. and Crawford, B. F. (1933). The luminous efficiency of rays entering the eye pupil at different points. *Proceedings of the Royal Society B*, **112**, 428–50.

Sturr, J. F. and Teller, D. Y. (1973). Sensitization by annular surrounds: dichoptic properties. *Vision Research*, **13**, 909–18.

Troland, L. T. (1915). The theory and practice of the artificial pupil. *The Psychological Review*, **22**, 167–76.

Weast, R. C. and Astle, M. A. (ed.) (1981). *CRC handbook of chemistry and physics* (62nd edn) CRC Press, Boca Raton, FL.

Westheimer, G. (1966). The Maxwellian view. *Vision Research*, **6**, 669–82.

Westheimer, G. (1986). The eye as an optical instrument. In *Handbook of perception and human performance* (ed. K. R. Boff, L. Kaufman, and J. P. Thomas), pp. 4.1–4.20. Wiley, New York.

Williams, D. R. (1985). Aliasing in human foveal vision. *Vision Research*, **25**, 195–205.

Williams, D. R., Brainard, D. H., McMahon, M. J., and Navarro, R. (1994). Double-pass and interferometric measures of the quality of the eye. *Journal of the Optical Society of America A*, **11**, 3123–35.

Wyszecki, G. and Stiles, W. S. (1982). *Color science: concepts and methods, quantitative data and formulas* (2nd edn). Wiley, New York.

Zalewski, E. F. (1995). Radiometry and photometry. In *Handbook of optics*. Vol. II: *Devices, measurements, and properties* (2nd edn) (ed. M. Bass, E. W. van Stryland, D. R. Williams, and W. L. Wolfe), pp. 24.1–24.51. McGraw-Hill, New York.

Zalewski, E. F. and Duda, C. R. (1983). Silicon photodiode device with 100% external quantum efficiency. *Applied Optics*, **22**, 2867–73.

Note

1. An exception can occur when a coil filament is imaged on an aperture. Thermal expansion of the filament as it warms can move one of the turns of the coil out of the aperture and change the light flux by as much as 25%.

2

Light sources

JOHN G. ROBSON

2.1 Introduction

Although the foundations of vision science were laid down in the days when it was necessary to rely on the light from the luminous flame of a candle or an oil lamp for all experimental observations that could not be made using sunlight, the modern vision scientist almost universally relies on light from some form of electrically excited light source. While incandescent filament and open arc lamps have been available for well over a 100 years, the continuous development of electric lamps based on these methods of light generation, as well as the introduction of lamps based on many other modes of excitation, has made an enormous difference to the way in which vision research can be undertaken today. Although many instruments for examining the eye or measuring some aspect of its function may make use of light in their operation, in this chapter we shall mainly be concerned with light sources used for providing visual stimuli and directing our attention primarily towards vision in humans and other mammals. In this context we shall look at some of the basic principles that govern the choice of a light source for a particular experimental project, discuss some of the types of lamp whose characteristics make them particularly useful in vision research, and also consider aspects of the operation of these lamps. We shall also look at some of the ancillary equipment that may be used in conjunction with these lamps (e.g. light guides, integrating spheres, shutters) to provide a light source with some functionality not inherent in the lamp itself.

This chapter is intended to be a simple introduction to a diffuse field and not a treatise. Other sources of general information are listed at the end of the chapter. It has not been thought to be very useful to provide detailed information about particular devices as this is subject to rapid change, and up-to-date information will usually need to be obtained from manufacturers' publications. The World Wide Web has already become a good place to find such information, or at least to discover how to get it. Although it will sometimes be appropriate to purchase specialized, laboratory light sources from optical equipment suppliers (see Chapter 1), light sources and lamps that are intended for other applications will often be more appropriate (and certainly cheaper). It is hoped that this introduction will enable vision researchers to utilize such devices in their work.

2.1.1 Wavelength range

The eye is sensitive to radiation having a relatively narrow range of wavelengths. Although normal human vision is usually said to respond to radiation whose wavelength lies between about 400 nm and 750 nm (this range of wavelengths constituting 'light'), it is worth noting that the short wavelength limit in primates is set by absorption in the lens of the eye and not by failure of the retina to respond to 'ultraviolet' radiation (Stark and Tan, 1982). In other vertebrate species (excluding squirrels) the optical media of the eye only absorb significantly below about 300 nm and visual sensitivity extends well into the near ultraviolet region. Moreover, it is now becoming clear that ultraviolet sensitivity may result not only from the response at these wavelengths of the photopigments that primarily respond to 'visible light' but also in some species from activation of retinal photopigments specifically sensitive in the near-ultraviolet as well (Jacobs and Deegan, 1994). While retinal sensitivity to ultraviolet radiation may not be of much importance when studying human or primate vision, those working with other animals should be aware of the visual effectiveness of ultraviolet radiation that may be present in the lights they use as stimuli but which they themselves cannot see.

It should also be noted that ultraviolet radiation may be converted into visible light by fluorescence within the eye as well as externally, and that both ultraviolet and infrared components of a stimulus may be potentially damaging to the eye, though invisible. It is therefore always good practice, where necessary, to use appropriate filters to limit the range of wavelengths present in a visual stimulus to those that are specifically required. As a caveat to this it should be noted that light that is intended to mimic the normal daylight or sunlight illumination of coloured objects may only provide accurate colour rendering if it contains ultraviolet radiation comparable to that of the natural light, thereby exciting the same fluorescence.

2.2 Photometric considerations

One of the most important considerations in choosing a light source will be the provision of sufficient light for the purpose in hand. This will necessarily require a knowledge of the photometric characteristics of different light sources—approximate information about typical light sources will be given later as a useful rough guide. Here we will consider the various characteristics that may need to be taken into account.

2.2.1 Total light output

In some circumstances the most important characteristic of a lamp will be its total light output. This may well be the case when the lamp is required to provide a specified level of diffuse illumination of a target, and the spectral composition of the light does not need to be modified by filtering it. This is the usual case for general lighting requirements; and for lamps that are intended for this purpose the manufacturer will normally provide information about the total output in *lumens*, the standard photometric unit of light flux. Even if this information is not available for any particular lamp, it is usually possible to estimate the output simply from a knowledge of the kind of lamp and its electrical power input, since the efficacy with which electrical energy is converted into light (the *luminous efficacy* of the light source) is characteristic for each different type of lamp (Table 2.1). However, there are several possible complications that should be borne in mind.

Table 2.1 Light output of different lamps in both phototopic and scotopic units

Source	Approximate luminous efficacy (photopic lumen.watt $^{-1}$)	Typical luminance (photopic cd.m^{-2})	Photopic/ scotopic ratio
Sun's disc	–	10^9	–
Tungsten halogen lamp (3000K)	20	1.2×10^7	0.67
Fluorescent tube (white)	70	10^4	–
Xenon arc (150W)	20	2×10^8	0.43
LEDs (high intensity)			
450–650 nm peak	4–20	2×10^6–10^7	0.1–45
white (Nichia)	7	3×10^6	0.42
Electroluminescent panel (green, 400Hz)	–	70	–
Xenon flash tube	10–30 photopic lumen.s.J^{-1}	(depends on current density, duration, etc.)	0.35

First, there is the fact that there are really two different sets of photometric units, *photopic* and *scotopic*, which take into account the relative effectiveness of lights of different wavelengths in producing a visual sensation under dark-adapted and light-adapted conditions, respectively (or rather under conditions in which vision is mediated by rods and cones, respectively). Invariably, the output of commercial lamps is specified in *photopic* lumens (lm) and luminous efficacy in *photopic* lumens per watt (lm W^{-1}), though this is rarely spelled out. While photopic units will often be appropriate, there will be occasions when scotopic units are necessary. Calculation of the scotopic equivalence of a photopically specified light requires a knowledge of its complete spectral distribution (see Section 2.2.5); it is not sufficient to know the colour of a photopic light to calculate its scotopic equivalence. Table 2.1 shows the light output of some lamps in photopic units together with some indication of the ratios of photopic and scotopic efficacies.

Second, there is the possible confusion of a lamp's *luminous efficacy* (its luminous flux *output* divided by its electrical power *input*) with the *luminous efficacy* of the lamp's radiation (the luminous flux divided by the power of the light *radiated*). Both these quotients are expressed in units of lumens per watt and it may require judicious interpretation of the context to know which is intended. An ideal lamp would convert all its input power into radiation and both measures of luminous efficacy would then be equal. Real lamps are much less effective than this, the best converting no more than about half their input power into radiation and the worst as little as 1 or 2%. Moreover, only a small proportion of the total radiated power may fall in the range of wavelengths that are visually effective and the overall luminous efficacy may therefore be very low.

Third, it should always be remembered that the definitions of all photometric units depend upon the application of either the photopic or scotopic luminous efficiency functions that describe the visual effectiveness of monochromatic luminous flux at different wavelengths relative to that at the optimum wavelength (the CIE V_λ and V'_λ functions, Table 1.3). Although these standard functions are based on averages of empirical observations on normal human observers, in reality they do not exactly describe the visual performance of individual observers even under the particular conditions in which they were made, while even less do they correspond to their performance under other conditions or the performance of other species.

Anyway, assuming that the total light flux from a lamp is known and that it has been possible to make some estimate of the proportion of the radiated light available for illuminating a surface of known area, it will be possible to calculate at least the average density (light flux per unit area) of the light flux falling on the surface. This quantity is the *illuminance* of the surface, and if light flux is measured in lumens and area in square metres it has units that may be called *lux* ($1\ \text{lux} \equiv 1\ \text{lm m}^{-2}$).

2.2.2 Source luminance

While in some cases it may be the total light flux from a lamp that is important, in others it will be the luminance of the radiating surface that is critical. This will usually be so when it is necessary to illuminate a small aperture (e.g. the entrance slit of a monochromator) with an image of the source formed by a lens or mirror, and where the emitting surface is larger than the aperture to be illuminated.

The source luminance of practical lamps varies much more widely than their total light output or their luminous efficacies; the source luminances of a number of lamp types are given in Table 2.2. Like total light output, the luminance of a lamp's radiating surface is always normally specified in standard *photopic* luminance units, photopic candelas per square metre (cd m^{-2}, sometimes called *nits*) or, for very high-luminance lamps, in cd cm^{-2} (sometimes called *stilbs*). In general, the luminance of a radiating surface is dependent upon the direction, relative to the surface, in which it is measured. In many instances, however, emitting surfaces are roughly Lambertian, that is their luminance is independent of direction (especially for directions not too far from the normal to the surface), and this makes estimation of their light output relatively simple.

For a uniformly radiating surface that is Lambertian, the total light flux Φ (in lumens, lm), the luminance L (in cd m^{-2}), and the area of the surface S (in m^2) are related by $\Phi = \pi LS$. In the general case where the radiating surface does not have a uniform luminance, this relationship will only apply to infinitesimal elements of the surface, and total light output must be calculated by integration over the surface. However, in many cases of practical interest where only a small area of the radiating surface is being imaged, the simple formulation is adequate for preliminary planning purposes. The total luminous flux from an infinitesimal element of a surface is known as its *luminous emittance* (or *exitance*) and, like illuminance, is measured in lumens per square metre (lm m^{-2}).

2.2.3 Source intensity

When the dimensions of a light source are small compared with the distance at which the effect of the light is significant (i.e. when it can be considered to be a 'point source'), it is only the product of the area of the emitting surface and its luminance that is important. This is the *luminous intensity* of the source and is measured in candelas (cd). A source of intensity I cd that radiates equally in all directions (into a total solid angle of 4π steradians) provides a total light flux of $4\pi I$ lumens. In many lamps in which the directional distribution of the radiated light is deliberately modified by the geometry of the radiating surface or by built-in mirrors or lenses so as to increase the intensity over some reduced range of angles, the manufacturer will usually specify the luminous intensity of the light (in *photopic* candelas) in the direction in which it is greatest (effectively defining the principal axis of the lamp), as well as either the angular width of the radiated beam or some more general indication of the way intensity depends upon direction relative to the axis.

The total amount of light that can be collected by a lens from a small source is set by the F number of the lens (F = focal length/diameter) as this determines the solid angle over which light can enter the lens. It must be appreciated, however, that the total light flux collected in any particular case will depend upon the actual distance of the source from the lens (this being set by the magnification that is required) as well as upon the angular distribution of source intensity. As a rough approximation it may be helpful to note that a lens of radius r at a distance R from an isotropic source of intensity I cd will collect about $Ir^2/4R^2$ lumens, and that for a lens set to produce a collimated beam of light (i.e. placed at a distance from the source equal to its focal length) the light flux in the beam will be about $I/16F^2$ lm. But it should also be noted that even if the source is truly isotropic (i.e. the intensity is independent of the point of entry into the lens), the illumination across the exiting beam will not be uniform. Similarly, in a Maxwellian view arrangement where the source is imaged in the pupil of the eye (see Section 1.2.2.1), the retinal illuminance across the visual image will also not be constant. The cosine fall-off in retinal illumination towards the periphery of the image of the Maxwellian lens, that results simply from the geometry of the illumination by the source, will be exacerbated not only by any general shaping of the angular distribution of intensity that further reduces the illumination of more peripheral parts of the lens but also by localized irregularities that may result from non-uniformities in the glass envelope of the lamp.

2.2.4 Retinal illumination

Vision research is normally concerned with the effects of light that are initiated by its action on photoreceptors in the retina, and thus it will often be necessary to take into account the strength of the stimulus that is available at this level. In the simplest case, i.e. direct viewing of an external illuminated surface, the illuminance of the retina will obviously directly depend not only upon the luminance of the target surface but also upon the area of the pupil. This has led to the introduction of the troland (named after Leonard Troland) as a measure of 'retinal illumination', 1 troland (1 td) being the 'retinal illumination' that is established when a surface with a luminance of 1 cd m^{-2} is viewed through a pupil with an area of 1 mm^2. This definition gives rise to both photopic and scotopic trolands according to the luminance measure employed. Even when visual stimulation is not produced in such a direct way, particularly, indeed, when it is not (e.g. in a Maxwellian view), it is desirable to express the effective retinal illumination produced by the stimulus as a troland value.

The actual retinal illuminance corresponding to 1 td depends, of course, not only upon the primary variables of luminance and pupil area but also upon various parameters of the particular eye, notably its dimensions and the losses in transmission to the retina. Moreover, the visual effectiveness of a stimulus will also depend upon other factors, including: the collecting area of the photoreceptors; the angular dependence of the entry of light into the photoreceptors (the Stiles–Crawford effect) which reflects the position of the entry beam in the pupil; the proportion of light absorbed by the photopigment; and the quantum efficiency of the isomerization of the photopigment that is required to initiate the visual process. The use of the troland goes some way towards simplifying the comparison of observations made with different pupil sizes or with different optical arrangements; but caution must always be exercised when making such comparisons, particularly between different individuals and even more so between species. While there is little dependence of rod sensitivity on the position within the pupil of light entering the eye, the Stiles–Crawford effect for cones is greater, and its possible significance should be considered whenever the entrance pupil is more than a millimetre or two in diameter or is not very well stabilized with respect to the eye.

2.2.5 Radiometric vs. photometric quantities (see also Section 1.12)

Even though our principal interest may be in light as a visual stimulus and we shall therefore normally wish to assess light sources in terms of their visual effectiveness (i.e. in *photometric* terms), it is inevitable that we shall sometimes have to consider the radiation from these sources in basic physical terms (i.e. *radiometrically*). This will usually be the case when we need to determine the visual effects of stimuli in circumstances in which the normal photopic spectral sensitivity is either not a sufficient description (i.e. when considering colour vision, but see Chapter 4 for a full discussion of this) or not an appropriate one (e.g. at low light levels, or in some non-human species), when we examine visual processes operating at light levels at which quantal (photon) considerations are necessary and whenever we become involved with instrumental measurements of light intensity.

Light sources are characterized radiometrically in ways that are equivalent to those already discussed photometrically. The basic radiometric measures are, however, effectively simpler than their photometric equivalents because they do not involve anything comparable to the visual luminous efficiency functions, radiant flux being measured directly in the normal units of power (i.e. in watts, W). But even in physical systems it is generally necessary to take into account the relative efficacy of radiations of different wavelengths in producing whatever effect is of particular interest, and it is therefore usual for radiometric measurements to be determined either over a specified range of wavelengths (within which the efficacy may be more-or-less constant) or at a number of different wavelengths (when they must relate to some specified small range of wavelengths around each quoted value). In this latter case a set of discrete measurements at different wavelengths provides an estimate of what is really a continuous *spectral density function*. Thus the flux density of radiation emitted from a surface (its *spectral radiant exitance*), or incident upon it (*spectral irradiance*), may be measured in W m^{-2} nm^{-1}, with *spectral radiance* being measured in W sr^{-1} nm^{-1}.

Although photometric measurements are still reported using many different units (lumens, candelas, lux, and others) these are all directly related, and only one basic unit is really necessary. Historically, photometric calibrations were based upon a physical standard of luminous intensity (a standard candle) and this is reflected in the adoption of the *candela* as the SI (Système Internationale des unités) basic photometric unit. However, the definitions of both the candela and the lumen are now both based upon radiometric measurements of spectral radiant flux. Thus one candela is defined as the luminous intensity of a source that has a radiant intensity of 1/683 watt per steradian and emits monochromatic light whose frequency is 540×10^{12} Hz (corresponding to a wavelength in air of 555.016 nm), while 1 lumen can be defined as the luminous flux provided by a radiant flux of 1/683 watt of this same light. These definitions are independent of the spectral luminous efficiency functions of human vision (the CIE V_λ and V'_λ functions), and can be used equally well as the basis for both photopic and scotopic measurements relating to monochromatic lights at other wavelengths as well as to polychromatic lights by employing V_λ and V'_λ as weighting functions. Because the wavelength chosen for the definition of the lumen is close to the peak wavelength of the photopic efficiency function (at 555 nm), the peak efficacy for photopic vision is about 683 lumens per watt. Since the scotopic luminous efficacy is also by definition 683 lm W^{-1} at 555.016 nm, it follows from the shape of the scotopic luminous efficiency function V'_λ that the peak efficacy for scotopic vision (at 507 nm) is about 1700 lm W^{-1}.

The complete luminous efficacy functions K_λ and K'_λ, which relate luminous and radiant fluxes at different wavelengths for photopic and scotopic vision, are obtained by multiplying the V_λ and V'_λ functions by 683 and 1700, respectively, and are shown in Figs 2.1(a), (b). The ratio of the areas under these curves is 4.7, this being the ratio of the scotopic to photopic intensities of a white light with an equal energy spectrum.

2.2.6 Photon fluxes

Neither photometric nor radiometric measures really provide the most relevant information about visual stimuli. Since vision depends upon a photochemical rather than a heating effect of light, the most appropriate measure will usually be based upon photon fluxes rather than power. To convert from photometric and radiometric measures to photon-based ones it is sufficient to know that a radiant flux of 1 watt of a monochromatic light with a wavelength in air of λ nm corresponds to a flux of $\lambda \times 5.04 \times 10^{15}$ photons per second. On this basis 1 lumen (both photopic and scotopic) is provided by a flux of 4.38×10^{14} photons per second at 555 nm and 1 scotopic lumen by a flux of 1.503×10^{14} photons per second at 507 nm. Photon fluxes corresponding to 1 lumen at other wavelengths can be calculated by adjusting the appropriate relative luminous efficiency at that wavelength (Table 1.3).

2.3 Lamp types

2.3.1 Incandescent lamps

Light is emitted from any body whose temperature is sufficiently high (incandescence). However, in most incandescent electric lamps the light-emitting surface is a

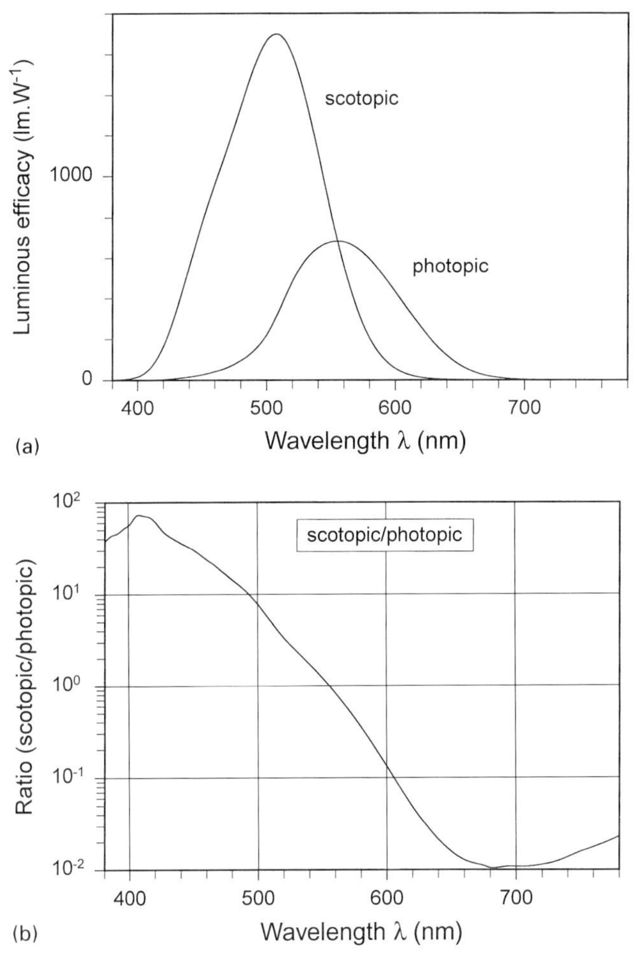

Fig. 2.1 (a) CIE luminous efficacy functions. (b) Ratio of CIE luminous efficacy functions (scotopic/photopic).

metallic wire (usually tungsten) raised to a high temperature by the passage of a current. The radiation from the heated surface of the wire is characterized by having a broad spectral distribution with a peak at a wavelength that becomes shorter as the current through the filament, and hence the filament temperature, is increased. While gases can be raised to temperatures high enough for the wavelength of the spectral peak to appear in the visible range (for example the sun's radiation peaks at about 450 nm), this is not possible with metal-filament lamps as the peak wavelength is still about 800 nm even at the melting point of tungsten (3680 K). Although a heated tungsten wire produces rather less radiation at any given temperature than that of an ideal 'black body' (having an *emissivity* of only about 45% at visible wavelengths), the spectral distribution of the energy in the radiation is very similar to that of a black body, extending from the long-wavelength infrared to some shorter wavelength that depends upon the temperature (see Fig. 2.2).

Table 2.2 Photopic and scotopic luminous efficacy

λ (nm)	Photopic (lm.W^{-1})	Scotopic (lm.W^{-1})	λ (nm)	Photopic (lm.W^{-1})	Scotopic (lm.W^{-1})
380	0.012	1.001	585	564.2	152.8
385	0.049	1.887	590	530.7	111.4
390	0.173	3.757	595	491.8	79.73
395	0.525	7.701	600	449.4	56.44
400	1.366	15.79	605	405.7	39.27
405	3.080	31.45	610	360.6	27.03
410	5.983	59.16	615	315.6	18.53
415	9.904	102.7	620	271.8	12.53
420	14.62	164.2	625	231.5	8.449
425	20.15	244.8	630	193.3	5.678
430	26.43	340.0	635	155.7	3.808
435	33.88	447.1	640	122.9	2.550
440	42.41	557.6	645	95.62	1.717
445	51.02	668.1	650	73.76	1.150
450	61.13	773.5	655	55.46	0.780
455	72.40	872.1	660	41.19	0.532
460	87.42	963.9	665	30.12	0.366
465	104.5	1054	670	21.72	0.252
470	126.4	1149	675	15.44	0.175
475	150.3	1248	680	10.86	0.121
480	173.6	1348	685	7.581	0.085
485	203.5	1447	690	5.293	0.060
490	231.5	1537	695	3.675	0.043
495	269.8	1613	700	2.541	0.030
500	314.9	1669	705	1.749	0.022
505	362.7	1697	710	1.209	0.016
510	414.6	1695	715	0.833	0.011
515	468.5	1658	720	0.578	0.008
520	520.4	1590	725	0.400	0.006
525	562.1	1496	730	0.278	0.004
530	597.6	1379	735	0.194	0.003
535	631.1	1246	740	0.136	0.002
540	657.0	1105	745	0.096	0.002
545	670.7	958.9	750	0.067	0.001
550	677.5	817.7	755	0.048	0.001
555	682.3	683.4	760	0.034	0.001
560	681.0	559.3	765	0.024	0.001
565	670.7	448.8	770	0.017	—
570	652.3	353.6	775	0.013	—
575	624.9	272.0	780	0.009	—
580	593.5	205.7			

Tungsten-filament lamps are made in an immense variety of shapes and sizes, and with wire filaments variously formed to suit them for many different purposes. Those most likely to be useful for vision research will probably have been designed for use in projectors or vehicle headlights in conjunction with focusing mirrors or lenses, or both. Such high-performance lamps have compact filament structures with the wire

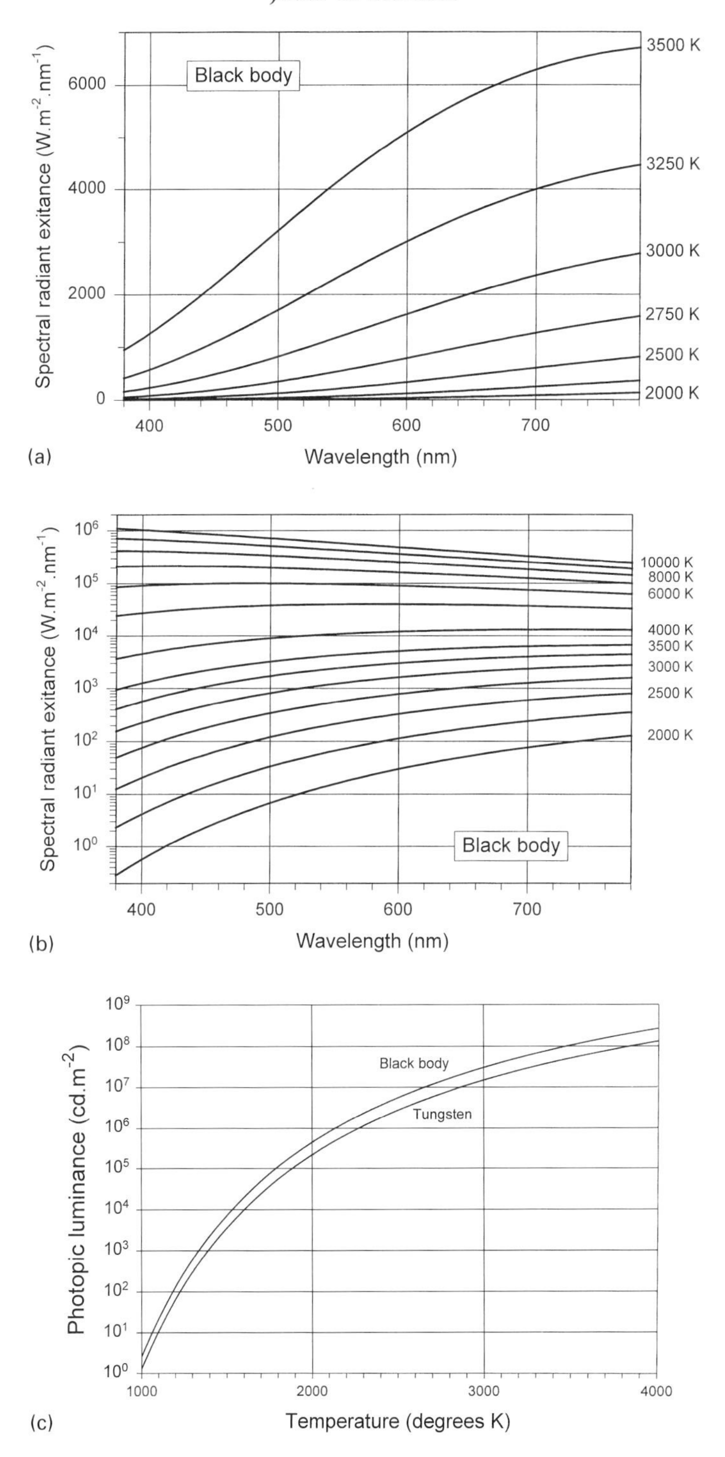

Fig. 2.2 Black body: (a) linear plot; (b) logarithmic plot; (c) luminance vs. temperature (K).

coiled or folded in such a way as to approximate a continuous surface. Most lamps now made for this service are of the tungsten–halogen variety, and these have so many advantages over the older types that they are likely to be the automatic choice for virtually all purposes and the only kind that we shall consider in detail. As an exception, there may still be some place for the ribbon-filament lamp which is a regular incandescent lamp that has a tungsten (or platinum) strip as its radiating element. This provides the nearest approach to a uniform high-luminance, Lambertian radiating surface that is conveniently available, making this a good secondary luminance standard and a good source for Maxwellian view systems.

2.3.1.1 Tungsten–halogen lamps

A major problem with simple, gas-filled, tungsten-filament incandescent lamps is that as the temperature of the filament is raised and it becomes an increasingly more effective radiator of light, the rate of evaporation of the tungsten also rises and seriously limits the life of the lamp at temperatures substantially less than the melting point of the tungsten. Although evaporation of the filament material eventually results in failure of the filament itself, the reduction in useful life of such lamps comes primarily from the reduced light output caused by absorption by the metal deposited on the glass envelope. To minimize this problem a small amount of a halogen (iodine or bromine) can be introduced into the lamp. At the lower temperature near the bulb, the halogen combines with vaporized tungsten to form a volatile compound which then diffuses towards the filament where, at the higher temperature there, it dissociates into its elements, the tungsten being deposited back on to the filament and the halogen becoming available for taking part once more in the regenerative cycle. For this cycle to take place the filament of a tungsten–halogen lamp must be run at a temperature of at least 1700 °C and the bulb must be kept at a much higher temperature than that of conventional tungsten-filament lamp. Compared with a conventional lamp the filament of a tungsten–halogen lamp is usually made more compact and the bulb made much smaller and either of fused silica (quartz) or of a much higher melting-point glass. The drive towards small size and compactness of the filament dictates that lower wattage lamps are run from low voltages, but, although mains voltage lamps are made, many of the easily available lamps with powers of less than 100 W are made to run from 12 V (even lamps intended for general lighting service). In addition to the higher luminance and efficacy of these lamps, their low voltage is often of direct advantage to the vision researcher who wishes to run such lamps from readily available, regulated power supplies.

While the regenerative cycle greatly increases the longevity of tungsten–halogen lamps, it remains the case that there will always be a trade-off between an increase in filament temperature (and hence luminance and efficacy) and a reduction in useful life (the tungsten is not redeposited on the hottest parts of the filament from which it mainly comes). Although this applies in the case of individual lamps and the user may choose to over-run a lamp and accept the consequential reduction in life, it should be noted that the detailed design of each type of lamp is optimized for use at some particular filament temperature. Tungsten–halogen lamps that are designed to run at colour temperatures between 2400 and 3400 K can have rated lifetimes that vary from 5000 to 50 hours. If a lamp with a high colour temperature is definitely required then it is sensible to choose one that is designed to run this way rather than over-running one that is not.

When a light source with a given intensity is needed, it may not necessarily be appropriate to choose a lamp whose filament runs at a very high temperature in preference to a higher wattage lamp running at a lower temperature and having a much longer life, unless it is important to minimize the electrical power consumed. Again, if light with a high colour temperature is required, it may be more satisfactory to choose a lamp that runs at a lower colour temperature and correct this with suitable filters than to use a higher-temperature lamp with a necessarily much shorter life. In any case, it is not possible to run any incandescent light at a high enough temperature to provide light corresponding to either natural sunlight or daylight. If this is to be provided by incandescent lamps they must inevitably be used in conjunction with colour-temperature correcting filters (e.g. Hoya light balancing filters from BES Optics).

The small size of tungsten–halogen lamps makes it easy to use them in conjunction with small ellipsoidal or parabolic reflectors to provide collimated or focused beams—a number of lamps are available that have integral mirrors. Some of these have dichroic mirrors that reflect visible light but allow a substantial proportion of the unwanted infrared to pass through, thereby producing, if not the 'cold light' of the 'blurb', at least a useful reduction in the heat load on the subsequent optical elements. A potentially useful variant of the integral-reflector lamp has a faceted mirror designed to create a uniform light distribution across its output plane.

Using tungsten–halogen lamps As indicated above, tungsten–halogen lamps are designed for use under quite specific conditions with some particular power input; if run at other powers (both lower as well as higher) their life can be seriously compromised. Thus, even if the spectral change that inevitably occurs when the temperature of an incandescent lamp is varied is not important, no attempt should be made to control the light from a tungsten–halogen source over any substantial range by changing the supply voltage or current. Rather, the output should be attenuated using neutral-density filters or wedges, or, if the spectral distribution is wide and must remain particularly constant, with variable apertures (see also Sections 1.4.4 and 4.3.9). Tungsten–halogen lamps may initially be under-run by a few per cent, but not much more than this, to allow for any subsequent increase in current needed to counteract the slight fall-off in efficacy as they age (perhaps under feedback control, see Section 2.5.1). Lamps may be equally well run from direct or alternating current supplies, though if run from 50 or 60 Hz supplies the light output will show some ripple at twice the supply frequency. The magnitude of this ripple depends upon the thermal inertia of the filament structure, and will be least for filaments made of the thickest wire (i.e. for high-current, low-voltage types) and greatest for lamps run at high colour temperature. If their light output is not stabilized by a feedback system that monitors the light intensity, tungsten–halogen lamps are best driven from a well-regulated supply. This can be a constant-voltage or a constant-current type, the latter providing immunity against variations in lead and contact resistance, though a more stable light output can probably be achieved by using a constant-voltage supply that monitors the voltage directly at the lamp. If calibrated tungsten–halogen lamps are to be used as sources of known illuminance or irradiance, a constant-current supply is preferred as calibration information will always specify the lamp current for which it applies.

Tungsten–halogen lamps should only be run within the range of positions specified by the manufacturers; some are made for horizontal use, some for vertical, while some can be used over a larger range than others. The envelopes of tungsten–halogen lamps can reach very high temperatures in use and appropriate precautions must be taken. Lamps should, of course, only be removed or inserted into sockets when cold; even so, they should not be handled directly as the surface of the bulb must remain clean and free of grease. If necessary, the bulb should be cleaned (when cold) with alcohol.

Tungsten–halogen lamps produce considerably higher intensities in the ultraviolet than do conventional incandescent lamps. This comes about not only because of their higher filament temperature but also because of the generally lower UV absorption of the material from which their bulb is made. For visual stimulation these lamps should always be used in conjunction with a suitable UV-absorbing filter.

2.3.2 Discharge lamps

Discharge lamps operate by passing an electric current through a gas or vapour between two electrodes. Electrons accelerated in the electric field between the electrodes collide with atoms of the gas causing them to become excited or even ionized. Relaxation of the atoms from the excited states to which they have been raised is accompanied by emission of radiation, while ionization produces positive ions and more electrons that are then available to collide with further atoms of the gas. If the pressure of the gas is low its temperature is not raised much by the discharge, and the radiation spectrum will have a well-defined line structure reflecting the energy levels of the excited states of the particular atoms involved. A typical low-pressure discharge is that through the mercury vapour of the normal fluorescent tube, whose primary radiation is in the ultraviolet at 253.7 nm, or in the low-pressure sodium lamp that radiates mainly yellow light at 589 nm. If the pressure of the gas is much higher, then the discharge may raise its temperature to a level (4000–6000 K) at which the spectral lines become much broader and less distinct and the spectrum becomes more nearly continuous and like that of a thermal radiator. High-pressure discharge lamps in which the location of the discharge is not constrained by the enclosing bulb are often called *arc lamps*.

2.3.2.1 Fluorescent lamps

The ubiquitous fluorescent tube used for general lighting purposes is a discharge lamp containing an inert gas at low pressure together with a little mercury that vaporizes as the tube warms up. At working temperatures the mercury is excited by electrons from the heated electrodes to radiate mainly in the ultraviolet (at 253.7 nm), though a number of weaker spectral lines in the near-UV and visible range (e.g. 365.0, 365.5, 404.7, 407.8, 435.8, 546.1, 577.0, 579.1, 623.4, 671.6, and 690.7 nm) are also produced (and may be used for wavelength calibration of monochromators and spectroscopes). Although the discharge does give some visible light, the high efficacy of these lamps results from the conversion of the much stronger primary ultraviolet radiation into visible light by the fluorescent phosphor with which the inside of the tube is coated. Most fluorescent tubes have phosphors, or mixtures of phosphors, that produce more-or-less white light, though tubes with phosphors fluorescing in relatively narrow bands to give strongly coloured lights are also available. Some of these latter are produced for decorative purposes, many for use in technical systems such as photocopiers. A wide variety of different 'white' phosphors are used to achieve different compromises between colour appearance, colour-rendering, efficacy, cost of manufacture, and lifetime. The surface luminance of standard fluorescent lamps is usually in the range $10\text{–}20 \times 10^3\,\text{cd m}^{-2}$. Traditional straight-tube fluorescent lamps are made in many sizes, the smallest being about 150 mm long (4 W), though more compact lamps with folded tubes have now become common as direct replacements for domestic incandescent lamps. A range of annular tubes is also manufactured that may be of particular relevance to vision researchers needing to illuminate the area surrounding a stimulus pattern generated on a cathode-ray tube.

Fluorescent tubes are commonly driven by mains-frequency alternating currents whose magnitude is limited by a reactive element or 'ballast' (typically an inductor called a *choke*) in series with the tube. Provision is also made for heating the electrodes sufficiently to obtain adequate thermionic emission of electrons. In some tubes the electrodes are heated by current from a transformer and remain powered during normal running; in others the filaments, after preheating by the starter current, are subsequently heated by the discharge current itself. The light output of fluorescent lamps can be controlled (i.e. reduced from the normal operating level) by reducing the voltage or by increasing the effective reactance of the series element, but satisfactory control over more than a small range is only possible if the electrodes are independently heated.

Problems with fluorescent lamps The light from fluorescent lamps driven at mains frequency has a large ripple component at twice the frequency of the mains (i.e. 100 or 120 Hz), because the discharge current goes to zero and then reverses twice in each cycle. The light generated directly from the discharge follows the variation in current with very little delay, though the fluctuations in the light generated by fluorescence may be somewhat smoothed out and delayed by the slower relaxation time of the excited phosphor molecules. Where mixed phosphors are used the light from different components may be differently delayed. This can result in variations in colour of the light during each half cycle. Such variation is the cause of the coloured fringes often seen when viewing a rotating spoked wheel illuminated by a fluorescent lamp. Because the discharge is not of uniform intensity along the length of a fluorescent tube (there is a dark region near the cathode) the light from near the electrodes may be significantly modulated at mains frequency and give rise to visible flicker (at least when viewed in peripheral vision). To counter these effects (and to provide an increased efficacy) fluorescent lamps can be run from electronically generated voltages alternating at much higher frequencies; commercial 'electronic ballasts' that generate such voltages from the mains supply often run at around 30 kHz. At these frequencies there is essentially no modulation of the light output, though there may still be some small degree of modulation at twice the line frequency because of incomplete smoothing of the internal DC supply to the high-frequency generator. It is also possible, for laboratory purposes, to run normal fluorescent lamps with a direct current, though it then becomes necessary to use either a resistive ballast to limit the current or a constant-current supply. Some arrangement to heat the cathode and to start the discharge will also be needed. With a DC supply, modulating the light output by modulating the current is possible, though the current cannot be reduced to very low levels without the discharge becoming extinguished or unstable. In fact the discharge may in any case exhibit oscillatory behaviour at frequencies of a few kHz even when running at normal currents. Such oscillations may result in substantial modulation of light output from local regions of a tube without there being much fluctuation in total light output. Whenever such effects might be of consequence, the nature of the light output from fluorescent lamps should be examined directly using a photodetector and an oscilloscope (see Section 2.4.1).

Although fluorescent lamps are in many ways cheap and convenient sources of large amounts of light when high levels of general illumination are required, they should be used with caution. First, there is the possibility that even though their light appears steady it is substantially modulated at frequencies at which the peripheral elements in the visual system certainly respond (e.g. Berman *et al.*, 1991). Second, the spectral composition of their light can be very different from that of natural or incandescent light even though its apparent colour may be similar (note that different fluorescent lamps with the same colour appearance can have very different spectra). Particular caution should be exercised when using fluorescent lamps in work on non-human vision as both the flicker sensitivity (e.g. Eysel and Burandt, 1984) and colour-rendering requirements of other species may differ significantly from our own.

Another problem with normal fluorescent lamps is that they are much more dependent than other lamps on the ambient temperature. Changes in ambient temperature are reflected in changes in pressure of the mercury vapour that produce changes in both light intensity and spectral composition. For the same reason it may take quite a long time after a fluorescent lamp is switched on for its output to reach a stable level because of the time required to establish thermal equilibrium.

2.3.2.2 Cold cathode lamps

Fluorescent lamps can be made with electrodes that provide adequate electron emission without being heated, so-called 'cold cathode' tubes. While these are not commonly used for general lighting because they are not sufficiently efficient for this purpose, small lamps of this kind made primarily for instrument panel and display illumination may be useful in vision research applications. The smallest of these miniature fluorescent lamps are no more than about 80 mm long and can be run from low-voltage supplies using a suitable small DC–AC converter to provide the 200 V or so that they require. The tubes are made with an auxiliary starting electrode near each main electrode. This enables a priming glow discharge to be set up in the inert gas by passing a small resistor-limited current between the auxiliary and main electrodes. The luminance of these small tubes is less than that of regular fluorescent tubes but can still be as much as 2000–3000 cd m^{-2}. Optical feedback stabilization (see Section 2.5.1) can be easily implemented by controlling the low voltage driving the inverter.

2.3.2.3 Xenon arc lamps

Although there are many different kinds of high-pressure discharge lamps available, the kind that is most generally useful in vision research applications is the xenon arc lamp. The usefulness of this lamp comes from the very high luminance of the arc coupled with the relative flatness and smoothness of its spectrum in the visible region (Fig. 2.3), as well as its fairly high luminous efficacy. High-pressure xenon lamps are made in a wide range of powers, from 50 W to hundreds of kW, though it is only low-power lamps of less than a few hundred watts that are likely to be of much interest for applications in the vision laboratory. The largest lamps provide enormous light outputs but do this mainly by having long arcs contained in long tubes. Lower power lamps are all of the 'compact' type with only a short arc. Although the arc in a low-power compact lamp may be quite small, the luminance of the brightest part (close to the cathode) may be only slightly lower than that of the arc in a much more powerful lamp (the brightest part may, however, be less than a millimetre across). The luminance of the arc in low-power lamps (up to 500 W) may be expected to be in the range 10^8 to 5×10^8 cd m^{-2} (the luminance of the sun's disc is about 10^9 cd m^{-2}).

Compact xenon lamps are normally designed to be run from a direct current supply and usually have solid tungsten electrodes—a large cathode and a smaller anode—that are mounted to enable heat transferred to them from the arc (which may run at a temperature exceeding 6000 °C) to be conducted away and dissipated. Such lamps must usually be run vertically with the cathode down. Because the arc in compact xenon lamps is so short the voltage drop across the lamp is often only about 20 volts, the current being in the range of 5 to 50 amps. Like all discharge tubes the xenon lamp appears to have a negative resistance and must therefore be run either from a constant-current (high impedance) supply or from a voltage source with a suitable series resistor. To start a xenon lamp a very high voltage pulse (several tens of kV) is applied between the electrodes; this requires a special pulse transformer that is often incorporated into the lamp housing or power supply. After a lamp has been running and the pressure inside is much higher than the cold level, restarting may be difficult or even impossible.

In the most common kind of low-power, compact xenon lamp the arc is formed between electrodes that are only a few millimetres apart at the centre of a relatively

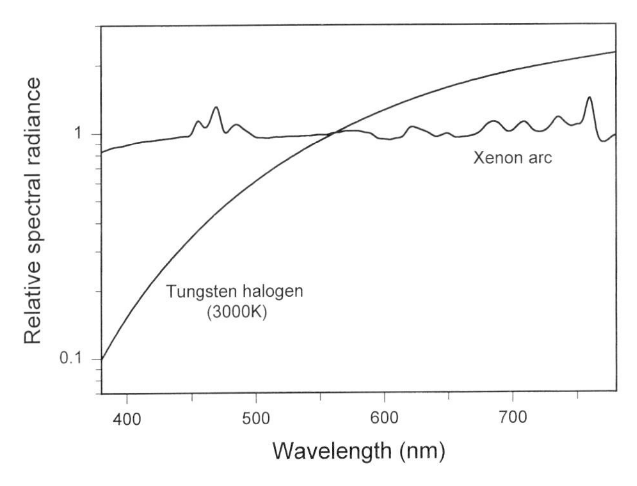

Fig. 2.3 Xenon arc spectrum (and tungsten–halogen).

large quartz bulb (though still no more than about 30 mm in diameter). Because the arc is not physically restrained by the envelope of the lamp its exact location may change from time to time as a result of small uncontrolled changes in the electrodes or convection patterns in the gas. Such instability can be troublesome; it is sometimes possible to improve the stability of an individual lamp quite considerably by slightly changing the operating current.

Compact xenon lamps of the usual form are particularly convenient in two respects. In applications requiring the highest possible illumination of a small area, such as the input of a fibre-optic light guide or the input slit of a monochromator, the small size of the bulb and the general form of these lamps make it possible to mount them along the axis of an ellipsoidal mirror of large aperture with the arc at one focus (as in the Photomax system from Oriel Instruments). In this way a large fraction of the light can be collected and made available at the second focus of the mirror for entry into a small aperture. On the other hand, where the lamp is required to provide light for several independently controlled components of a visual stimulus (as in the several channels of a three- or four-colour mixing system), a vertically mounted lamp can be surrounded by several separate condenser lenses, each providing the light for one channel. Inevitable fluctuations in arc intensity will thus be common to all channels and their visual effect thereby lessened (though arc wander can still be a problem). A single arc lamp with a number of condensers can also be used to feed several fibre-optic light guides.

Using arc lamps The pressure in a compact xenon lamp is usually several atmospheres when it is cold and can rise substantially when it is running. This poses a potential risk of explosion, therefore compact arc lamps should always be handled with care when cold and run within a suitable enclosure. Xenon arc lamps mounted in suitable housings and provided with suitable starting gear and power supplies can be obtained from several optical supply companies, and it is strongly advised that such complete systems should be used. The high temperature of the arc and the nature of the envelope result in there being a substantial short-wave ultraviolet component in the emitted light. This should always be filtered out when the light is to be used as a visual stimulus. If for some reason it is necessary to view the unshielded lamp then UV-absorbing goggles should be worn. The high levels of UV emitted by the arc can produce ozone in the air around the lamp and it is usually desirable for this to be vented to the outside. Commercial housings for larger lamps generally make provision for ozone venting or absorption.

2.3.2.4 Flash tubes

Flash tubes are discharge lamps that are intended to produce occasional brief flashes of very high intensity. They are mostly made either for photography or as warning beacons; they are also commonly used in stroboscopes. In the context of vision research, flash tubes are useful for generating after-images, particularly during eye movements, as well as for rapidly bleaching photopigments in the retina. They are commonly used (see Section 8.2.8) for evoking the electroretinogram or flash VEP (visual evoked potential).

Flash tubes are usually narrow-bore tubes that have an electrode sealed into each end and are filled with xenon. The tube may be straight, U-shaped, or coiled. The flash is produced when a capacitor, that is connected between the two electrodes and is initially charged to a high voltage (usually in the range 200 to 1000 V), rapidly discharges through the tube after conduction in the xenon has been initiated by a very high voltage-trigger pulse (several kV) applied to an external electrode on the tube. During the discharge very large currents (hundreds or even thousands of amps) flow through the tube. The duration of the flash depends upon the dimensions of the tube, the pressure of the gas, the capacitance of the capacitor, and the voltage to which it is charged. By using a small capacitance the flash duration may be reduced to about 1 μs, but 100 μs would be more usual. Following a flash it takes some time for the residual ions in the gas to disappear, and this can result in an untriggered discharge if the voltage across the tube subsequently rises too quickly. In practice, however, the time to recharge the capacitor is often limited more by the power available than by the need to prevent premature discharge. Flash tubes made to be used in stroboscopes, where they may be flashed at intervals of a millisecond or two, have lower pressure fillings, may include traces of other gases, and operate at lower energies per flash.

About half the electrical energy delivered to a xenon flash tube is converted into radiation. However, the spectrum of this radiation is very broad, extending from well below 200 nm to far into the infrared, and only a only a small proportion of the radiation is in the visible range. The smaller tubes that are likely to be of most interest to vision researchers are made of glass that will filter out most of the very short-wavelength UV. Although the spectrum displays a number of lines associated with the electronic excitation of the xenon, at normal current densities the majority of the radiation forms a continuous spectrum (Fig. 2.4). It should be noted that the ordinate in this figure shows the spectral density of the efficiency with which electrical input energy is converted into radiant energy in a single flash. When the duration of a stimulus flash is much less than the integration time of the photoreceptors, the instantaneous intensity or luminance of a source are of little consequence; it is the total energy in the flash (i.e. the time integral of these quantities) that determines the visual effectiveness of the stimulus. Small flash tubes can handle input energies up to about 25 J per flash (1 joule ≡ 1 W s), though

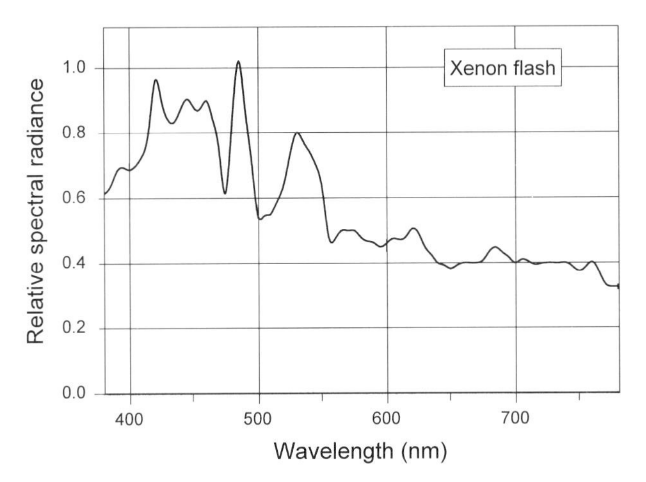

Fig. 2.4 Xenon flash spectrum.

there will also be a power limit (related to the ability of the electrodes and bulb to dissipate heat and typically a few watts) that will limit the frequency of flashing. The visual efficacy of such tubes is around 20 lm s per joule.

The exact position and nature of the discharge within a flash tube can vary from flash to flash, giving rise to random fluctuations of a few per cent in the light output. Special tubes that have extra electrodes to guide the arc are available (e.g. from E G & G Optoelectronics and Hamamatsu Photonics) but these are expensive and, having only a short arc length, can handle only quite small energies (less than 1 J per flash). Flash tubes that have been in the dark for some time may show triggering irregularities because of the total disappearance of gaseous ions within the tube; if possible, some continuous external illumination of the tube should be provided.

The large amount of ultraviolet radiation in the output of xenon flash tubes (even glass ones) makes it imperative to use an external UV filter in conjunction with any xenon flash source that is to be placed at all close to the eye. Polymethylmethacrylate sheet, intended for glazing pictures and display cabinets and which absorbs strongly below 400 nm (e.g. Perspex type VA), can conveniently be used for this purpose.

2.3.3 Electrically excited sources

2.3.3.1 Light-emitting diodes (LEDs)

Although many semiconductor junction diodes emit some light when an electric current is passed in the forward direction, light-emitting diodes (LEDs) are made of materials specially formulated to increase the proportion of the electrical energy that gives rise to radiation, rather than being dissipated as heat, and are constructed to allow as much as possible of the generated light to be radiated. LEDs normally produce radiation with a single spectral peak and a bandwidth sufficiently narrow for them to appear strongly coloured; many different semiconductor materials and dopants are now used in the production of LEDs and can provide a wide range of colours from the violet into the near-infrared (see Fig. 2.5).

Although the luminous efficacy of even the brightest LEDs is less than that of most other light sources (the best LEDs currently available produce around 2 photopic lumens per watt at 425 nm and 670 nm with a maximum of about 20 lumens per watt at 600 nm), LEDs can be at least as efficient as other lamps in producing light near their peak wavelength. While their high efficiency in producing coloured light accounts for the ubiquity of LEDs as signal lamps in portable equipment, their robustness and long life-expectancy are also very important factors in their widespread commercial use. The usefulness of LEDs in vision research applications, however, depends mainly on the fact that their light output can be rapidly varied over a wide range (by varying the applied electric current) without much change in spectral distribution. This makes LEDs almost ideal light sources where the electrical control of light output is required, particularly as the low voltages and currents at which LEDs operate are very easy to provide and control with simple circuitry (see below).

LED chips are typically intended to operate with a current of about 20 mA. The voltage required to produce this current depends upon the wavelength of the emitted radiation, typically being about 2.0 V for LEDs operating in the far red and 4.5 V in the violet, i.e. at a voltage little higher than is minimally necessary to provide the

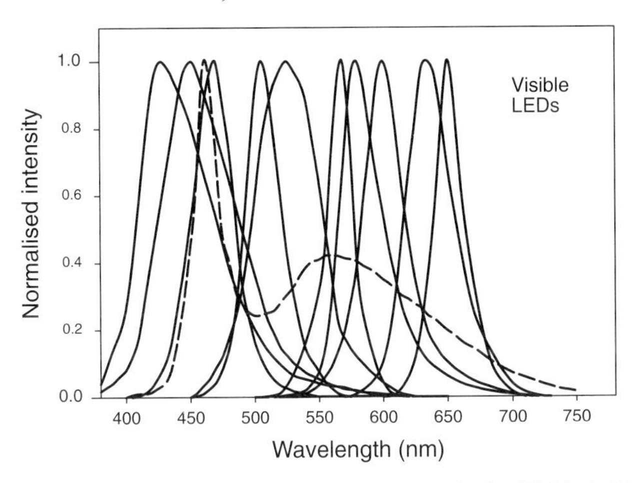

Fig. 2.5 Light-emitting diode (LED) spectra. The dashed line is the Nichia 'white' LED (redrawn from data sheets of Nichia Chemical Industries, Hewlett Packard Ledtronics, and Cree Research).

energy of the emitted photons (photon energy at 700 nm is 1.8 eV and at 400 nm is 3.1 eV, equivalent to 2.83×10^{-19} J and 4.96×10^{-19} J, respectively). The relationship between current and voltage is very non-linear, the current rising extremely rapidly with voltage in the operating range; this means that LEDs can only be run from a constant-voltage supply if a suitable resistor is connected in series with the LED. In this case reasonable stability will be achieved if the constant voltage is at least twice the nominal running voltage of the LED. Alternatively, LEDs may be run from an appropriate constant-current source. Multiple LEDs may be connected in series, but each string must have either its own constant-current supply or its own series resistor. When run with a steady current the maximum light output from an LED is normally limited by the maximum permissible power dissipation of the chip. This will depend upon the way that the chip is mounted, but is typically about 100 mW. While it is usually possible to over-run LEDs without damaging them, it is generally undesirable to do so—not only because of the reduction in lifetime, but because the increase in temperature will cause the spectral distribution of the light to change. If LEDs are to be turned on infrequently for short periods of time only (not more than 2–3 ms) the maximum current can be increased several-fold. The maximum permissible current limits set by manufacturers take into account the effects of high currents on the lifetime of the LEDs, but this is not usually of much significance in vision research applications and it is often possible to exceed the manufacturers limit substantially without serious effect. LEDs are cheap enough for the practical limit to be determined experimentally. Unless the pulses are very brief, the limit on total light output in a flash may, in practice, be set by the effects on the spectrum of the transient temperature rise at the junction; this will depend upon the thermal inertia of the junction and must be investigated experimentally if it is likely to be of consequence (see Section 2.4.2).

The intensity of the light emitted by an LED is generally related non-linearly to the current, the actual relationship depending upon the nature of the semiconductor materials and the construction of the junction. The light intensity usually rises more rapidly than the current at low currents but more slowly at high ones. The intensity–current relationship is somewhat dependent upon temperature and may be slightly different at different wavelengths (i.e. the spectral distribution may be somewhat different at different currents, an effect that mainly depends on temperature). These characteristics are of little importance in most technical applications of LEDs and are not well controlled by manufacturers; individual LEDs of a given type may show significantly different behaviour, particularly at currents that are much lower than those at which they are expected to be operated. While this does not mean that the light output of LEDs cannot usefully be controlled over a very large range by altering the current, it does mean that attention should be paid to possible spectral changes (the effects will be minimized if narrow-band filters are used), and each system must be individually calibrated if a measurement of the current is to be used as a substitute for a measurement of the light intensity actually achieved. Given the exceptional linearity, high stability, and low cost of silicon photodiodes, it is often better to make a direct measurement of relative light intensity whenever it is actually required rather than rely on time-consuming calibrations made at other times (see Section 2.4). Alternatively, a feedback system that uses such a photodiode to monitor the light from the LEDs can be used. Feedback systems are easily set up to produce a voltage-controlled light source that can equally well provide either a steady light intensity proportional to some constant voltage, or a light whose intensity is modulated with any waveform that can be provided as a varying voltage.

While the intensity of light from an LED can be altered (non-linearly) by changing the strength of the current flowing through it, it is also possible to control the effective light intensity by switching a fixed current on and off at a high frequency and then controlling the fraction of the time for which the current is switched on. With suitable switching parameters the average light output can be closely proportional to the on-time. When the switching frequency is sufficiently high (at least well above the flicker-fusion frequency) the train of light pulses that are produced in this way will be indistinguishable from a continuous light, so long as the light subtends only a small angle at the eye and its image does not move rapidly across the retina. If there is rapid image motion, maybe as a result of large eye movements, the switching frequency will have to be very much higher than the fusion frequency to prevent the temporal variations in intensity being visible as spatial variations.

In fact there are two rather different ways in which the proportion of time for which an LED is switched on can be altered: either the duration of the current pulses can be kept constant and their frequency changed (pulse-frequency modulation, PFM) or the frequency of the pulses can be kept constant and their duration changed (pulse-width modulation, PWM). Both forms of pulse modulation are easily implemented using switching signals that are generated either by analog-timing circuits that can be voltage controlled or by digital counter-timers that can be directly computer controlled. Both forms of pulse modulation can provide linear control of effective LED intensity very precisely, and may equally well be used to adjust the intensity of steady lights or to generate any desired modulation waveform unaccompanied by any significant change in spectral composition. A practical PWM system might switch the LEDs on 1000 times per second (i.e. at 1 kHz) and control the on-time between 1 μs and 1 ms to give a range of 1:1000 in effective luminance.

Alternatively, a PFM system could use 0.9-μs pulses at frequencies between 1 kHz and 1 MHz to give about the same range of luminances. Switching signals generated by standard digital integrated circuits can be very simply interfaced to LEDs using transistors (most simply by logic-level MOSFETs as current switches (see Fig. 2.6).

LED chips are commonly made into lamps by encapsulation in a transparent plastic material, which not only protects the semiconductor junction but can also be shaped to form a lens to concentrate the radiated light into a beam. LED lamps are made with diameters between 2 and 20 mm and with various beam widths; angular beam widths as narrow as 3 or 4° are possible for the larger lamps. Lamps intended for general illumination have no lens and radiate over about 160°.

Although the high luminance and small size of LED junctions make them good sources in Maxwellian view systems, it is necessary to place a diffuser in front of the LED if good uniformity of the field is required. To minimize the reduction in retinal illumination, this diffuser should be thin and as close as possible to the LED junction; many types of white translucent plastic materials make satisfactory diffusers for this purpose. Good results can be obtained either by using a wide-angle LED with no integral lens or by carefully cutting off the lens of a normal LED lamp close to the junction.

Although most LED lamps incorporate a single semiconductor chip, they can also be obtained with two or three independent chips which may be red and green or red, green, and blue, making it possible to vary the colour of the emitted light by differentially altering the current through the chips. It is also possible to obtain LED lamps that contain many chips connected together (either all of the same type or red, green, and blue to give a white light) and mounted on a ceramic base. Such lamps are intended for more general illumination purposes and can have a much larger light output than a single chip, though the maximum available light output may be less than expected from the number of chips because of heat-dissipation limitations; these lamps do not form the light into a beam, but radiate over a full 180°. Of course, it is also possible to obtain more light by using a number of single-chip lamps connected

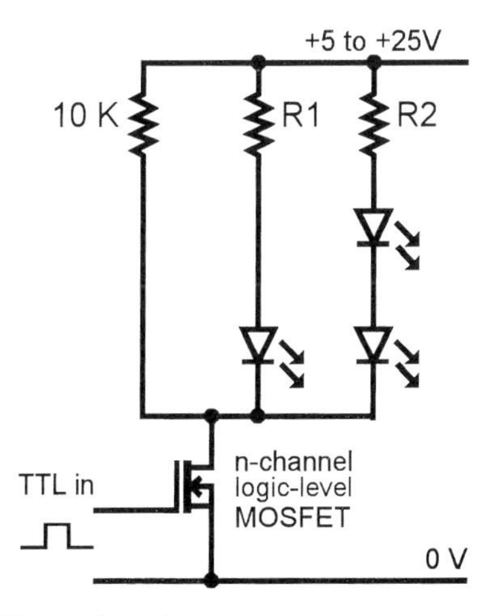

Fig. 2.6 A logic-level MOSFET transistor interface.

together externally, though again this will usually only be appropriate if they are to be used to provide diffuse illumination. Similarly, the light from several LEDs of different colours can be combined (most satisfactorily in an integrating sphere) to produce a diffuse source whose colour can be controlled electrically. Where large numbers of LEDs are used to provide distributed illumination it may be convenient to use surface-mount devices that are no more than 2 × 3 mm and can be fixed adjacent to each other.

LED chips are also incorporated into various packages that themselves incorporate diffusers. While at one extreme this may be no more than a single, uniformly illuminated rectangular area a few millimetres on a side, at the other there are packages that comprise a large number of individual elements that can be illuminated in various combinations to display different patterns or characters. These include bar-graph displays that have a row of 10 abutting rectangular elements as well as matrix displays with up to 8 × 16 elements, the familiar 7-segment digital displays, and more elaborate character displays. The luminance of the elements of these displays can be controlled in exactly the same way as the intensity of separate LED lamps. Uniformly illuminated areas can also be provided by assemblies of LEDs around the circumference of a transparent plastic sheet coated with a diffuser on one surface. These are made in sizes suitable for back-illumination of instrument legends and liquid-crystal information displays.

2.3.3.2 Electroluminescent sources

Electroluminescent panels generate light by direct electrical excitation of phosphors or fluorescent dyes that form a thin layer between two extended electrodes, one of which is transparent. These electrodes can be deposited on either rigid panels or flexible films and light is emitted when an alternating voltage is applied between them. The colour of the light depends upon on the nature of the phosphor, and its luminance upon both the voltage and the frequency. While most rigid panels are custom-made for large-scale applications, thin and flexible electroluminescent sheets that can be readily cut with scissors are now available in many different colours for small-scale uses (e.g. from Seikosha via RS Components). The maximum luminance available from these panels is in the range 50 to 200 cd m^{-2} and is obtained by driving them with up to 200 V at frequencies up to 3 kHz. This current can be obtained from small DC–AC converters that run from modest DC supplies. The light output has a substantial alternating component at twice the frequency of excitation, but as this can be much above the flicker-fusion frequency the fluctuation may be of little consequence.

2.3.3.3 Lasers

Although both solid-state and gas lasers providing continuous outputs at various wavelengths in the visible range are available, there are not many circumstances in which they will be an obvious choice as a light source for providing visual stimuli. Semiconductor lasers that operate at wavelengths shorter than 600 nm are not yet commercially available, but following the introduction of practical gallium nitride LEDs operating in the blue region of the spectrum, it seems likely that they soon will be. Lasers with an output power of more than a few milliwatts are rather expensive, and because their effective intrinsic luminance is very high they can easily give rise to retinal damage if accidentally viewed directly—so great care is needed in their use.

Lasers generate extremely monochromatic light with a high degree of coherence, but the only visual stimulus situation in which these characteristics are really essential is that in which the light is used to generate extended grating patterns by interference (see for example He and McLeod 1996; Sekiguchi, Williams, and Brainard 1996). In other contexts, the coherence of the light may be more of a disadvantage, giving rise to unwanted 'speckle' phenomena.

2.4 Photodetectors and light measurement

2.4.1 Silicon photodiodes

There are many different types of radiation detector, but the only kind that is now of much practical significance in vision research applications is the silicon photodiode. This is a semiconductor junction diode in which absorbed photons generate electron-hole pairs in the silicon, that then migrate to the electrodes and give rise to a current that flows in the external circuit to which they are connected. Although silicon photodiodes are usually described as having a sensitivity that peaks near 900 nm and then steadily declines into the ultraviolet and infrared (in terms of the current produced by a given amount of radiant energy at different wavelengths, this is correct), in fact they are best described as having an almost constant quantum efficiency (the ratio of electrons to photons) over the whole range of visible wavelengths and into the near-UV and IR (the quantum efficiency is most nearly constant in 'blue-enhanced' diodes). Moreover, not only is the actual quantum efficiency high (typically about 70%) and independent of light intensity over a very wide range, the speed of response is much faster than that of the eye. Such photodiodes are therefore almost ideal devices for measuring and monitoring lights that are to be used as visual stimuli.

While a silicon photodiode may be used directly whenever a photodetector with a more-or-less constant quantum efficiency over the visible range is appropriate, there are many cases in which it is more relevant to have a photodetector with defined spectral characteristics corresponding to those of the visual system. For this purpose, several manufacturers (e.g. UDT, International Light) provide photodiodes paired with glass filters that are designed to have an overall spectral sensitivity corresponding to either the photopic or scotopic luminous efficiency functions as specified by the CIE. While such diode/filter combinations are normally supplied as part of a complete photometric measuring system, they can also be obtained as separate items and effectively used with simple current-measuring circuits.

Silicon photodiodes are usually used in conjunction with amplifier circuits that convert the small photocurrent (at 70% quantum efficiency 1 nA is equivalent to 9×10^9 photons per second) into voltages that are large enough to be easily measured or satisfactorily used as control signals. Figures 2.7(A) and (B) show two simple circuits that can easily be constructed. In both circuits an operational amplifier is used to convert the current from the photodiode into a voltage, and at first sight it may not be obvious what the difference is between the two. Basically, it is that in A a substantial voltage is maintained across the photodiode (the diode is operated in the 'photoconductive' mode), while in B the voltage across the photodiode is always kept at zero by the feedback (the diode is operating in the 'photovoltaic' mode). Operation in the

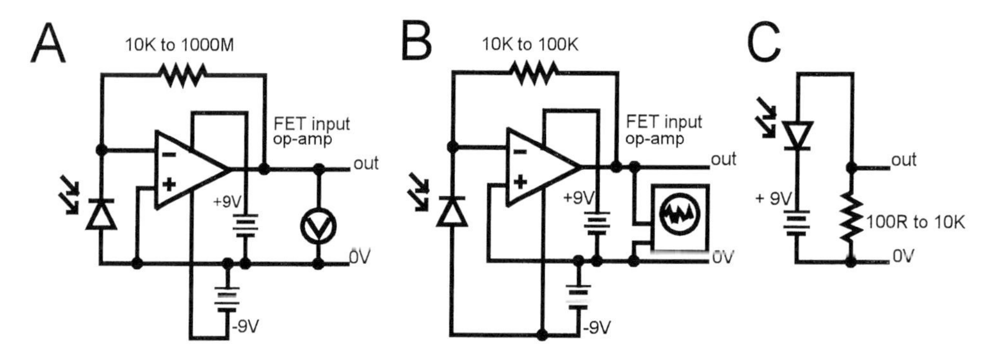

Fig. 2.7 A typical circuit for observation of modulation waveforms.

photoconductive mode results in a faster response and the best possible linearity, but is not so good when very low light levels have to be measured because of the existence in this mode of a leakage current through the diode in the absence of any light. The problem with this is that the 'dark current' is very sensitive to temperature and other changes and makes the zero unstable. For most vision research applications silicon photodiodes are probably best used in the photovoltaic mode. The linearity in this mode is quite good enough for most purposes, and adequate sensitivity can usually be obtained by using an operational amplifier with a low input current in conjunction with a feedback resistor of sufficiently high resistance. Use of the photoconductive mode usually only becomes necessary if the exact time-course of brief flashes (say less than 1 ms) or high-frequency modulation (more than a few hundred Hz) is to be observed. For simple observations of modulation waveforms it is normally quite satisfactory to use the circuit shown in Fig. 2.7(C) in which the voltage developed across a resistor in series with the photodiode and a battery is fed directly to a suitable oscilloscope. A resistance of 0.1–10 kΩ will often be appropriate.

Although silicon photodiodes with quite large photosensitive areas (up to 1 cm^2) can be obtained, it will often be perfectly acceptable to use a smaller (and therefore cheaper) one, unless very low levels of illumination are to be measured. When weak lights are to be measured, it should be noted that, if at all possible, it is always better to focus the light on to a small photodiode rather than to use a larger one. Although, if the light is diffuse, increasing the area of a photodiode will result in a proportional increase of the photocurrent, this will also result in a greater dark-current noise. If the diode is operating in the photoconductive mode the dominant noise will probably be that from the photodetector dark-current and the resulting signal-to-noise ratio will only improve as the square-root of the diode area. However, if the diode is being used in the photovoltaic mode the dominant noise source may well be the amplifier itself, and a greater improvement will be obtained by using a larger diode.

2.4.2 Spectroradiometers

Photodiodes are not only available as individual photodetectors but are also manufactured in large arrays. A linear array of photodiodes placed at the exit of a grating spectroscope can be used to monitor the intensity of the light entering the instrument at many different wavelengths simultaneously. Modestly priced spectroradiometers, based on this principle and utilizing the computational capabilities of a personal computer, have recently become available (e.g. from Ocean Optics) and will make it possible for far more vision scientists to check the spectral characteristics of their stimuli than hitherto. Light can be conveyed to these spectroradiometers by a fibre-optic light guide, and this can easily be arranged to pick up from a small area of a stimulus. Unlike scanning spectroradiometers (which are in any case mostly prohibitively expensive), spectroradiometers of this kind can be used to examine the spectrum of single brief flashes as well as steady lights.

2.5 Stability of light output

Unless it is deliberately intended to modulate the intensity of a light source, it is usually desirable that this should remain absolutely constant, but all lamps display some instability or variation of light output over time. This may result from:

(1) fluctuations in the electrical supply to the lamp (including variations in lead resistance);
(2) progressive changes that occur after a lamp is switched on as it reaches its working temperature;
(3) uncontrolled changes in the lamp's environment (e.g. temperature changes produced by air currents, movement of a filament resulting from mechanical disturbance);
(4) effectively random changes within the lamp (notably arc-wander in some discharge lamps); and
(5) effects of ageing (usually resulting in steady loss of output).

While instability is normally most obvious as a change in intensity, it is also important that due attention be paid to the possibility of changes in spectral distribution that may remain even if changes in intensity are corrected.

2.5.1 Feedback stabilization

Most of these instability effects and ways of reducing them, or their significance, have already been noted in describing the various types of lamps; here we consider a generally applicable method of minimizing such disturbances. This is the use of negative feedback to control lamp intensity using a signal from a photodetector monitoring the light output. Commercial systems of this kind are available for many types of lamp, but it is in fact quite easy to set up a simple system in the laboratory for controlling either low-voltage filament lamps or LEDs. The use of feedback control to provide a stable output is really no more than a variant of its use to provide a modulated output with the same prime requirement, a stable photodetector of adequate sensitivity. A silicon photodiode (see Section 2.4.1) is entirely suitable for this purpose, though if a lamp's luminous output is to be stabilized, the spectral sensitivity of the photodetector should have been corrected with a suitable filter to match, or at least to approximate, the visual luminous efficiency function (unless, of course, the spectral output of the lamp is known not to change). Another requirement in this

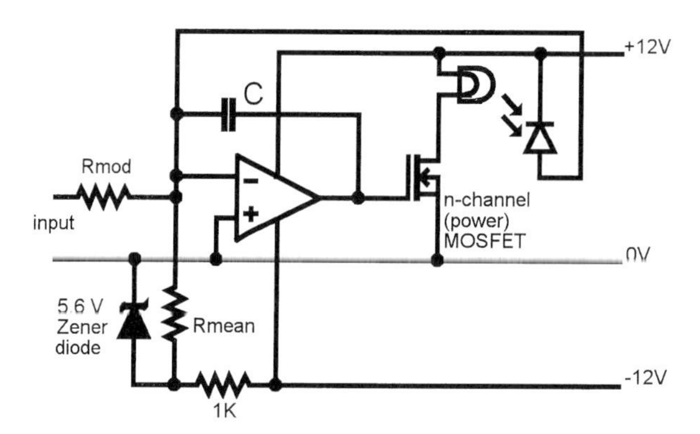

Fig. 2.8 Simple circuit for a negative feedback circuit to control lamp intensity using a silicon photodiode and a stable current source.

case is a stable current source as a reference with which the output of the silicon cell can be compared, any difference being used to change the drive to the lamp. A simple circuit is shown in Fig. 2.8. The resistor determining the reference current should be initially adjusted so that the lamp is run at somewhat less than its rated power, to allow for subsequent reductions in efficacy with ageing that must be compensated for by raising the lamp current.

Although it is not necessary to use a highly stable primary supply to power such a feedback circuit because the feedback will make up for deficiencies in its stabilization, it is desirable that the supply should be of such a voltage and well enough regulated that the lamp will not be damaged if the feedback control should fail. A feedback circuit of this kind will not only greatly improve the long-term stability of light output but can also greatly reduce the amplitude of short-term fluctuations, both ripple and 'noise'. The same basic circuit can be used to stabilize the output of fluorescent tubes (most easily if they are of the cold-cathode type) and electroluminescent panels by using it to drive an unstabilized 'converter' which converts low-voltage DC to the higher DC or AC required by the lamp. The output of arc lamps and lasers can also be stabilized in the same general way, but the circuitry required is more elaborate and home construction is less practicable.

2.6 Uniform luminance surfaces

2.6.1 Diffusers and integrating spheres

A common requirement for vision research is a uniformly illuminated field. The provision of such a field using a Maxwellian view system has been mentioned previously here and discussed extensively in Chapter 1. When a Maxwellian view is inappropriate and a subject is required to view a uniformly illuminated screen directly, the experimenter has the choice of illuminating a translucent diffuser from the back or an opaque diffuser from the front (or a combination of the two). In either case it is difficult to get a perfectly uniform field if the diffuser is directly illuminated by a lamp, both because of the operation of the inverse square law (unless the lamp is a long distance from the diffusing screen the different parts of the screen will be at significantly different distances) and because of variations and irregularities in the angular distribution of the light from the lamp. In both cases the uniformity of illumination can be greatly improved if two or more diffusers are used, and can quite readily be made almost perfect if the viewed surface is illuminated by multiple diffuse reflections as in an *integrating sphere*.

An integrating sphere is a spherical enclosure whose inner surface has a diffusely reflecting white coating and whose wall contains two (or sometimes more) relatively small holes (or *ports*). Light enters the sphere through one (or more) entrance ports and is then repeatedly reflected from the diffuse white surface. Light leaving the exit port comes from all directions and makes this port act as a diffusely illuminated surface. So long as the light source itself and the region of the sphere's surface that receives its light directly is not visible through the exit port (this may require white baffles to be suitably positioned within the sphere), and in so far as the white surface reflects a large fraction of the incident light, the exit port will have the characteristics of a uniformly illuminated Lambertian surface. Although very good integrating spheres can be obtained from optical suppliers, they are expensive; but it is worth noting that remarkably good uniformity can be achieved even when the enclosure is not spherical, the exit port is quite large (with a diameter up to say one-third the dimensions of the enclosure), and the inner surface is simply painted with normal high-quality matt white paint. If a translucent diffuser is placed over the exit port, very good luminance uniformity can be obtained with an enclosure of almost any shape so long as it is reasonably large. The efficiency of an integrating sphere (the proportion of the total ingoing light that leaves the exit port) depends upon the reflectance of the inner surface and the size of the ports relative to the size of the enclosure. It can be roughly estimated, by assuming that the whole inner surface is evenly illuminated, that the exit port 'absorbs' all the light falling on it, and that the rest of the surface absorbs $(100 - r)/100$ of the light, where the reflectance of the white surface is r% (perhaps 95–97% for a good paint).

Integrating spheres make very good light mixers. The excellent uniformity across the output field can be very useful when coloured lights have to be mixed because of the high visibility of even small changes in spectral balance in different regions of an extended field.

2.7 Shutters

2.7.1 Ferroelectric shutters

Although the light output from some types of lamps may be satisfactorily controlled electrically (both modulation and complete switching have already been discussed), in other cases this is not possible and it is necessary to use ancillary shutters or attenuators (see Chapter 1). Most such devices are either basically mechanical (e.g. simple shutters and apertures) or require mechanical operation (e.g. neutral-density wedges, cascaded polarizers). While these devices can incorporate electromechanical actuators whenever it is necessary to control them electrically, it is not possible to make such mechanisms operate fast enough for some purposes, and it is also often difficult to prevent all signs of the image motion that tends to result from having moving elements in the light path. In these cases it will sometimes be preferable to use electro-optical devices to effect the necessary control. Of the several types of electro-optical shutter available, it is probably the ferroelectric type that is most likely to be useful in this regard.

Ferroelectric shutters (e.g. those from Displaytech) contain a very thin layer of a ferroelectric, birefringent, liquid-crystal material sandwiched between two glass plates, each of which has a transparent conductive coating to enable it to function as an electrode. This assembly is in turn sandwiched between crossed polarizers. If the axis of polarization of the light passing through the first polarizer is unchanged by the liquid crystal, it will be stopped by the second polarizer; only if its polarization axis is substantially changed will the light be transmitted. Application of a voltage between the two electrodes causes the molecules of the liquid crystal to be aligned in one of two directions according to its polarity, and in this way the polarity of the applied voltage determines whether the liquid crystal will rotate the axis of polarization of the light or not. The shutter thus operates as a voltage-controlled light gate, having only on and off states.

Ferroelectric shutters have the advantage over the much commoner nematic liquid-crystal types of being much faster (they can switch completely in less than 100 μs) and having a much better extinction ratio (on-to-off transmission ratio can be more than 1000:1). They operate on low voltages and their optical quality is high.

Although the binary nature of ferroelectric light shutters is obviously well suited to switching lights on and off (they are good for viewing frame-sequential stereoscopic cathode-ray tube displays), it is perfectly possible to use them to modulate light intensity by employing a variable on-to-off ratio technique in much the same way as was described earlier for controlling the light output of LEDs. For example, the speed of ferroelectric shutters makes it possible to control the effective intensity of a light source whose output is being gated at 200 Hz over a range of about 1:100 by adjusting the on-time between 50 μs and 5 ms. As with the control of LEDs, the effective light intensity is closely proportional to the fraction of time that the shutter is open. Ferroelectric shutters are relatively expensive and quite easily damaged. Careful attention should be paid to the instructions for their operation.

Small pixellated ferroelectric devices suitable for acting as spatial light modulators and intended for use in virtual reality displays are now available (also from Displaytech), though they are still rather expensive. These reflective devices offer substantial advantages over other liquid-crystal displays, since they are fast enough to be used in colour-sequential modes in conjunction with switched light sources (e.g. LEDs) and thus provide visual stimuli having arbitrary spatial and spectral characteristics. They are likely to become much cheaper after further development.

2.8 Light guides

Although the light from a light source can be transmitted from place to place in a beam using relay lenses and mirrors or prisms (see Chapter 1), it is often convenient to use fibre-optic light guides instead. The commonest light guides are round bundles of randomly arranged, flexible glass or plastic fibres whose function is simply to convey light from one place to another; they are often used in illuminators. Each individual fibre has a cylindrical core of high refractive-index material surrounded by a sheath of lower refractive index. Light that enters the fibre at an angle not too far away from the axis of the fibre is constrained to travel in the core by repeated total internal reflections at the boundary with the sheath material. Once the light has entered a fibre, it may be transmitted over long distances with little loss. As well as simple circular light guides it is also possible to obtain guides in which the shape of the entrance or exit aperture is not round. A guide with a rectangular aperture at one end and a round one at the other can make an efficient coupling between a light source and the entrance slit of a monochromator. It is also possible to purchase light guides in which the fibres start as a single bundle at one end but are separated into several bundles at the other. These might be used where a single, controlled light source was required to illuminate several separated targets. Liquid-filled light guides can provide the highest transmission, but the improvement is not great at visible wavelengths. Other fibre-optic devices are also available. These include both rigid and flexible optic-fibre bundles in which the topological arrangement of the fibres remains fixed along the length of the bundle. A coherent light guide of this kind can transmit an image from place to place.

As well as manufactured fibre bundles the vision experimenter may well find a use for very simple light guides made from glass or polished methylmethacrylate (e.g. Perspex, Plexiglass) rods. Although not very efficient, such a rod can be a convenient way of transferring light to a photodiode or of providing a fixation light in some position where it is inconvenient to place a lamp. For this latter purpose a glass rod that has been pulled to a relatively fine thread before being heated to form a small ball on the end can be particularly useful. Single optic fibres made from plastic materials can also be used for this same purpose. These have the advantage of leaking much less light and may therefore not need to be shielded along their length.

While optic-fibre light guides may be fed by focusing the light from any lamp on to the end of the guide, the experimenter may find it convenient to use one of the numerous fibre-optic illuminators that are manufactured for many different purposes such as medical endoscopy or microscopy. Although many commercial illuminators use tungsten–halogen sources, illuminators are also made with xenon arc lamps (e.g. using the ILC Cermax lamp) and metal halide lamps (high intensity discharge lamps containing mixtures of metal halides designed to give a broad spectrum but having many narrow lines). Illuminators intended for stage- and decorative-lighting purposes may incorporate computer-controlled colour filter wheels.

References

Berman, S. M., Greenhouse, D. S., Bailey, I. L., Clear, R. D., and Raasch, T. W. (1991). Human electroretinogram responses to video displays, fluorescent lighting, and other high-frequency sources. *Optometry and Vision Science*, **68**, 645–62.

Eysel, U. T. and Burandt, U. (1984). Fluorescent tube light evokes flicker responses in visual neurons. *Vision Research*, **24**, 943–8.

He, S. and MacLeod, D. I. A. (1996). Local luminance nonlinearity and receptor aliasing in the detection of high-frequency gratings. *Journal of the Optical Society of America A*, **13**, 1139–51.

Jacobs, G. H. and Deegan, J. F. (1994). Sensitivity to ultra-violet light in the gerbil (*meriones unguiculatus*). *Vision Research*, **34**, 1433–41.
Sekigichi, N., Williams, D. R., and Brainard, D. H. (1993). Aberration-free measurements of the visibility of isoluminant gratings. *Journal of the Optical Society of America A*, **10**, 2105–17.
Stark, W. S. and Tan, K. E. W. P. (1982). Ultraviolet light—photosensitivity and other effects on the visual system. *Photochemistry and Photobiology*, **36**, 371–80.

Useful sources of information

Cayless, M. A. and Marsden, A. M. editors (1983). *Lamps and lighting: a manual of lamps and lighting prepared by members of the staff of Thorn EMI Lighting Ltd.* Edward Arnold, London.

Wyszecki, G. and Stiles, W. S. (1982). *Color science: concepts and methods, quantitative data and formulae.* Wiley, New York, Chichester. An essential handbook for vision researchers (not only those working in colour vision).

Horowitz, P. and Hill, W. (1989). *The art of electronics* (second edition). Cmbridge University Press. An invaluable companion for anyone having a direct involvement with electronic devices and circuits.

Barss, M. (editor-in chief) and van Stryland, E. J. W., Williams, D. R., and Wolfe W. L. (associate editors) (1996). *Handbook of Optics* (second edition). Sponsored by the Optical Society of America and published by McGraw-Hill. The Vision section is particularly relevant.

Oriel Instruments' catalogue of light sources. This is free of charge and a mine of useful information and practical advice.

Photonics Directory published annually by Laurin Publishing Co. Inc. comprises four volumes:

(1) *Photonics Corporate Guide to profiles and addresses*
(2) *Photonics Buyers' Guide*
(3) *Photonics Design and Applications Handbook*
(4) *Photonics Dictionary.*

World Wide Web. Information from all the manufacturers mentioned in this chapter (as well as very many others) are available on the internet (*URLs for companies mentioned in Chapter 2 are given below*). It is also possible using ordinary search techniques to find much useful (if not always authoritive) information from enthusiastic amateurs (e.g. on flash tubes, stage lighting equipment).

Bes Optics	`www.besoptics.com`
Cree Research	`www.cree.com`
Displaytech	`www.displaytech.com`
E G & G Optoelectronics	`www.egginc.com`
Hamamatsu Photonics	`www.hamamatsu.com`
Hewlwtt Packard	`www.hp.com`
ILC Technology	`www.ilct.com`
International Light	`www.intl-light.com`
Ledtronics	`www.ledtronics.com`
Nichia Chemical Industries	`www.mesh.ne.jp/nichia`
OceanOptics	`www.oceanoptics.com`
Oriel Instruments	`www.oriel.com`
RS Components	`rswww.com (sic)`
UDT	`www.udt.com`

3

Topics in computerized visual-stimulus generation

TOM ROBSON

3.1 Introduction

In the last 20 years or so computers and other electronic equipment have become increasingly important for vision research. This is because they provide a flexible way to generate many kinds of stimuli which may be altered or adjusted in reply to an external input such as a subject's response. Although popular, these systems have many limitations from an engineering point of view, and it is therefore important that the vision researcher should have a good understanding of the way they work, what the potential problems and pitfalls are, and how they may be minimized. It is also necessary to have a quantitative knowledge of the deficiencies so that the researcher can decide whether or not they are significant in the desired situation. In many cases, it is possible to augment a particular parameter of the system by trading it off against another, so an idea of relative priorities is also needed.

A basic system for vision research consists of a computer, which is often an IBM-type PC or MAC, and a cathode-ray tube (CRT) display monitor. The monitor may be the same one as that used to control the computer but is usually a separate and more fancy device that can be sited remotely, in a darkened room for example. In addition to the basic system, the user can add devices for capturing images from the real world such as CCD cameras or CD-ROMs, devices for producing hard copy such as laser printers, devices for producing special effects such as stereo glasses, and devices for calibrating the system (very important). Video monitor configurations are not the only possible uses for computers in vision research, but since they are certainly the most popular and cause the most trouble they are the ones we shall consider here.

In this chapter we shall start by reviewing some of the fundamental aspects of using digital systems in an analog world and discuss a few of the engineering terms and concepts used by equipment designers and suppliers. After that, we will describe the principle of operation of some of the commonly used equipment and show how they can be used to generate different stimuli. Finally, we will mention some of the other miscellaneous aspects not covered elsewhere. In order to keep things simple, the scope of the chapter is mainly confined to monochromatic stimuli; throughout the approach will be largely non-mathematical and the reader referred to one of the many books available for a more rigorous discussion.

3.2 How do computers represent images?

All digital computer-based systems represent real-world images as a series of discrete picture elements which are commonly known as *pixels* or sometimes *pels*. Each of the pixels is used to record the luminance and colour of the image at a particular point in space (x, y) and time (t), the idea being that if the samples are closely spaced enough when they are subsequently displayed on an output device they will appear to coalesce and form a continuous display.

The pixels can be loaded into the computer's memory in one of two ways: either a real-world scene can be converted into pixels using a device known as a scanner (so-called because it 'scans' the image line by line), or they can be generated mathematically using a knowledge of the desired stimulus. In many respects the mathematical description has advantages over using 'real life' images as it allows the researcher to know and control the important parameters of the stimulus, such as the frequency content or the average luminance. Sometimes though, particularly when researching higher order cognitive functions, stimuli showing the sort of variations found in real life may be preferable. Once inside the computer, the image must be processed (if necessary) and then presented to the subject using a display device. No one has yet invented a display device that can take a digitized stimulus and transform it back into a continuous light source such as you might see from your window, so we are reliant on systems that reproduce the pixels in a similar manner to that in which they are stored within the computer. In other words, display devices can be considered as being made up of thousands of little individual light sources each of which is modulated by the stored pixel data. Each of the light sources should be thought of as having a uniform emittance across its whole emitting surface. Luckily, as everyone knows from watching television, under certain conditions the observer can't resolve the individual pixels and the image appears indistinguishable from one generated by a light source that is truly continuous in space and time. This of course begs the question: 'How many pixels are needed to achieve this effect and how frequently must they be changed?' The density of pixels per unit area is known as the spatial resolution, whilst the rate at which the pixels can be changed in time is known as the frame rate. This term is derived from CRT display technology where one complete scan of the display is called a frame, but it should not be confused with a similar term used by TV engineers for one-half of a pair of interlaced scans called a field. Note that input devices, computers, and output devices can all possess resolutions in exactly the same way, and in a well-designed system these will tend to mirror each other in order not to have unused (but paid for) capabilities in any particular component.

3.2.1 Some theory

What is spatial frequency? Everyone is familiar with the idea of Fourier analysis of a time-domain signal; that is to say, taking a continuous time signal such as a speech pattern in which pressure is a function of time, and decomposing it into the sum of numerous sinusoidal harmonics each of which has its own frequency, amplitude, and phase. This is a very useful technique, not only because it allows us to consider signals in frequency space which is normally more intuitive but also because, for linear systems at least, we can work out the effect of complete systems by considering their influence on each of the input frequency components separately and adding them all up at the end. The same concept can be applied to two-dimensional images where the luminance is a function not only of time but also of position in space. Starting with the case of a static image such as a photograph we can decompose our picture into the sum of spatial harmonics, each of which has a spatial frequency, amplitude, and phase. Unlike temporal frequency which can be understood as individual notes played on a piano, spatial frequency is less intuitive, but it can be useful to think about it as a measure of feature size. Things with big dimensions contain low spatial frequencies, while small things, or those with sharp edges, have a large, high spatial-frequency content. Obviously the apparent size of something depends on the position of the observer relative to it, so spatial frequencies are measured as the angle subtended by one cycle of the waveform at the observer in units of cycles per degree (cpd). For example, the fence posts (spacing 2 m) at the bottom of the garden (distance 30 m) have a fundamental spatial frequency of 0.3 cpd. Of course, there are also many higher frequencies present which are needed to define the exact shape of the posts, the surface texture, etc. It is convenient to remember that with a 57-cm viewing distance one cycle of a 1-cpd waveform measures 1 cm.

3.2.2 What is meant by bandwidth

Once we have the idea that images or stimuli can be represented in the frequency domain (both spatial and temporal) it is natural to consider what are the highest and lowest frequencies present, as any system that will have to display or manipulate these stimuli will need to work equally well for all these frequencies. In other words, in order not to introduce unnecessary distortion into our stimulus, the reproduction system should have a *flat* frequency response across this range. The bandwidth of the system must exceed the bandwidth of the signal. In practice, the observer of the stimulus will place upper and lower bounds on the required bandwidth as there is no point in reproducing signals if they are can't be utilized. For reference, the frequency response of a human, very roughly, is about 0.1–40 cpd in the spatial domain and 0.1–70 Hz in the temporal domain. It is difficult to characterize these bandwidths precisely as they depend significantly on the viewing conditions, including such things as the luminance and contrast of the stimulus itself and what else is happening in the neighbourhood. Please beware that the frequency response of a component of a visual system, such as a retinal cell, may be significantly higher than the frequency response of the visual system as a whole and this should be considered when probing internally. It is much easier to measure the bandwidth of a machine such as a CRT display and this is often quoted by the manufacturer on his data sheets; but more of this later.

The spatial bandwidth of a device is normally determined by its physical size at the low-frequency end and the density with which it can produce dots at the high-frequency end (the dot pitch). It is worth noting here that the bandwidth of a CRT display, but not a scanner or laser printer, also depends on the ability of the video circuitry to turn the electron beam on and off quickly. In fact it is not just the monitor that has this problem but the visual-stimulus generator as well, so it is convenient to lump them together and talk about the bandwidth of the system as a whole. This electronic bandwidth limits the ability of the system to reproduce horizontally adjacent pixels and therefore has the effect of reducing the resolution in this direction. Vertically adjacent pixels are unaffected by electronic bandwidth as they are really separated, as far as the electronics is concerned, by the time it takes to scan one line. A high-quality, visual-stimulus generator/display system will have a video bandwidth of at least 100 MHz, which corresponds to rise and fall times of about 3.5 ns. (A good way to measure the bandwidth is to generate a stimulus of parallel dark and light bars and measure the average screen luminance. Do this for horizontal and then vertical bars and for different separations. When the bars are horizontal, changing the spacing between them shouldn't affect the total luminance, but when the bars are vertical the bandwidth reduction caused by the video system combined with the non-linearity of the display will create a reduction in overall luminance.) (See Fig. 3.1.)

3.2.3 What is a sampled data system?

As explained above, computerized visual-stimulus generators choose to store continuous images as a series of closely spaced pixels or samples, each of which is derived by measuring the value of a real (or imaginary) continuous time (and space) image on a regular spatial and temporal matrix. This is called a sampled data system. In order to understand how a computer generates a stimulus consisting of a single spatial harmonic on a CRT-based display, let us start by considering the illustration shown in Fig. 3.2. The sine wave depicts the luminance profile that we wish to reproduce from left to right on our monitor, while the spaces between the thin vertical lines are supposed to indicate the phosphor dots on the screen that will emit the light. For the purposes of simplicity the screen is covered in uniform areas of phosphor without any gaps in between. The computer has been programmed to control the voltage to the display such that the luminance rises instantaneously to the commanded value and stays constant until instructed to change for the next pixel. The luminance of each whole pixel is chosen to be the luminance of the desired waveform at the start of the pixel. From the picture it can be seen that the resulting luminance on the display has a distinctly stepped appearance; but with a little imagination it can at least be seen to have the correct spatial frequency, albeit with a slightly different

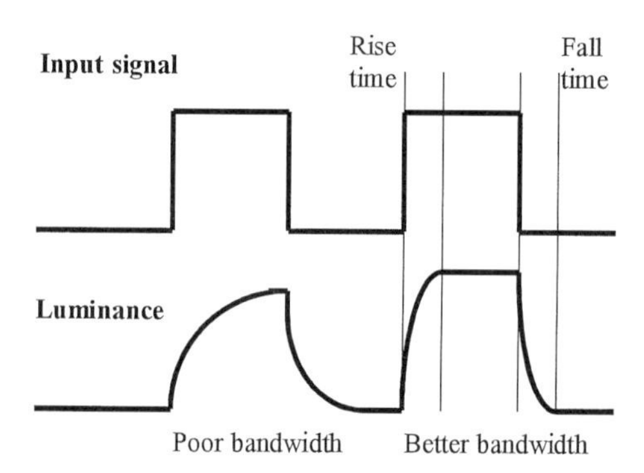

Fig. 3.1 What happens to the luminance when the bandwidth is poor?

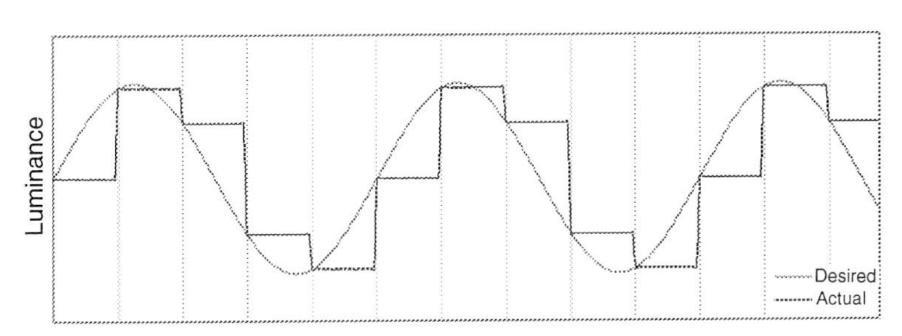

Fig. 3.2 Single frequency reproduction.

phase to the desired one. The next stage in the process is to take the stepped profile and pass it through a low-pass filter to remove all the higher harmonic components and which, if suitably adjusted, will reconstruct the desired waveform exactly, although still with a slight phase shift. But how do we implement this low-pass filter? Conveniently, the optical system of the eye will perform this function for us and, even though we have no control of the pass band, by placing our samples at an appropriate frequency we can achieve the required frequency response.

Figure 3.3 shows what happens if we try and use this system to reproduce a luminance profile of a higher spatial frequency than that shown in Fig. 3.2. The pixels still have the same spacing on the screen but, as can be seen, their luminance values have been chosen from different cycles on the waveform which results in the stepped profile shown. This is very similar to, and in fact has the same frequency as the example given in Fig. 3.2, so that when the low-pass filter is applied the two waveforms are indistinguishable. By not using enough samples to reproduce our desired waveform we have, as far as the observer is concerned, transformed one frequency into an unintended lower one. This phenomenon is called *aliasing*. Another way of looking at aliasing is as a beat frequency; if two frequencies f_1 and f_2 are mixed together the result will have components at the sum frequency $(f_1 + f_2)$ and the difference frequency $(|f_1 - f_2|)$. Looking again at Fig. 3.3, and considering the periods rather than the frequencies, we can see that the desired waveform has about 10 cycles across the diagram while there are about 13 pixels in the same space leading to an actual waveform that has $13 - 10 = 3$ cycles in the same space.

To put this yet another way, consider trying to reproduce a waveform of spatial frequency f cpd with a sampling or pixel frequency of s cpd. When low-pass filtered

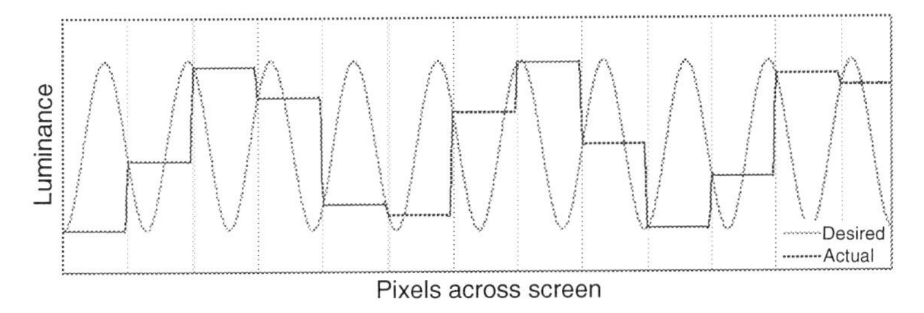

Fig. 3.3 Single frequency reproduction showing aliasing.

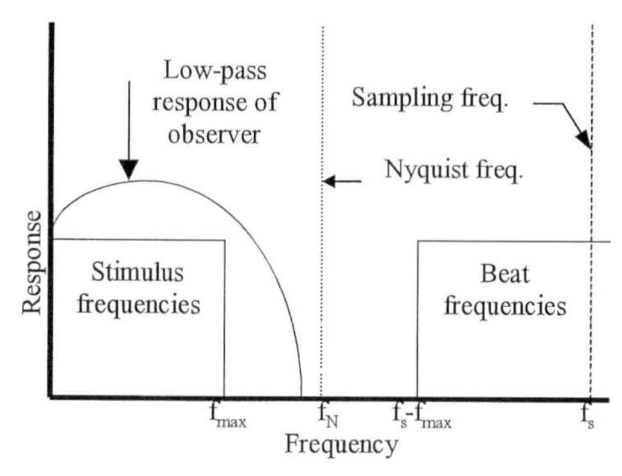

Fig. 3.4 The important frequencies in a sampling system.

there will be four frequencies present in the output, one at f, one at s, one at $s-f$, and one at $s+f$. If the beat frequency at $(s-f)$ is not to interfere or alias with the desired output at f then $(s-f)$ must be greater than f, or in words, the sampling frequency must be greater than twice the maximum frequency that is to be reproduced.

The important conclusions to be drawn from this discussion are these (and see Fig. 3.4):

1. To prevent aliasing, the sampling frequency (f_s), which in this case is the pixel frequency, must be at least twice as high as the highest frequency (f_{max}) that is to be reproduced. The corresponding temporal condition is that the frame rate of the display must be at least twice as high as the maximum temporal frequency that is to be reproduced. This is called the *Nyquist criterion*. The frequency $f_s/2$ is called the *Nyquist frequency* (f_N):

$$f_s > 2 \times f_{max}.$$

2. To prevent observable beat frequencies, the pixel frequency minus the highest spatial frequency ($f_s - f_{max}$) must lie outside the pass-band of the observer.

3.2.4 A real example

Having discussed sampling and the Nyquist frequency we are in a position to see what properties a display must have in order to be able to reproduce signals that span the complete range of human visual capabilities, say from 0 to 80 Hz temporally and 0 to 40 cpd spatially. To satisfy the Nyquist criterion in time we should have a sample frequency of at least 2 × 80 = 160 Hz; on a CRT monitor this translates to a frame-rate of the same value, while to satisfy the Nyquist criterion in space we need a sampling frequency (f_s) of at least 2 × 40 = 80 cpd. A standard colour monitor will have a mask pitch (d_{mask}) of about 0.28 mm, so this implies a viewing distance (d_{view}) of :

$$d_{view} = f_s/2 \times d_{mask}/\tan(0.5) = 1.3\text{ m}.$$

Remember that with modern colour displays increasing the frame rate will cause a corresponding decrease in the number of lines on the screen, and therefore the vertical sampling frequency, without a commensurate change in the horizontal sampling frequency.

If the resolution on the display is rather low and therefore a large viewing distance is needed for recreating high spatial frequencies it is probably easier to use a mirror than find a bigger lab. The software can easily be adjusted to correct the image reversal that this will introduce.

3.2.5 Square waves are nasty

It follows from the statement of the Nyquist criterion above that it is only possible to reproduce signals that are band-limited by using a digitally based stimulus generator. If an attempt is made to generate a non-band-limited stimulus, however, all the energy in the signal that is supposed to be above the Nyquist frequency (f_N) will be aliased into the pass-band of the system and a distorted waveform will be produced. The group of signals that most commonly fall into this category is square-waves. Fourier analysis of a square wave (f_s) with a fundamental frequency f_0 shows that it is composed of the sum of the odd harmonics (f_n) with decreasing amplitude to infinity:

$$f_s = f_0/1 - f_3/3 + f_5/5 - f_7/7\ldots, \text{etc.}$$

Even if the sampling frequency is arranged such that the fundamental harmonics, or even the lower ones, pass the Nyquist criterion, the relatively slow convergence of the series will mean that significant energy is available for aliasing.

To avoid this problem and ensure that the stimulus is as faithfully generated as possible without artefacts, think in terms of sine waves only and shun square waves or anything with sharp edges. This applies equally to spatial and temporal waveforms.

3.3 A typical computer-based, visual-stimulus generator

A typical system for use in vision research will consist of three principal parts: something for capturing an image, a computer for processing the image, and a display device for presenting the image to a subject. In a large number of cases the capture stage is avoided by devising a stimulus that can be described in simple mathematical terms and using the computer to generate it, but all of the situations discussed here will have a computer and some kind of display device. Figure 3.5 shows such a typical system.

3.3.1 How to get data into the computer

Most visual researchers will choose to generate their stimuli entirely within the computer without reference to the outside world at all, but sometimes it is necessary to work with real scenes that contain natural selections of colours, luminances, and

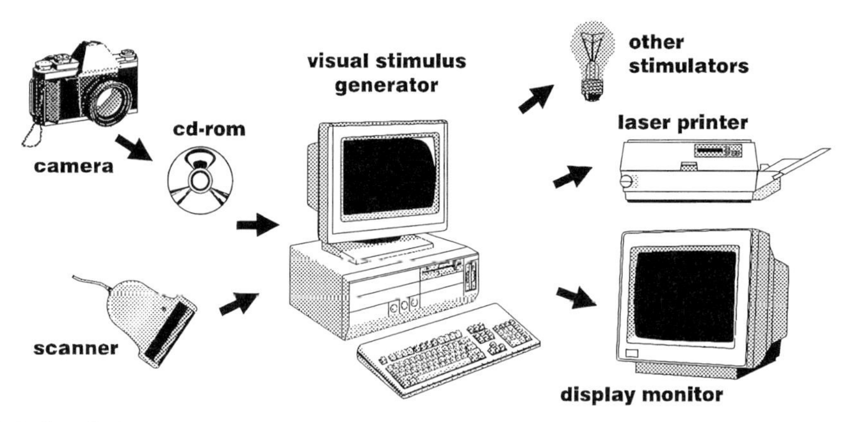

Fig. 3.5 A typical computer-based, visual-stimulus generator.

spectral composition. In this case an input device of some description is needed. Exactly which one to choose depends on which parameters are important, but the questions to ask are: 'Do I want colour?'; 'Do I want moving or static images?'; 'What resolution is needed?'; and, of course, 'How much money is available?'. Here is a short summary of the available options:

1. *Scanners.* Scanners are very freely available now, and often come in a flatbed format in which you insert the original material to be scanned, close the lid, and press the button. Typical resolution is 1200 dots per inch and 8 bits of greyscale or 24 bits of colour. Scanners will only work with paper or flat originals and obviously can't handle moving pictures.
2. *CD-ROM.* This is a very popular format now, particularly for high-resolution images where the amounts of data are very large. Prerecorded images can be purchased from many suppliers, or a better solution is to capture your own on 35-mm film and have them scanned and recorded on to CD-ROM by almost any high-street processor (very cheap). Scanned photographs have a very high resolution, typically 3000 × 2000 pixels with 24 bits per pixel, so the amount of data is enormous. You will definitely need to subsample these and reduce the quantity of data before using them.
3. *Video or CCD.* This is the only way to go if you want any kind of moving image or sequence of frames. It normally involves adding a card to your computer to digitize the image and a source of video data such as a CCD camera or VCR. Resolution is medium (256 × 256 pixels) and colour quality is poor, particularly when the image is obtained from a television system.
4. *Internet.* Many researchers like to use the same images as their colleagues in order to compare results more easily. These can be obtained from many sites on the Internet and have the benefit of being free. Indeed it has become something of a game among the image-processing fraternity to try and establish pictures of children or students as *de facto* standards in this area.

3.3.2 Data rate

From the preceding discussion it should be apparent that real-world images take up a lot of memory in the computer, so something should be said about moving this data around. For example, a photographic-quality image of the kind obtainable from a 35-mm colour print has a resolution of 3000 × 2000 (or 6 megapixels) and if stored with 24 bits per pixel would need 18 megabytes! To give this figure some kind of perspective, the maximum data transfer rate of the ISA bus found in ageing PCs is about 2.4 megabytes/second, while even that of the PCI bus is only 33 megabytes/second. If you are considering showing photographic-quality movies from disk on even a high-end PC don't be surprised if it doesn't work. Much work has been put into developing data-compression techniques that allow the display of at least TV-quality images in real-time, which means a resolution of about 320 × 200 at a frame-rate of 50 Hz, with the sort of data rates that can be sustained by a CD-ROM drive and a PC—you will often see the acronym MPEG (Motion Picture Experts' Group) in this context—but so far these techniques are of little benefit to vision researchers. Of more interest are the data-compression algorithms designed for still images, such as JPEG or PCX which can be used to compress images for storage on disk. They are often used as a common-file interchange format between graphic and image-processing programs, but the computation required in compression and decompression renders them unsuitable for real-time use.

3.3.3 How are images represented inside a computer?

As indicated in an earlier section, computer systems store images in their memory as a matrix of pixels. If the image changes with time, then a succession of these matrices is stored; one for each time frame. The data recorded for each pixel must contain its position on the screen (indicated by its location in the matrix), its luminance, and, in the case of colour images, its colour. In the simplest case, such as a laser printer, each pixel can only be black or white, being a presence or absence of toner, so only one bit of information need be stored for each point. In a typical personal computer graphics system each pixel normally has 8 bits, or one byte, allocated to it. This number can be thought of as a code that is used to select the final displayed colour and luminance from an array of preselected ones. The analogy commonly made is with an artist's palette. Although theoretically able to mix combinations of paints and thinners to make millions of different colours, the artist is only able to have a small selection of them on a palette at one time. Note that numerically consecutive pixel values need not translate to adjacent luminances or colours, and that the palette may contain colours and luminances selected from the entire gamut of the display device. In this respect, the pixel value acts only as a place-holder. It is worth pointing out that *nothing* can be inferred about the resolution of luminances and colours with which the palette can be filled by knowing that each pixel is only represented by 8 bits. What can be known though is that it is only possible to have 256 different luminances (and colours) displayed simultaneously on the same frame. As computer memory becomes cheaper we are seeing the rise in popularity of so-called *true-colour* graphics systems. These typically use 24 or 32 bits in which to store each pixel, some of the bits (often 14) being used to store the luminance information and the remainder used to store the colour information—the idea being that all the data is contained within the pixel so that the palette is superfluous. True-colour systems are superficially attractive, but they have the drawback that each image occupies three or four times more memory than in the 8-bit systems; which, while not of itself a problem as memory is now fairly cheap, means that it takes three or four times longer to load or construct thus limiting the dynamic capabilities of the system.

3.3.4 The importance of output resolution

The output resolution of a visual-stimulus generator is the number of distinct levels that can be used to reproduce the stimulus. On a colour CRT display this corresponds to the resolution with which the voltages to drive the red, green, and blue guns can be created. (Laser printers are given greyscale capability using a method of super-pixels which is discussed elsewhere.) A typical personal computer graphics card will use one 8-bit digital-to-analog converter (DACs) for each of the three outputs, resulting in a resolution of 1 part in 256 for each colour. Imagine using this resolution to represent a spatial waveform with a contrast of 0.5%—a not untypical requirement—and you will see that there aren't very many luminance levels left to define it (one or two). The situation is made worse by the fact that CRTs have a non-linear transfer function between voltage and luminance (approximately x^2), so that voltage steps near the top end of the voltage range produce a far greater change in luminance than those near the bottom. In a colour system the situation is worse still as most of the luminance comes from the green gun with very little being contributed by the blue, yet they have the same voltage resolution for both. A good visual-stimulus generator therefore should have high resolution for the blue output, very high resolution for the red output, and ultra resolution for the green output. Failing this though, high resolution for all three guns will do. Look for at least 12–15 bits per gun. Do not confuse these with true colour graphics systems that still only have 8-bit DACs on their outputs, and remember that it is perfectly possible to generate high-quality stimuli with only an 8-bit framebuffer provided that it has high-resolution DACs at the output.

3.3.5 Manipulating the data inside the computer

Once the stimulus is inside the computer it may need to be manipulated in some way or other. This is often done as part of an experiment in which one parameter of the stimulus, such as contrast, is changed as a consequence of responses from an observer. There are many processes encompassed by the idea of manipulation and only a few can be discussed here. Many of the operations that are performed on real-life images, such as filtering and feature extraction, are normally classed as image processing and are well dealt with in other literature. Of the principal techniques used in visual-stimulus generators, one involves the use of look-up tables and the other the summation of two components of a stimulus using interleaving.

3.3.5.1 Look-up tables

Look-up tables, or LUTs, form a basic technique of visual-stimulus generation. The idea is simple (Fig. 3.6): one number is converted to another by using it as an index into a table of other numbers. Imagine a set of pigeon-holes each consecutively numbered and spanning the range of the input set, 0–255 for example. Into each of these boxes place a piece of paper on which is printed another number selected from an output set. To use the look-up table and convert one number to another, just go to the box whose address is the number you currently have and extract the new number on the piece of paper and use that one instead. Note that the numbers in the look-up table (or the paper in the boxes) need have no relationship to the address of the box, nor the numbers in the other boxes, nor need they be unique. Indeed it is quite possible, and useful, for all the numbers in the look-up table to be the same as each other.

What can we do with LUTs? A look-up table can be implemented easily both in hardware and software. A software LUT is just an indexable array of numbers, the index being the input set and the contents of the array the output set; while a hardware LUT is just a random access memory (RAM) where the address lines are driven by the input number and the data lines provide the output number. Some method has to be found of writing data in the RAM in the first place but this is usually easy.

A perfect use for an LUT is as a method for correcting non-linearities in a display device, often called gamma correction. The input set is the desired luminance expressed on a linear scale from 0 to 255, while the output set comprises those numbers which must be sent to the digital-to-analog converters (DACs) to provide the voltage that will cause the display to generate that luminance. In this case, the contents of the LUT are usually derived from prior measurements of the display transfer function. Of course, there is no reason why the LUT contents need to span the available dynamic range of the display, so if the user needs to reduce the contrast of the stimulus or change the mean luminance then this can be done just by changing the contents of the LUT and without change to the original stimulus at all. In many cases a stimulus is presented to an observer using a defined temporal envelope such that the display is initially blank, the contrast is increased slowly to the desired maximum and then decreased again until the stimulus has been removed. This is done to reduce artefacts caused by transient responses and can be implemented entirely using LUTs. Before the experiment begins, a sequence of LUTs, one for

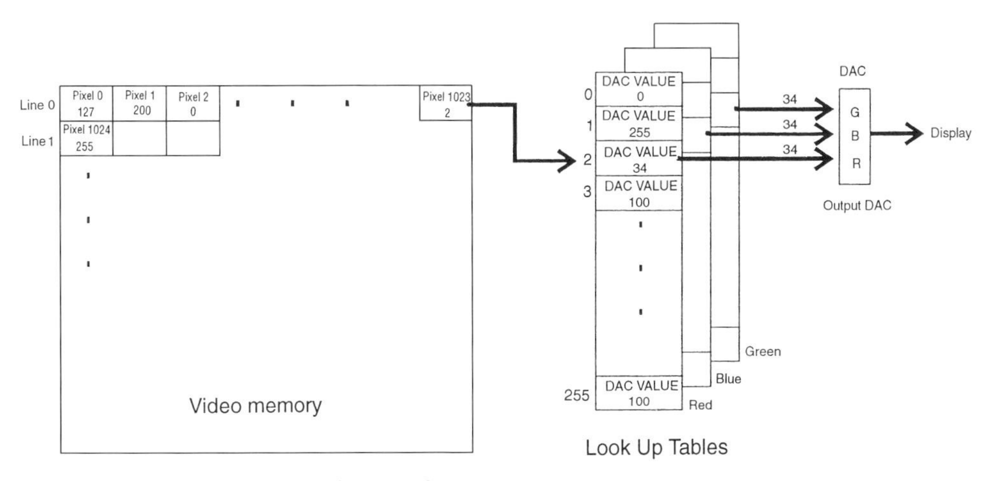

Fig. 3.6 Palette and look-up table operation.

every frame of the stimulus, is precalculated and loaded on to the visual stimulus generator. Then on receipt of a start command the look-up table is loaded from the precalculated values on a frame-by-frame basis.

LUTs are also commonly used in the generation of the stimulus itself. Consider the problem of trying to generate a dynamic noise field on a CRT display. One technique that can be employed is to draw a complete noise field into the framebuffer of the stimulus generator. Then after a predetermined time (say one frame) to draw a completely new field and display that instead, repeating the procedure for as long as required. The drawback of this scheme is that it is rarely possible using conventional equipment to achieve both the recalculation and drawing of the new frame in the time available. A better technique is to draw a noise field where each little spot of noise is given a value from the available range of pixel values (normally 0–255). The pixel values are then converted to real luminances using a look-up table. Simply by changing the contents of the look-up table between frames, which may only involve rewriting 256 numbers, a completely new field can be created. If binary noise is desired, i.e. the pixels are either black or white, then different combinations of pixel value can be mapped to one of two luminances in the LUT.

Another example of this kind of technique is the generation of a drifting sinusoidal grating. As with the noise example above this could, in theory, be generated by either precomputing successive frames of the stimulus and showing them sequentially, which needs a lot of memory, or by drawing each new frame on the fly, which needs a lot of processing power. A neater method involves drawing a saw-tooth waveform into the framebuffer with the desired orientation and spatial frequency in such a way that the numbers from the framebuffer form the input set of a look-up table. If the look-up table is filled with luminances taken from a table of sines then the resulting output will also have a sinusoidal function. If, in addition, the look-up table is also reloaded between frames with numbers from the same sine table but in a different phase then the grating will appear to drift. Figure 3.7 illustrates the idea.

3.3.5.2 Temporal and spatial summation

Some common stimuli, such as plaids, are composed of two (or more) independent components summed together. One way to do this is to calculate the two components

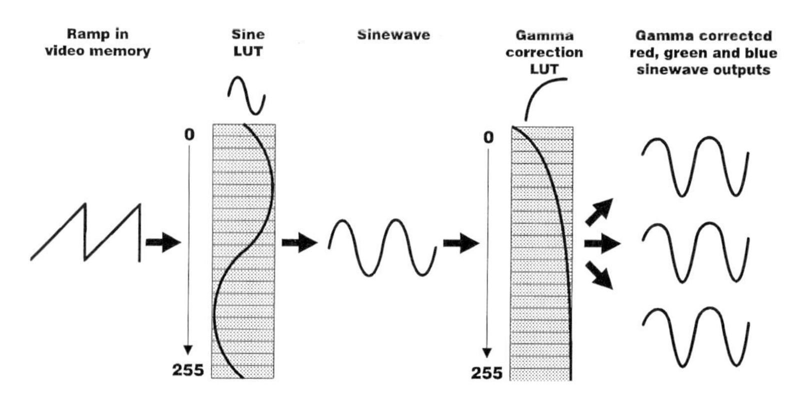

Fig. 3.7 Generating a gamma-corrected signal using ramps and LUTs.

separately, add them together numerically, and then display the result. If the parameters of one of the components then needs to be changed the whole operation needs to be repeated, which can take a long time. A second way of doing it is to divide the framebuffer into two sections, each of which contains one of the components of the plaid, and arrange the display controller to display each of the sections on alternate frames of the display. If the display rate is fast enough (> 140 Hz), the observer will add the two components together optically and perceive the same result. The beauty of this scheme is that each component can be manipulated individually. A similar thing can be done by interleaving the components on alternate columns of the display, which if narrow enough will give spatial summation.

3.4 Tricks and tips to improve aspects of the system

In most real-world situations a vision researcher is presented with some particular apparatus left from a previous era or a finite sum of money with which to purchase new equipment. In the latter case, the researcher examines the available suppliers and rapidly finds that it is easy to meet most of the requirements within the specified budget, but that straying outside the scope of commodity products causes the cost to increase alarmingly. What can be done to squeeze that extra bit of resolution from the system? In many cases it is possible to trade an excess of one parameter for an improvement in another, although if the system only just 'cuts the mustard' on all accounts then almost certainly nothing further can be done.

3.4.1 Spatial

The most obvious symptom of the lack of spatial resolution in a display is the appearance of jagged edges or 'jaggies'; that is to say those disconcerting little jumps that occur in lines that are drawn close to the vertical or horizontal. As might be expected, jaggies can be explained in terms of a sampling process. A straight line or edge is not a band-limited phenomenon, but contains frequency components that extend to infinity which, as we already know, is forbidden in a sampled data system if aliasing is to be avoided.

As jaggies are an aliasing phenomenon, the obvious way to improve or remove them is to increase the sampling rate, i.e. the spatial resolution of the display. However, in many systems this is not possible as it will already be running at its limit anyway, so another scheme is called for. If the sample rate cannot be increased then the high-frequency content of the waveform must be reduced by filtering it, and the simplest way to do this is to utilize the ability of the stimulus generator to reproduce *greyscales*. For an easy way to do this, look at the illustration in Fig. 3.8 which shows a section of a dark bar drawn across a grid of pixels (the larger squares) to which has been added a grid of smaller imaginary pixels in the ratio 1:64. Instead of allocating one of two luminances to the real pixels in the traditional way, just count the number of smaller pixels covered and use this number to allocate one of 64 different luminances instead. Although this tends to blur the edge, because we are in effect low-pass filtering it, it also allows the edge to be positioned spatially with far greater resolution than before; a situation analogous to hyperacuity in the human visual system.

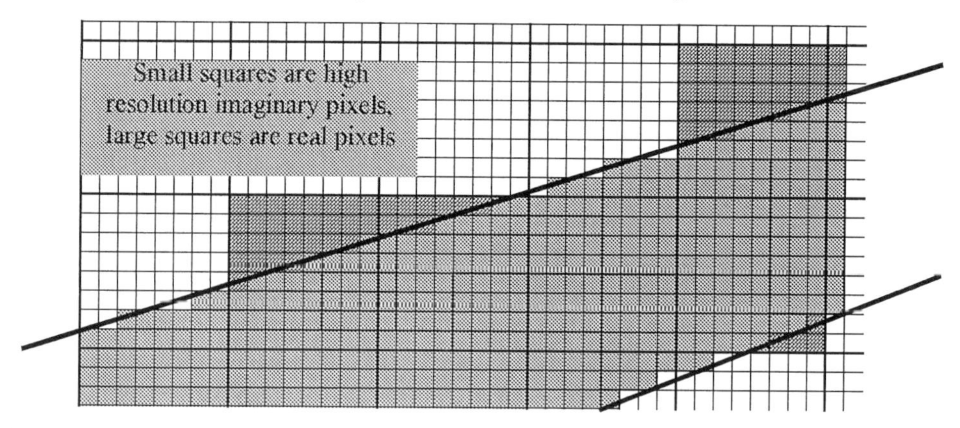

Fig. 3.8 Removing jaggies.

However, it is an interesting fact that many observers state a subjective preference for images with a significant, if spurious, high-frequency content over those without, even if the high frequencies come from the raster of a CRT display or random lines drawn on the image.

3.4.2 Temporal

Techniques for increasing temporal resolution are exactly analogous to those that can be used for increasing spatial resolution. For example, consider presenting a stimulus to an observer by turning it full-on on one frame of the display, leaving it at the desired luminance for 10 frames, and then turning it off again. If the display is operating with a 100-Hz frame-rate this corresponds to viewing the stimulus through a 100-ms square time-window. As usual, the word square should be a warning that bandwidth limiting is not operating and that the frequency components of the underlying stimulus are liable to be modified when viewed in this way. The preferred technique in this case is to utilize available luminance resolution by viewing through something like a raised cosine or Gaussian window, whereby the stimulus intensity is slowly increased from zero over several frames before reaching a plateau and then being ramped off again. The desired stimulus waveform should be seen as coming from a continuous function, or at least a very long list of numbers, which is subsequently sampled by the display process once per frame. In this way, the temporal resolution of the system is determined only by the available luminance resolution and the accuracy with which the function can be calculated, and not by the display's frame-rate.

3.4.3 Luminance

Increasing the available luminance resolution of display devices can be brought about in two ways: software and hardware.

3.4.3.1 Software

The software schemes are the easier to understand and implement and are particularly important when trying to generate images using laser printers. In a laser printer, a small laser is used to apply a dot of electric charge to the print drum which either attracts or repels the grains of toner material. Toner is black, paper is white, and the toner is either there or it isn't; grey levels are intrinsically unreproducable in this process. The saving grace of the laser printer though is its high dot resolution; 600 dots per inch is common today and 2500 dots per inch is available on some systems. The soft solution to giving these devices greyscale capability is to assemble dots together as superpixels in the form of a matrix of say 8 rows by 8 columns and fill it in according to the desired grey level; no dots for white and 64 dots for full black, or in other words 6 bits. Of course, the theory is better than the practice and you can't really get the full 6 bits hoped for, principally because the dots are approximately circular so they don't tessellate well.

Similar techniques can be used to increase the luminance of CRT displays, although the improvements are less dramatic than with the printers as the amount of excess spatial resolution is less. It is easy to overlook the temporal domain here. If your display is capable of running at 150-Hz frame-rate (many modern ones are), then an extra bit of luminance control can be gained by displaying similar images on alternate frames and allowing the eye to integrate the two optically.

3.4.3.2 Hardware

The hardware solutions to stealing a few extra bits of luminance control are really only possible with CRT-type displays, as even the most accomplished home engineer will have trouble fiddling with a laser printer. The most common hardware-based scheme is used with monochrome displays when one has the benefit of colour-capable display hardware. Assuming that the outputs are already 8 bits then the theory is this: take the green output, divide it by 256, and add it to the red output using the combined signal to drive the monochrome monitor. In this way, the levels generated by the green signal will fit nicely between those from the red output giving $256 \times 256 = 65\,536$ distinct levels, or the equivalent of 16-bit luminance control. Unsurprisingly, anybody expecting to achieve this type of improvement may well be disappointed, but a very useful gain can be made this way. The principal problem is that the video DACs found on graphics cards are only accurate to 0.5 LSB. This means that the output levels may be as much as one part in 512 from their ideal level. Resistors for the dividing circuit are not available in exactly the required values with the necessary precision either. And what is more, it is very difficult to build a circuit that works well at video frequencies of over 100 MHz. And if that's not enough, who has a monochrome monitor anymore? Nevertheless, 12 bits can be obtained using a technique like this.

3.5 Miscellaneous

3.5.1 What is gamma?

Most input and output devices used in vision research have a non-linear transfer function. That is to say that in the case of input devices the number or voltage produced at the output is not linearly related to the luminance being measured, and in the case of output devices the luminance is not linearly related to the driving voltage or number. Hot-electron tubes such as CRTs are characterized by a transfer function of the form:

$$\text{Luminance} = k\,(V - V_0)^{\gamma}$$

where V is the applied voltage, V_0 is the brightness level, and γ is about 2.5. For vidicons (TV cameras) γ is about 0.5 and for silicon diodes (photocells) γ is about 1.

This is an approximate relationship and doesn't apply to other devices such as laser printers and photography where the underlying mechanism of the non-linearity is different. Despite these differences, however, they are all considered undesirable as they distort the final stimulus and it is the goal of the vision researcher to remove their influence as far as possible; a process that is known in all cases as gamma correction.

Before a piece of equipment can be gamma corrected the non-linearity must be characterized in such a way that an inverse transfer function can be concocted. The best way to do this will depend on the equipment to hand and how much effort can be spared. If it takes two minutes to perform each measurement one is inclined to devise a system that makes few readings, whereas if the system can be fully automated a blunderbuss approach may be more appealing. In either case, sufficient measurements should be taken to ensure that the users' error criteria can be met. Gamma correction of CRTs can be performed very adequately by making a dozen or so measurements across the full luminance range and fitting a function of the general gamma form using a standard minimization technique. For good results the readings should be taken in a dark room with the brightness control on the monitor adjusted such that a zero voltage input gives a just-black screen and the contrast control adjusted so that with the maximum voltage the display achieves peak output but doesn't saturate (Fig 3.9). Once set, these controls should be anchored to prevent the chance of inadvertent adjustment by visitors to the lab. Careful use of a hand-held photometer mounted on a tripod and a general purpose spreadsheet will enable gamma correction within the errors of the original readings. If using a colour display, the three guns should be characterized individually and checked using a white stimulus. If the display is a good quality one, the three guns will add together independently.

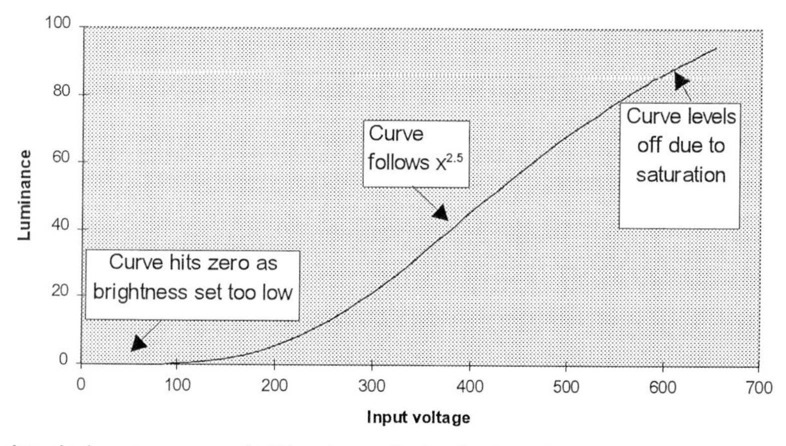

Fig. 3.9 A typical gamma curve and the signs of misadjustment.

3.5.2 Stereoscopic displays

To achieve stereopsis it is necessary to generate two images and present them separately to each eye. Depending on the purpose of the exercise, the images may or may not be correlated. In an ideal world there should be no timing skew between the two images, as the processing paths from the two eyes are more sensitive to timing disparities than would be suggested by considering the temporal bandwidth of the overall system. There are several ways to do this, each of which has its own disadvantages.

3.5.2.1 Two-stimulus generators

The simplest way to create two stimuli is to have two distinct stimulus generators and display devices, and use an optical system to ensure that the image from each impinges only on the intended eye. This works well if the subject can be fixed rigidly in a chair and a septum is used to prevent crosstalk. However, it may be difficult to arrange if the display devices are large as it may not be possible to get them very close together. In addition, there remains the problem that the two displays have to be adjusted to have the same geometry, brightness, luminance, and colour and some mechanism needs to be found to synchronize the two frame scans, not to mention the two generators.

3.5.2.2 Split screen

A variation on the two-stimulus generators' method above is to use one stimulus generator with a large display and split the screen into two areas, again using a septum to prevent crosstalk. The advantage here is that there is no need to solve the synchronization problems and that any adjustments for geometry and luminance will apply equally to both images. Even though this technique halves the available viewing area you can be quite sure that there is no timing difference between the two eyes.

3.5.2.3 Shutters

The third method of simulating stereopsis, and that most often practised, is to use a pair of shutter goggles synchronized with the display. The idea is to present the images destined for each eye on alternate frames of the display and then prevent the light reaching the wrong eye by closing a shutter in front of it. The ideal characteristics of these shutters should be instantaneous switching from open to closed and vice-versa, 100% distortion-free transmission in the open state and 0% transmission in the closed state, and they should be light enough to be worn either as goggles or made into spectacles. In the past, these have been constructed of black rotating discs that have apertures cut out of the sides driven synchronously with the display, rather similar in concept to the shutter systems on movie projectors. While having very good optical characteristics their portability qualities leave room for improvement: not everyone likes the idea of something akin to a circular saw blade 5 cm from their face. In recent years the emergence of liquid-crystal materials has provided an appropriate non-moving alternative. These rely on the property that some liquids have of being able to rotate the plane of polarization of light when a small electric field is applied. When sandwiched between a pair of crossed polarizers and connected to a suitable voltage source they can form the basis of a shutter system light enough to be worn on the head. Currently, the best available system uses a ferroelectric material that switches state in less than 100 μs and can be driven from a simple 5-V signal source. In the open state, transmission is 25%, while when closed it may be as much as 3-log units lower. The closed state transmission relies heavily on the accuracy with which the polaroid sandwich has been aligned, and shows very substantial degradation at elevated temperatures and off the principal axis. When expressed as a contrast, an open to closed transmission ratio of 1000:1 sounds very impressive, but will undoubtedly fail to impress a dark-adapted observer who expects to be able to see nothing through them when they are closed. The primary sources of error when using this stereoscopic system are a guaranteed eye-to-eye temporal disparity of one display frame time and crosstalk from one eye to the other caused by the finite extinction ratio of the goggles and the decay time of the display phosphor. It is possible to buy monochrome CRTs with a short persistence phosphor for this purpose but, in practice, the deficiencies of the goggles are the dominant factor. Whether the crosstalk is invisible or important will depend on the exact stimuli being used. Remember that the eye is sensitive to local contrasts rather than luminances, so that if one eye is presented with a high-contrast pattern while the other eye is shown a black image crosstalk is almost inevitable. (To check the crosstalk in a system display a stimulus, such as a plaid that is constructed by interleaving frames, and close one eye and see how much of the image intended for the other eye is visible.) Also, don't forget when using this system that the net frame rate of the image presented to each eye is half the total display frame rate. If the display is operating at 140 Hz each eye will see an image presented at only 70 Hz and consequently it may seem somewhat flickery.

3.5.3 Using a CRT as a tachistoscope

A tachistoscope is a device that has been used by visual psychologists for many years to present stimuli of differing durations to the subject. Its three essential elements are a constant (non-pulsed) light source, a slide containing the desired image such as a shape or word, and a shutter mechanism to control the illumination of the slide by the light. By adjusting the shutter, the image on the slide can be exposed to the subject for variable lengths of time. In a typical experiment, the tachistoscope is used to show the subject a word for increasing amounts of time until he or she can just read it. Because of the inherent inflexibility of these mechanical devices it would obviously be very useful to be able to reproduce the functionality of the tachistoscope using computer-generated images which could be presented for computer-controlled lengths of time. This would even allow more complicated experiments involving such techniques as masking to be designed.

At first sight it seems that a CRT-based display would not be suitable for such a system as the frame rate inherently limits the time resolution to about 10 ms, which is probably ten times too low to be useful. As we shall see, however, this is an erroneous analysis based on several misunderstandings—a computer system can be used as a very successful tachistoscope. To help the explanation it is useful to know a little about the response of the eye to pulses of light such as those emitted by a CRT display, but there is no need to consider the brain or assume anything about higher order cognitive functions. Many people have described and modelled the *impulse response* of the eye: which technically is the response to a pulse of light of infinitesimal duration and unit area, but is effectively the response to the short pulse (in the order of 1 ms) produced by a CRT display as each area of the screen is scanned. There is general agreement that it consists of a positive response followed by a slightly delayed negative response, with the whole event happening over a period of about 150 ms which we shall call the *natural time*. In addition to the shape of the response, it is generally accepted that the eye can be considered, under most conditions, as a linear system so that the response to an arbitrary input is simply the convolution of the input with the impulse response. In other words, to find out what happens when we expose the eye to two successive pulses of light from two consecutive frames from a display we can simply add together two impulse responses, with the second response being delayed by one frame time. The situation is shown in Fig. 3.10 below. In this figure, one curve shows the response to a single displayed frame, while the second curve shows the response to two successive frames.

The most obvious feature about the two curves is not that one is twice as long as the other, which might have been expected from the fact that it was created by displaying two frames rather than one, but that the response to two pulses is nearly twice as large as that caused by the single pulse. In fact, it can be shown that the response to twin pulses depends little on the width of the pulses and little on their separation (i.e. frame rate) and a lot on their aggregate area which is a measure of the stimulation energy. Even the small difference in phase between the two curves shown can be removed if time is measured from the centroid of the stimulation rather than the start. In situations like this in which the stimulation time is

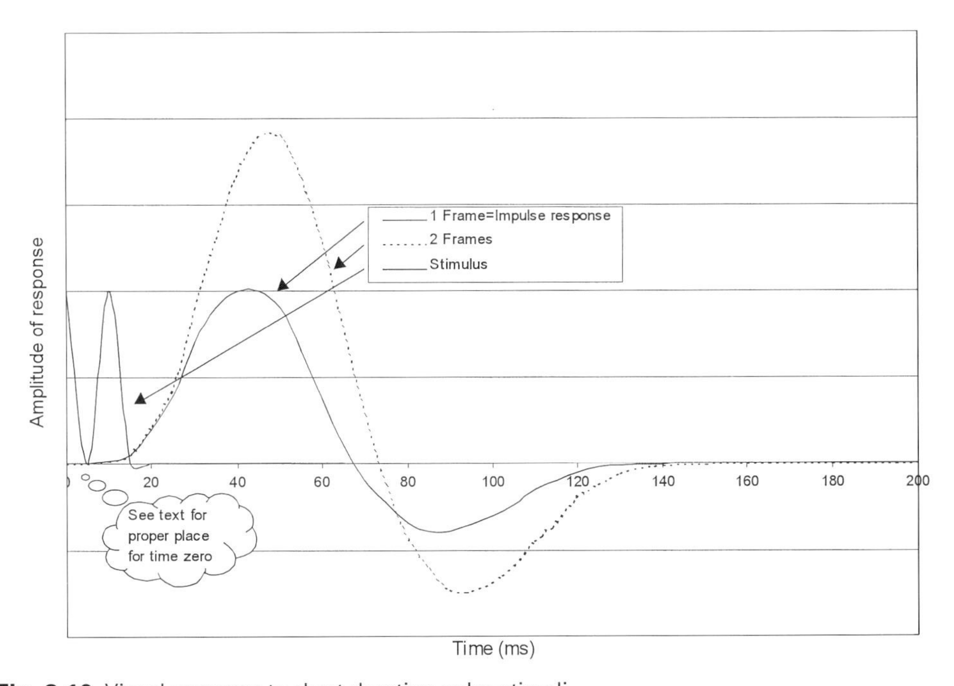

Fig. 3.10 Visual response to short-duration pulse stimuli.

significantly shorter than the *natural time* of the eye, all we achieve by increasing the stimulation time is to increase the magnitude of the response, and it makes little difference if the stimulation is provided by a uniform light source such as found in a tachistoscope or a pulsed light source, e.g. that provided by a CRT. If you still think that the stimulus duration is important in this situation then the energy input can be maintained at a constant level simply by adjusting the luminance downwards as the stimulus duration is increased.

Let us now consider a different condition where the stimulation time is fairly long compared to the *natural time* of the eye. This situation is illustrated in Fig. 3.11 below for stimulations of 14 frames, 15 frames, and 15 frames where the last frame is at half luminance. A frame rate of 100 Hz was used, so this represents the response to stimuli of about 150 ms. As can be clearly seen, the shapes of these curves are very different to those produced by short-duration stimuli, and they show three distinct regions: the initial turn-on transient, a steady-state regime, and a turn-off transient. The turn-on and turn-off transients are the same for all three stimulations, while only the length of the steady-state region depends on the stimulation time. The ripple that can be seen in the steady-state region is the only legacy of the pulsed input, and this can be shown to be reduced as the frame rate is increased beyond 100 Hz. The most interesting fact to notice is that the turn-off transient starts at a time determined by the length of the stimulation: increase the stimulation time by one frame or 10 ms and the turn-off transient begins 10 ms later. More interesting even than that perhaps, is the fact that making the 15th and final frame half-luminance produces a response that occurs halfway in time between that caused by 14 and 15 frames at full

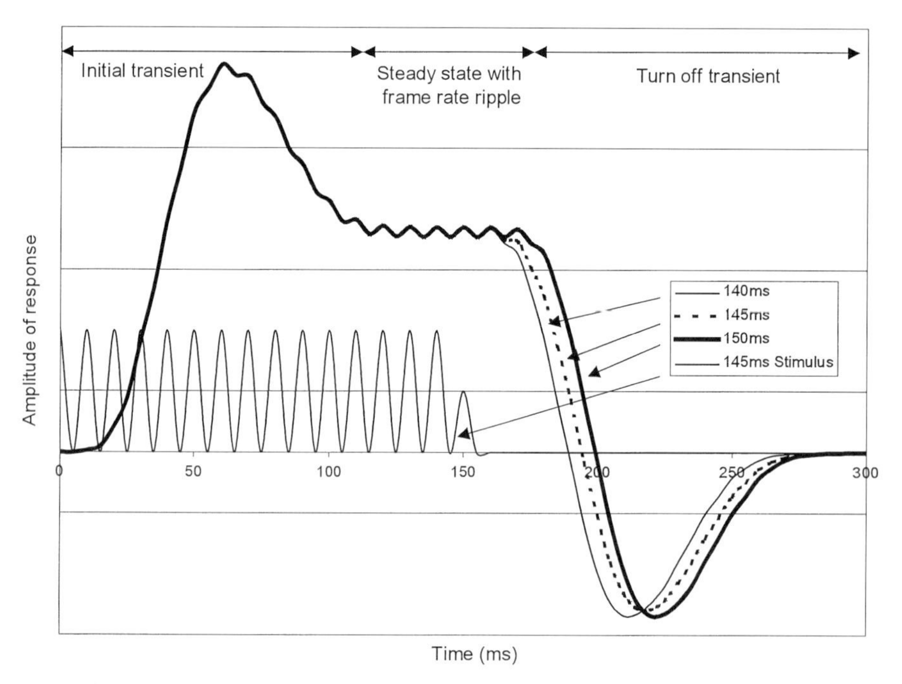

Fig. 3.11 Visual response to long-duration stimuli.

luminance. This means that by controlling the luminance of the final frame in the sequence we can effectively change the stimulus duration smoothly between 140 and 150 ms. If we use a monitor with a 100-Hz frame rate and have 256 linearly spaced luminance levels available, our temporal resolution can be increased from 10 ms to 40 μs and 'hey presto' we have a high-resolution tachistoscope.

A question that is often asked is, 'How can I present a mask followed by my special image with intervening times of less than one frame on a CRT display?' The answer is simple: adjust the luminance of the last frame of the mask before the onset of the experimental stimulus as described above. By varying the luminance from nothing to full, the last frame time can be traversed with high resolution. When doing this there may be one frame that has information from both the mask and the stimulus on it and these can just be summed, but remember that the result must still be within the linear range of the display (see Fig. 3.12).

3.5.4 When is the light emitted from a CRT?

After the discussion about using CRT displays as tachistoscopes it is probably worth saying something about when the light is actually emitted. As described earlier in the chapter, the face of the screen is scanned by the electron beam from left to right and top to bottom. As the beam impinges on each small area of phosphor a pulse of light is emitted, which typically rises quickly to a peak value before decaying to nothing over a period of about 1 ms. Different phosphors have different properties, but this represents a typical value. Therefore in a typical display used for vision research, the light that is emitted from the top left-hand corner of the screen happens about one frame time, or 10 ms, before that emitted from the bottom right-hand corner. The implication here is that any experiment that looks at the temporal response to a full-screen spatial stimulus is effectively convolving it with a 10-ms time window, with the effect being to limit the upper frequency response. This is especially obvious when making evoked potential recordings where the electrical potential being recorded is the sum of the contributions from the retinal cells, all of which have been stimulated at a slightly different time. However, help is at hand for those interested in the ultimate recordings simply by using the techniques described in the section about tachistoscopes. Consider displaying a chess (checker) board that counterphases at 1 Hz on a 100-Hz frame rate monitor so that each of the squares comprises 100 pulses of light followed by 100 no-pulses. In an uncorrected system the white squares at the top and bottom of the screen produce their pulses of light on the same frame and hence suffer a one-frame, or 10-ms, phase delay between them. To perform the time correction simply divide the screen into horizontal bands, any convenient number will do but think along the lines of eight, and advance, in time, the stimuli in the lower bands by reducing the luminance of the last frame in the 'ON' sequence and adding a corresponding amount to the last frame in the 'OFF' sequence. Do this in proportion to the distance of the band from the top of the screen. In this way, a square that lies in a band halfway up the screen will have a frame sequence consisting of 99 fully 'ON' frames, followed by one frame at half luminance, followed by 99 fully 'OFF' frames, followed by another frame at half luminance. The total energy of the stimulus remains at 100/200, but the time has been shifted by half a frame.

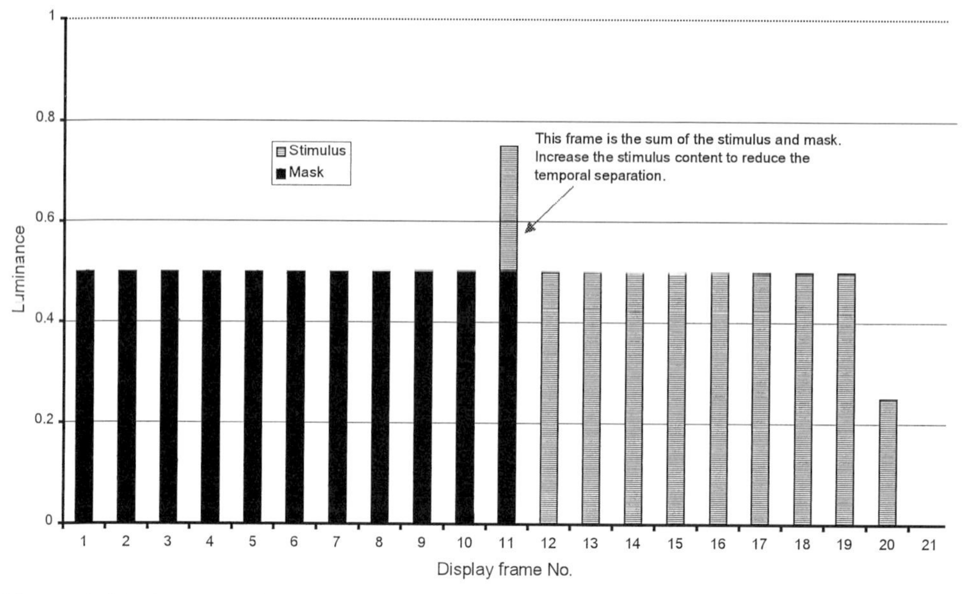

Fig. 3.12 Moving a stimulus closer to a mask.

3.5.5 Measuring reaction times

Many psychophysical experiments make use of the subject's reaction time to a specific event or events as a way of measuring his response, and it would seem that a computer system is an ideal way to do this. For example, the keyboard could be used as a convenient way for the subject to indicate that he had seen something appear on the screen. While good in theory this is more difficult than it seems. Most modern computers run multi-tasking operating systems, such as Windows 95, which means that a user's program may not in fact have control of the processor all of the time; if another task such as an autobackup wishes to access the hard disk it may steal quite large amounts of time from the other users, which can mean that a key press will go unrecognized by the user's program until control is subsequently returned—this then renders the reaction time invalid. A further source of error lies in the keyboard itself, which operates by scanning the rows and columns on a sequential basis. Typing is intrinsically a slow business so the scan speed may also be slow. This means that there is a variable time between hitting the key and the computer recognizing it that depends just where it was in its cycle when the event happened. Unfortunately, similar arguments apply to other convenient input devices such as mouse buttons. The advice here really has to be that if you need accurate reaction times and are not sure exactly how the computer works you should devise some other external hardware to perform the measurement instead.

3.5.6 Raster scanning—a brief description

Anybody who has ever tried to connect a CRT-based display to a computer or stimulus generator knows how difficult it can be to a find an acceptable combination of line-rates, frame-rates, back-porches, and other arcane parameters to provide the required resolution. So here is a résumé of some of the terminology used in the video world.

Raster scanning is the technique whereby an electron beam or spot of laser light is swept slowly from one side of the display area to the other and then returns quickly to the first side at a point slightly below where it started on the sweep above. The whole area is swept like this in a zigzag or sawtooth fashion until the spot reaches the bottom right-hand corner from where it is rapidly taken back to the top left-hand corner, and the process begins again. As the beam is scanned, it is turned on and off, or modulated, to cause the presence or absence of a dot on the image plane. The grid of lines so-produced is called the *raster*, the short sweeps from left to right form the lines of the display, and the slower sweep from top to bottom forms the frame. The quick return of the beam at the end of a line is called the *line-flyback*, while the return from the bottom to the top of the screen is called the *frame-flyback*. In order not to disrupt the display during the flyback the beam is always turned off, or *blanked*, at this time. Well-designed systems keep the flyback time as short as possible so as not to waste valuable display time. In a laser printer the laser is scanned by rotating a mirror and only one complete scan is needed to draw the image on to the photosensitive drum, while in a CRT the electron beam is scanned by a magnetic field and repeated scans are needed to keep the display refreshed; each time the electrons impinge on the phosphor a small packet of light is produced which decays to nothing in about 1 or 2 ms. From a technological point of view, the scanning in a laser printer is easier to achieve than that in the CRT as it doesn't need to be performed in real-time. That is to say, it is not important (within reason) how long it takes to scan the image as that will only affect the delay period between the computer sending the data and the first page being printed. In a CRT, however, the scan time is dictated by the requirement to perform at least 60 or 70 complete scans per second in order to give the impression of a continuous display, and this is much more difficult.

For historical reasons the evolution of CRT displays and laser printers have taken different paths; a typical laser printer comes complete with its own computer and framebuffer and needs only be sent the data (very easy), whereas CRT monitors are very simple and need a fully featured video signal to modulate their electron beams and synchronize to the raster. This is normally generated by the host computer or graphics card. The requirement to ensure that the raster on the display is locked to the video signal being supplied necessitates some extra synchronization signals. It is conventional that the video generator, in this case the host computer system, should supply two additional bits of information to the display: a pulse to indicate the start of a scan line and a pulse to indicate the start of a frame. These are the line and frame *synchronization signals* and they happen during the line flyback and the frame flyback, respectively. It is then up to the display electronics to ensure that it scans in synchrony with the generator, even though there is no method of feeding back if this is really the case. In an effort to keep the wires between the display and the computer to a minimum, these are often combined together to form a combined (or H/V) sync signal or combined and added to the green video signal (composite sync). This should not be confused with composite video used in television where all the video

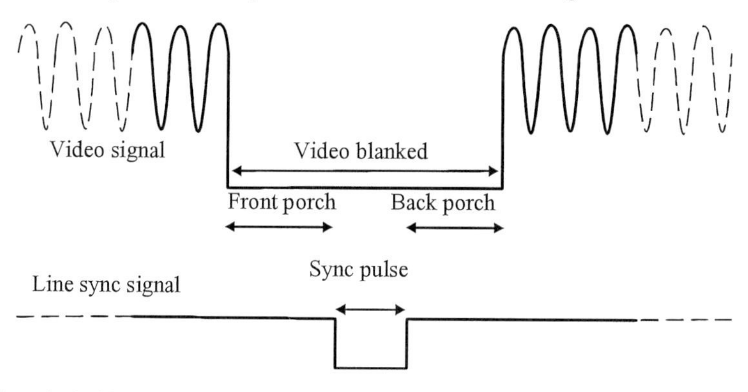

Fig. 3.13 A typical video signal.

and sync information is modulated on to one channel. A further term is needed in relation to the signals, and that is *porch*; the time between the end of the active video and the start of the synchronization pulse at the end of a scan line is called the *horizontal front porch*, while the time at the beginning of each line during which the display is blanked between the end of the synchronization pulse and the start of the active video signal is called the *back porch*. The equivalent times at the end and start of each frame are called the vertical front and back porches, respectively. (See Fig. 3.13.)

3.5.6.1 Monitor timing—a rough calculation

What are the important parameters for a display running at 120 Hz with a resolution of 800 pixels by 600 lines?

Assume that the line and frame flyback times are 10% of the line and frame display times. Then:

Frame scan rate = 120 Hz;
Line scan rate = 120 Hz × 600 lines × 110% = 79.2 kHz;
Pixel rate = 79.2 kHz × 800 pixels × 110% = 69.7 MHz.

Further Reading

Books

Travis, D. (1991). *Effective colour displays—theory and practice*. Academic Press. This book is really about 'human factors' but has good chapters on computer display systems and easy to understand overviews of colour spaces and colour space conversions.

Boff, K., Kaufman, L., and Thomas, J. (ed.) (1986). *Handbook of perception and human performance*. Wiley, New York. This book contains a chapter by A. B. Watson which gives a general description of the temporal response of the human visual system and indicates the problems with Bloch's Law. Read this if you are interested in tachistoscopes. This chapter is also available on the internet at the NASA vision group site.

Horowitz, P., and Hill, W. (1989). *The Art of Electronics*. Cambridge University Press. It is impossible to resist recommending this book which, although slightly long in the tooth, is probably the best book about real life electronics ever written. It contains more actually useful information about electronics than it would be possible to accumulate in a lifetime. Read it if you need to build *any* experimental equipment. (N.B. it is not too heavy on mathematics.)

The internet

The really up-to-date material is now available on the internet and this often includes books or chapters of books that are made available by their authors. Because of its dynamic nature, a list of all the interesting sites would soon be obsolete but once you have made one connection, the web like nature of the system will soon link you into other relevant areas. Good places to start are the NASA vision group site *http://vision.arc.nasa.gov/* and the visionscience site *http://www.visionscience.com/*.

One highly recommended site is run by Charles Poynton who is a forthright video engineer. Amongst other things, he has informative pages on gamma and colour in television and computer systems as well as links to other sites mainly of an engineering rather than scientific nature. Access his home page at *http://www.inforamp.net/~poynton/*.

4

Specifying, generating, and measuring colours

J. D. MOLLON

4.1 Introduction

In setting up an experiment on colour, the experimenter faces a fundamental choice: should the coloured stimuli be specified in physical terms or only in terms of the colour perception of a standard human observer? A specification of the former kind is called 'spectroradiometric', whereas a specification in terms of the human eye is called 'colorimetric'.

A *spectroradiometric* specification of a stimulus gives a 'spectral power distribution', that is a description of how radiance varies with wavelength over the range of the visible spectrum. However, normal human colour vision depends on the relative rates of quantum catch in only three classes of retinal cone, and so a normal observer can make a colour match by adjusting just three primary lights. Thus any spectral stimulus can be specified by three variables. Typically this *colorimetric* specification is given in terms of an internationally agreed luminance value and two *chromaticity coordinates* (see below). For many purposes (e.g. an experiment on visual search where colour is one of several attributes that distinguish the stimuli), a colorimetric specification is adequate; and it is clearly less clumsy than a full spectral power distribution.

However, if the subjects are unlikely to have the spectral sensitivity of the standard observer, if, for example, they are animals or are human observers with anomalous colour vision, then a conventional colorimetric specification is inappropriate. For the experiment to be replicable, the experimenter must in such cases use monochromatic stimuli of specified radiances and wavelengths or, if the stimulus is broad-band, must report its full spectral power distribution. One new factor that favours a spectroradiometric specification is the polymorphism of normal human colour vision: the colour matches of individual subjects may differ considerably from those of the standard colorimetric observer, and it is now known that much of the variance in such matches derives from inherited differences in the nucleotide sequences of the genes that code for the long- and middle-wave photopigments of the retina (Winderickx *et al.*, 1992). A radiometric specification may also recommend itself when mesopic conditions are being studied, that is when the subject's responses depend on the rods as well as the three types of cone: for two lights that have the same chromaticity and photopic luminance may differ greatly in their effects on the rods, and so a colorimetric specification would be inadequate.

In this chapter we discuss how to specify, generate, and measure coloured stimuli. Where appropriate we consider separately the spectroradiometric and colorimetric approaches. In experimental practice, of course, the two approaches are often mixed and many authors, although making their calibrations spectroradiometrically, report their results in terms of visual mechanisms, e.g. the contrast visible to an individual class of cones (Stromeyer *et al.*, 1985).

4.2 How to specify coloured stimuli

4.2.1 Spectroradiometric specification

If the stimuli are monochromatic (of bandwidth less than, say, 10 nm), a radiometric characterization should give the wavelength, bandwidth, and radiance of the stimuli. If the stimuli are broad-band, then the spectral power distribution should be reported, that is the variation of radiance with wavelength.

4.2.1.1 Spectral limits

In the case of man, the visible spectrum can conventionally be taken to extend from 400 to 700 nm.[1] However, in the case of insects, birds, fish, and even some mammals (Jacobs *et al.*, 1991), sensitivity may extend much further into the ultraviolet or infrared, and the human spectral limits would then be inappropriate. Photopigments with peak sensitivities as low as 360 nm are proving to be quite common.

4.2.1.2 Wavelength vs. wavenumber

For a time it was fashionable amongst colour scientists, especially those influenced by W. S. Stiles, to specify not the wavelength of visible radiation but its frequency or wavenumber. One valid ground for giving actual frequency is that frequency, unlike wavelength (or wavenumber), is independent of the optical medium; but people were probably most influenced by the belief that all photopigments had similar absorption spectra when the spectra were plotted on a frequency abscissa. This is now known to be quite wrong at the level of the photoreceptors (Barlow, 1982) (Mansfield, 1985). In fact, *log* frequency gives the most constant shape for pigment spectra (Baylor *et al.*, 1987), but wavelength is nowadays again the conventional abscissa except in specialist discussions of photopigments.

4.2.1.3 Quanta vs. energy

The energy of a light quantum varies with its frequency. The variable actually thought to determine the response of a cone cell is the total number of quanta absorbed: although the frequency of a given quantum will determine the *probability* that it will be absorbed by the photopigment, no information about the energy or frequency of the quantum is thought to be preserved in the physiological signal. For this reason many colour scientists prefer to specify the radiance of monochromatic lights in quanta rather than in energy units. The energy per quantum is given by the relationship $\varepsilon = hc/\lambda$, where h is Planck's constant (6.626×10^{-34} J s), c is the speed of light (3×10^{8} m s^{-1}), and λ is wavelength expressed in metres. So in laboratory practice the relationship that is most often needed is:

$$N = \lambda \times 5.03 \times 10^{15};$$

where N is the number of quanta per joule, and wavelength (λ) is now expressed in nanometres. Since a watt is a joule per second, one watt of monochromatic radiation at λ corresponds to N quanta per second.

4.2.2 Colorimetric specification

It would seem natural to have a way of expressing colours in terms of the relative excitations of the three cones of the normal retina, and the MacLeod–Boynton diagram, discussed below in Section 4.2.2.2, offers such a representation; but in 1931, when the Commission Internationale de l'Eclairage (CIE), in response to commercial and scientific needs, adopted a system of specifying colours in terms of three variables, the spectral sensitivities of the actual photoreceptors were not securely known. What were available were colour-matching functions, derived from the careful experiments of Guild and of Wright: these allowed any stimulus to be specified in terms of the proportions of three (arbitrary) primary lights needed to match it and they became the basis of the CIE 1931 chromaticity diagram (Wyszecki and Stiles, 1967; 1982).

4.2.2.1 The CIE system

Already in 1972 William Rushton was regretting that the CIE system was too often used 'to instruct the young and bewilder the old'; but the system is well understood by the trades that need to communicate about colours, it has useful features, and it is not really all that difficult to understand. Any light can be described in terms of three 'tristimulus values', X, Y, Z, which could be thought of as the photon catches of three imaginary photodetectors.[2] Their 'distribution coefficients', or relative sensitivities for an equal energy spectrum, are shown in Fig. 4.1. Although these distribution coefficients (designated $\bar{x}$, $\bar{y}$, $\bar{z}$) are not the human cone sensitivities, they are approximately linear transformations of the cone sensitivities of an average observer ($\bar{z}$ is in fact very close to the spectral sensitivity of the short-wave cones). Y was chosen to have the same spectral sensitivity as V_λ, which had already been adopted as the standard function for the luminous sensitivity of the eye (see Chapter 1); so $\bar{y}$ has a maximum value of 1.0 at 555 nm.

To specify the colour of a stimulus in the CIE system, one first finds the values of X, Y, Z by multiplying the power spectrum $P(\lambda)$ of the stimulus with $\bar{x}$, $\bar{y}$, and $\bar{z}$ in turn:

$$X = K \int P(\lambda) \times \bar{x}(\lambda)\, d\lambda;$$

$$Y = K \int P(\lambda) \times \bar{y}(\lambda)\, d\lambda;$$

$$Z = K \int P(\lambda) \times \bar{z}(\lambda)\, d\lambda;$$

where K is a scaling constant. One then gives the corresponding 'chromaticity coordinates' x, y, where:

$$x = X/(X + Y + Z), \text{ and}$$

$$y = Y/(X + Y + Z).$$

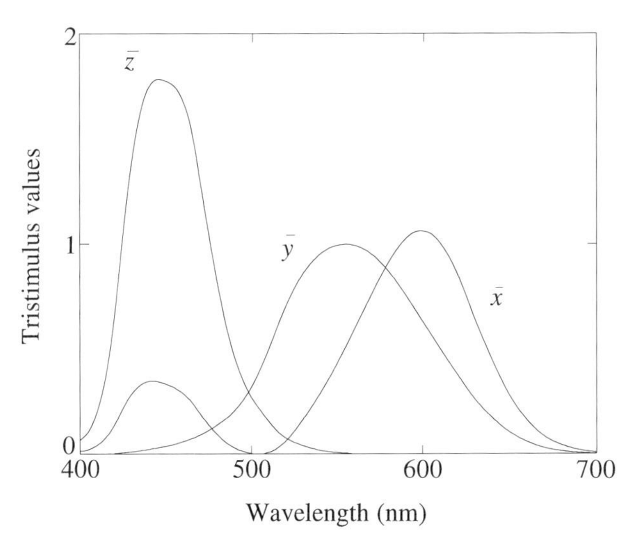

Fig. 4.1 The variation with wavelength of $\bar{x}$, $\bar{y}$, $\bar{z}$ in the system introduced in 1931 by the Commission International de l'Éclairage (CIE). Formally, the curves show the amount of each of three imaginary primary lights required to match the colour of one unit of radiant power of the indicated wavelength; but informally, it may be helpful to think of them as the sensitivity curves of three photodetectors. They are not the sensitivities of the human cones but are linearly related to those of the cones—or at least, the cones of some average observer.

Figure 4.2 shows the CIE chromaticity diagram, which has x and y as its axes.[3] The locus of monochromatic lights is shown as a solid line and the chromaticities of some common stimuli are indicated. To think physiologically about the CIE diagram, it is useful to know that lines radiating from $x = 1$, $y = 0$ correspond to loci of equal excitation of the short-wave cones when luminance is held constant (Fig. 4.3). It is also useful to know that lines radiating from approximately $x = 0.171$, $y = 0$ correspond to sets of chromaticities that are are confused by a tritanope (someone lacking the short-wave cones) and thus are chromaticities for which the ratio of long- and middle-wave cone signals is constant.

4.2.2.2 MacLeod–Boynton diagram

The CIE diagram is now so entrenched in commercial practice that it will not be quickly displaced. However, many visual scientists are adopting a physiologically realistic alternative, the chromaticity diagram of MacLeod and Boynton (1979), shown in Fig. 4.4. The vertical ordinate of the diagram corresponds to the relative excitation of the short-wave cones, expressed as a proportion of its maximum value near 400 nm; the abscissa represents the proportion of the total luminance contributed by the long-wave cones. In this diagram the results of colour mixing can be simply predicted by applying a centre-of-gravity rule using the luminances of the lights as weightings—in contrast to the CIE diagram where the weightings must be in terms of the sum of X, Y, and Z.

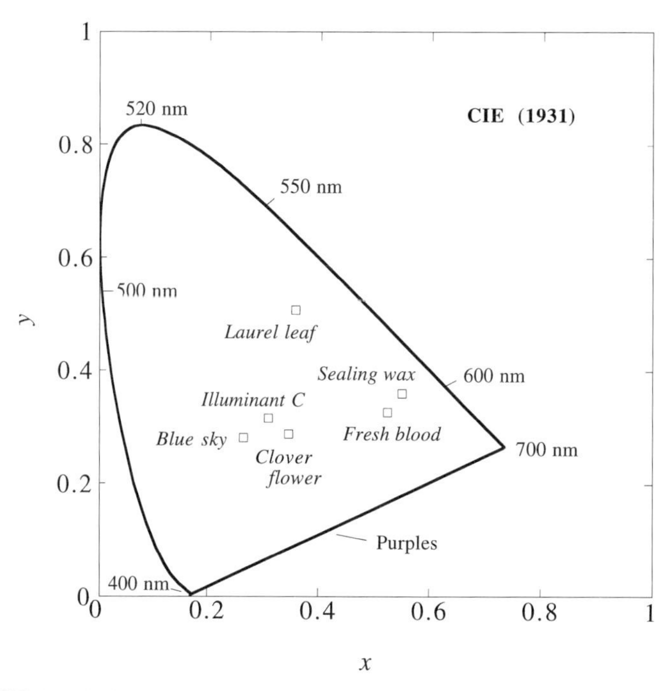

Fig. 4.2 The CIE (1931) chromaticity diagram. The solid line represents the spectrum of monochromatic lights from 400 to 700 nm plus the purple lights that consist of mixtures of long- and short-wave light. The chromaticities of all other physically realizable stimuli lie within this boundary. Coordinates are plotted for some illustrative stimuli. In the case of the reflective materials, the values shown are ones measured in Illuminant C, an approximation to northern daylight (see Section 4.3.8). Notice that quite strongly coloured stimuli may lie well within the interior of the diagram.

The MacLeod–Boynton diagram recommends itself because its axes correspond to the two chromatic channels that have been identified physiologically in the early visual system (Derrington *et al.*, 1984). The vertical axis corresponds to the phylogenetically older subsystem, which compares the signal of the short-wave cones with some combination of the signals of the long- and middle-wave cones. The abscissa corresponds to the phylogenetically recent subsystem, which compares the quantum catches of the long- and middle-wave cones (Mollon and Jordan, 1988). Horizontal lines in the diagram represent lights that give equal excitation in the short-wave cones. Vertical lines are tritan confusion lines, i.e. sets of chromaticities that represent a constant ratio of long-wave to middle-wave cone excitation. So the two physiologically significant sets of converging loci in the CIE diagram (identified in Section 4.2.2.1 and Fig. 4.3) become two sets of parallel loci in the MacLeod–Boynton diagram.

The MacLeod–Boynton diagram is derived from the cone sensitivities given by Smith and Pokorny (1975). The latter are not exact linear transformations of the CIE $\bar{x}$, $\bar{y}$, and $\bar{z}$ functions but rather are derived from the slightly different 1951 $\bar{x}_j$, $\bar{y}_j$, $\bar{z}_j$ functions of Judd (tabulated by Vos, 1978). The functions of Judd are preferred for work in visual science in that they incorporate a more accurate estimate of luminosity at short wavelengths. Vos (1978) gives a formula for transforming between chromat-

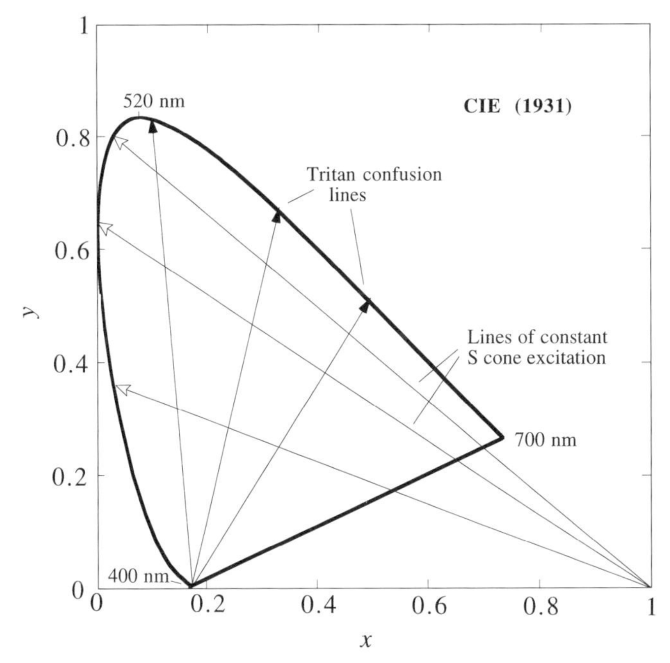

Fig. 4.3 The CIE (1931) chromaticity diagram showing some physiologically significant features. Any line radiating from $x = 1.0$, $y = 0$ (examples are shown with open arrow heads) represents a set of stimuli for which the excitation of the short-wave cones is constant for lights of constant luminance. A special case of such a line is the diagonal running from $x = 1.0$, $y = 0$ to $x = 0$, $y = 1.0$, which corresponds to zero excitation of the short-wave cones; between 700 nm and approximately 550 nm, this line effectively coincides with the spectrum locus, indicating that the standard observer is dichromatic in this part of the spectrum. Any line radiating from $x = 0.171$, $y = 0$ (examples are shown with solid arrow heads) is a tritan confusion line: along such a line the ratio of excitation of long- and middle-wave cones is constant and such stimuli would be confounded by a tritanope (i.e. a subject lacking short-wave cones). These relationships are shown here only to illustrate properties of the CIE chromaticity diagram: for research purposes it will often be desirable to establish them experimentally for a particular subject.

icity coordinates in the CIE (1931) and Judd (1951) systems, but this transformation is valid only for monochromatic lights. There is no unique transformation between the two systems for non-monochromatic lights. Thus, to use the MacLeod–Boynton diagram you strictly must know the spectroradiometric properties of your stimulus; you cannot simply measure the light with a colorimetric instrument that gives CIE (1931) chromaticity and luminance.

4.2.2.3 The limitations of chromaticity diagrams

To know the chromaticity coordinates of a stimulus is equivalent to knowing the relative signals it will produce in the three cone types of a (standard) human observer, but the chromaticity coordinates do not tell us how the stimulus will *look*: by manipulating simultaneous and successive contrast, we can generate almost any hue and any brightness from a given luminance and chromaticity.

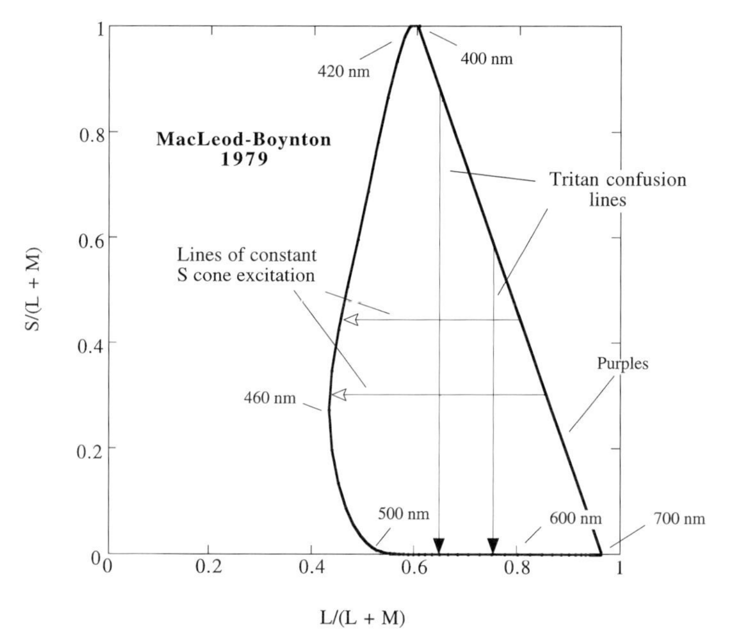

Fig. 4.4 The chromaticity diagram introduced by MacLeod and Boynton (1979). The abscissa represents the relative excitation of the long- (L) and middle-wave (M) cones: it is scaled to indicate the proportion of the total luminance that is accounted for by photons absorbed in the L cones. The ordinate represents the excitation of the short-wave (S) cones, arbitrarily scaled to have a value of 1.0 at the wavelength at which it is maximum at constant luminance. As in Figs 4.2 and 4.3, the solid line represents the coordinates of monochromatic spectral lights, plus the purples that consist of mixtures of short- and long-wave light. Also shown, using the same conventions as in Fig. 4.3, are examples of tritan confusion lines and lines of constant short-wave cone excitation. In the MacLeod–Boynton diagram, these two sets of physiologically significant lines become orthogonal.

Nor does distance in a chromaticity diagram tells us how salient will be the difference in colour *between* two stimuli. The philosophers' stone of colour science has been a *uniform colour space*, that is a representation of all chromaticities such that stimuli equally separated in the space are equally distinct. It is clear today that a space of this kind can be valid only for the stimulus conditions under which it was generated: its structure will vary with the spatial parameters of a display (Regan and Mollon, 1997), with the size of the colour differences (Mollon and Cavonius, 1986), and with the range of colours in the array (Webster and Mollon, 1995). Nevertheless, the CIE (1976) u', v' uniform chromaticity diagram, a linear transformation of the x,y diagram, has an honest use in choosing, say, chromaticities that are roughly equally separated in an experiment on visual search or in choosing a starting set of constants in a discrimination task intended for patients with varying degrees of colour loss. The relationship between the two diagrams is given by the following equations:

$$u' = 4x/(-2x + 12y + 3) \qquad v' = 9y/(-2x + 12y + 3)$$

$$x = 9u'/(6u' - 16v' + 12) \qquad y = 4v'/(6u' - 16v' + 12).$$

A rough working rule, for use only in the privacy of the laboratory, is to reduce the factor 9 in the formula for v' if the stimuli are brief (less than 200 ms) or small (less than 0.6 deg). These are the conditions under which the eye tends to tritanopia, and compressing the vertical ordinate of the u', v' diagram is a rough way of allowing for this.

4.2.2.4 Materials

Tables of chromaticity coordinates for the CIE and MacLeod–Boynton diagrams are available from a web site: *http://www-cvrl.ucsd.edu/*. Blocks of graph paper overprinted with the CIE chromaticity diagram are available from Beuth Verlag (see the Appendix at the end of this chapter for this and other addresses).

4.3 Generating coloured stimuli

4.3.1 Monochromators

The monochromator remains the workhorse of the colour laboratory. In almost all modern monochromators a diffraction grating is used to form a spectrum, and the desired wavelength is selected by rotating the grating until the required wavelength coincides with the exit slit of the instrument (Fig. 4.5). The surface of a diffraction grating consists of evenly spaced, parallel grooves. When white light falls obliquely on the grating, each groove can be thought of as a thin individual source. Consider light that is diffracted in a given direction from one groove. A wavelength will exist such that light diffracted in the same direction from the next groove has to travel a path that is longer by that wavelength, or by an integral multiple of that wavelength. The light of this wavelength will thus be in phase and will constructively interfere. Other wavelengths, out of phase, will destructively interfere. The favoured wavelength is different for different directions of diffracted light, and thus the spectrum is formed.

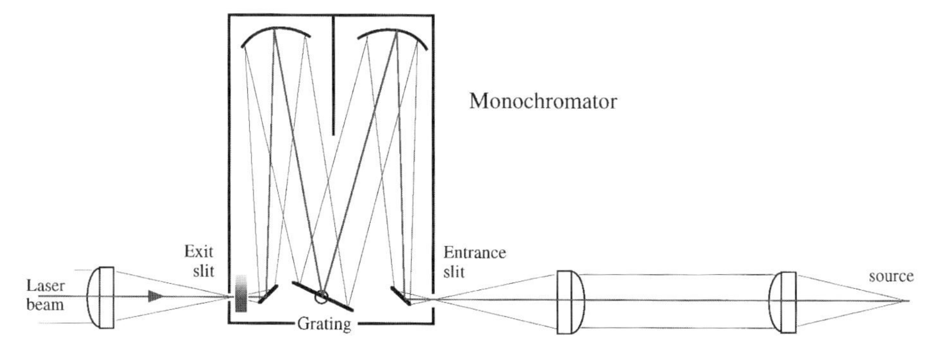

Fig. 4.5 A schematic diagram of a monochromator built into an optical system. White light is drawn from a source to the right and illuminates the diffraction grating, which is mounted on a rotable shaft. As the grating is rotated, different wavelengths are selected from the spectrum formed at the exit slit. For the purposes of lining-up, it is convenient to pass a helium–neon laser beam through the system from the position of the subject's pupil to the source (see Section 4.3.1, and Chapter 1).

The grating surface is exceedingly delicate: felt gloves are recommended when a grating is to be exchanged; there is absolutely no question of cleaning gratings with tissue or cloth; and compressed-air cleaners should be avoided since droplets of liquid may be deposited on the grating surface. Any dust on the grating will scatter white light indiscriminately and impair the purity of the spectral band selected by the exit slit: the best approach is to prevent dust entering the monochromator in the first place.

The bandwidth of the monochromator—the range of wavelengths transmitted—is controlled by the width of both the entrance and the exit slits. If the entrance slit is set to near its minimum and the exit slit is increased, then the width of the selected spectral band will increase linearly. If the exit slit is fixed at some small value, and the entrance slit is increased, a range of overlapping spectra are effectively superposed at the output, again increasing the bandwidth. Most monochromators are designed to be used with the entrance and exit slits set to the same width. Necessarily, the total radiant output will be lower the narrower the slits, and one almost always finds oneself compromising between bandwidth and throughput. One way to maximize the latter without increasing the former is to orient the monochromator (or possibly the source, if its operation will not be impaired) so that the longer dimension of the filament or the arc has the same orientation as the entrance slit of the monochromator.

Monochromators are available with integral stepping motors, allowing the selected wavelength to be changed under computer control. Such instruments make it possible to measure wavelength discrimination by forced-choice psychophysics, although such studies are still surprisingly rare. A suitable design is to present a bipartite field in two successive intervals: in one interval the two half-fields are identical in wavelength and in the other a wavelength change of $\Delta\lambda$ is introduced in one of the half-fields; the subject is required to say in which interval the half-fields differed and $\Delta\lambda$ is titrated according to a staircase routine (Mollon and Cavonius, 1987). In a forced-choice experiment of this kind, the threshold may be as little as 0.3 nm; so an instrument with a step of < 0.1 nm is required if the experimenter wishes to study the limits of human wavelength discrimination. However, this does not require that the *bandwidth* of the stimulus should be equally small: you would need a very powerful source to study photopic discrimination using a bandwidth of 0.1 nm, and the important (and remarkable) fact to remember is that the discrimination of centre wavelengths 0.3 nm apart is being achieved with retinal filters that have bandwidths of 100 nm. So a stimulus bandwidth of 1–2 nm is quite acceptable. A major problem to watch for is play in the coupling between the stepping motor and the grating: this may be equivalent to several nominal steps. The problem can usually be solved by programming your monochromator always to make its final approach to the desired wavelength from a single direction. Even with manual controls, this is good practice.

Manoeuvering a heavy monochromator in the way described in the box (opposite) is much easier if it is mounted on a fully adjustable table. Curiously, suitable tables do not seem to be available commercially: if you ask your workshop to make one, the table should ideally have three legs (each with a small adjustment in height), a lateral adjustment, and a small rotatory adjustment that allows the instrument to be rotated around the centre of the entrance slit. But when the monochromator is aligned in this way, what do you do if you find, as well you may, that the beam from the source does not emerge concentrically with the entering laser beam? If the

Building a monochromator into a Maxwellian-view system can often be a devil of a task. For aligning a monochromator that has entrance and exit slits in-line (Fig. 4.5), the following procedure has been found to be helpful in the writer's laboratory:

1. Pass a laser beam from the position of the subject's pupil to the intended position of the source, using a centre spike to ensure that the beam runs horizontally above the centre of the optical bench.
2. Introduce the source and ensure that the laser beam falls on the centre of the filament (or, in the case of an arc source, close to the tip of the appropriate electrode).
3. Introduce those lenses that are to lie between the source and the monochromator, making sure that the laser beam continues to fall on the centre of the source (see Chapter 1 for detailed guidance on the centring of lenses).
4. Turn on the source, check that its beam is everywhere concentric with the laser beam, and introduce the monochromator so that the source is focused on the centre of the entrance slit.
5. Now gently rotate the monochromator about the latter point until the laser beam passes into the exit slit at its centre.

optics of the monochromator are sealed, then give it back to the manufacturer. If internal adjustments are provided, then you will most probably need to adjust the mounting of the grating to ensure that the grating is perfectly vertical relative to the incoming beam. Another problem to watch for is a wavelength-dependence in the direction of the output beam of the monochromator: any small effect of this kind will mean that the stimulus beam will traverse the subject's pupil as wavelength is changed during the experiment, and the Stiles–Crawford effects will then produce changes in the effective radiance and chromaticity of the beam. If you find this fault, challenge the manufacturer with it.

An unattractive feature of many monochromators, even rather expensive ones, is that a perfectly circular entry beam emerges as elliptical. This distortion arises because spherical mirrors are used to collimate an obliquely incident beam before it is presented to the grating (Fig. 4.5). In principle, you could correct the problem by reflecting the exit beam off an external spherical mirror and adjusting the angle of the mirror to restore the circularity of the beam. The problem is avoided in 'imaging monochromators' by using toroidal mirrors, and in several Jobin–Yvon monochromators by using the same corrected holographic grating to form the spectrum and refocus the beam.

When using a monochromator, it is always necessary to consider the effects of stray light, especially if the selected wavelength is near one extreme of the visible spectrum. Suppose one wishes to use a violet stimulus of 430 nm and suppose the nominal level of stray light is 0.1% throughout the spectrum. When the latter is integrated across the middle of the spectrum, where the eye may be 100 times more sensitive than at 430 nm, it may well be the stray light that the subject detects in a threshold experiment. The problem can be easily eliminated by placing relatively broad-band, gelatin blocking filters in series with the monochromator.

Remember that the output of the monochromator will be partially polarized as a result of reflection at its internal surfaces and that this polarization may interact with later components in the system.

How should the wavelength calibration of the monochromator be checked? I have suggested using a laser to line up the monochromator in a Maxwellian-view system, and so you will necessarily be able to check that the scale reading is 632.8 nm when the helium–neon laser light is maximally transmitted. But the careful experimenter will wish to check at least one other wavelength. Special lamps, emitting one or more spectral lines, are sold for the purpose, and the manufacturers usually imply that you should place these at the position of your usual source. This is the very last thing you would want to do if you had just built the monochromator into a Maxwellian-view system. Instead, you could place the spectral source at the position of the subject's pupil and rely on the principle of reversibility of optical paths. Another, quite cheap, solution is to use a didymium filter. Such filters appear almost neutral in colour but actually absorb in certain, relatively narrow, spectral bands: if you scan your monochromator through the spectrum you can check (with a photodiode or other detector) whether these bands are in their proper places.

4.3.2 Narrow-band interference filters

Interference filters make it possible to select spectral bands only a few nanometres wide. Such filters typically consist of superposed layers of optical materials of different refractive index. Light is multiply reflected to and fro between the internal interfaces and the separations are chosen such that normally incident light of the desired wavelength constructively interferes with itself and is transmitted, whereas other wavelengths are attenuated by destructive interference.

If an interference filter is tilted relative to the incident beam, the favoured wavelength becomes shorter. This property is occasionally useful, but it does mean that interference filters are best used in a well-collimated beam. If the beam is convergent at the filter and rays pass through it at different angles, the selected wavelength in the centre of the beam will be higher than that at its circumference. For precise work, temperature should be controlled, since expansion of the filter may change the separation of the critical surfaces and thus the wavelength selected. Many modern interference filters pass little stray light outside the chosen band, but the warning given above for monochromators should be heeded when working at the extremes of the spectrum.

It is possible to buy 'interference wedges', interference filters that vary linearly in wavelength across their surface. Sometimes they are made in the form of a disc (with wavelength varying circumferentially) so that they can be conveniently mounted on a stepping motor. Such devices may seem seductively cheap by comparison with a monochromator or a set of fixed interference filters, but remember that they are intended to be placed at the focus of a beam and so the considerations of the preceding paragraph mean that they are suitable for use only when a small central area of the beam is to be used as the final stimulus. On the other hand, Palmer and Whitlock (1978) have ingeniously shown how narrow-band stimuli can be produced by projecting on to a single linear interference wedge an inverted image of itself.

Electronically tuneable interference filters of relatively large aperture have recently become available. A device of this kind has been evaluated in the writer's laboratory, but transmission in different regions of the aperture was not uniform enough for the intended purpose (spectral imaging with a digital camera.)

4.3.3 Broad-band filters

A large variety of broad-band colour filters are available in either gelatin or glass. Two standard series are the Kodak Wratten and the Ilford sets, and a booklet of transmission curves can be purchased for each. They are available from photographic dealers. A subset of the Ilford filters (nos 602–626) cover the spectrum in relatively narrow pass-bands. Very cheap filters are available in large sheets from suppliers of stage lighting, e.g. Stage Electrics of Exeter. Such suppliers often give free swatches of sample filters that are themselves big enough to be useful in the visual science laboratory.

Many gelatin filters (especially the yellow, orange, and red ones) are high-pass filters, in that they pass wavelengths above a certain value. The opposite—a low-pass filter that blocks transmission above a certain wavelength—would often be very handy but is not available in gelatin. Interference filters with this property can be bought (e.g. from Melles Griot) but are much more expensive than gelatin filters and usually have a small residual ripple in the low-pass part of their transmission spectrum. 'Dichroic' mirrors are available that transmit part of the visible spectrum and reflect the other half: a mirror of this kind is useful when you are combining two different wavelengths and want to lose as little light as possible when mixing them.

To remove ultraviolet radiation from the beam a Wratten 1A or an Ilford 025 filter is suitable. Infrared radiation can be cheaply removed with HA3 glass (but when you receive it, examine it between crossed polaroids to check for dichroism: otherwise it may interact with polarization in your system to yield striations across the stimulus field). Also available are 'hot' and 'cold' mirrors that reflect or transmit infrared, respectively. If you are working in Maxwellian view and making your calibrations with an unfiltered silicon photodiode, it is very important indeed to remove any infrared radiation that might contaminate your measurements.

4.3.4 Electronic template colorimeters

Potentially the most versatile instrument for producing spectral stimuli is a template colorimeter, in which a spectrum is formed (by means of a prism or a diffraction grating) and an opaque template is used to select the amount of each spectral region to be transmitted. If the light is then recombined (by an integrating sphere, for example), any arbitary spectral power distribution can be offered to the eye. Newton was perhaps the first to use such a device when he filtered his spectrum with a moveable comb, and template colorimeters have often been used in research on colour vision (Ives, 1915; Stiles, 1955; Holtsmark and Valberg, 1969). But what gives them new potential is the availability of electronic masks that can be rapidly varied under computer control. Bonnardel *et al.* (1996) placed a black-and-white liquid-crystal screen (such as sold for use with overhead projectors) in the plane of a spectrum. The device was used to measure the eye's contrast sensitivity for comb-modulated spectra with varying frequency and phase of comb-modulation, but could in principle be used to generate any spectral stimuli. Until now, two disadvantages of liquid-crystal displays have been that the 'ON'–'OFF' contrast ratio has been relatively poor (as little as 15:1) and that the contrast ratio is wavelength-dependent, being lowest for short-wavelengths. But such devices are improving in contrast, in speed, and in price. Moreover, it is now possible to buy the raw array without the packaging but with the driving electronics (e.g. from CRL Ltd).

The digital micromirror device (DMD), such as recently introduced by Texas Instruments, has great promise as the basis for a template colorimeter. The DMD consists of an array of tiny, hinged mirrors: each individual mirror is about 16 μm across and can be rotated rapidly between two states (+ 10 degrees and − 10 degrees, corresponding to 'ON' and 'OFF'). Arrays of 800 × 600 micromirrors are currently available, and each mirror can be thought of as an independent pixel. The effective luminance of the pixel is controlled by the proportion of time the corresponding mirror is in the 'ON' state. A spectrum could be formed by means of a prism or diffraction grating and a DMD placed in the plane where the spectrum is focused before recombination. Each column of pixels would then correspond to a narrow spectral band. At any instant, wavelengths required in the shaped spectral distribution would be reflected back towards recombination, while other wavelengths would be reflected out of this path. Particularly suitable for such an application might be the long thin DMDs (7056 × 64 pixels) that have been developed for hard-copy devices.

4.3.5 Three-primary colorimeters

If the experimenter requires only colorimetric, and not spectroradiometric, versatility, a range of stimulators can be designed on the basis of three primaries. The latter may be provided by colour filters, by light-emitting diodes, or by lasers.

Economical to build are colorimeters of the Burnham type. A beam is drawn from a single source and is passed through an array of three colour filters (four filters may be used if the experimenter requires access to a larger area of colour space). The array is mounted in the beam and can be mechanically adjusted in its own plane so that more or less of each colour of light is passed. The light is then mixed to give the chromaticity desired. The mixing can be achieved by a glass integrating bar, as in Burnham's original design (Wyszecki and Stiles, 1967; p. 369), by an integrating sphere, or by lenses that form an image of the source. If suitable neutral filters are mounted over each colour filter, then the experimenter can arrange that all the chromaticities generated are of equal luminance. Stepping-motor driven stages may be used to adjust the position of the filter array under computer control. Anyone who contemplates preparing a colorimeter of the Burnham type will find it useful to consult the description of the La Jolla colorimeter by Boynton and Nagy (1981, 1982).

4.3.6 Computer-controlled colour monitors

A special case of the three-primary colorimeter, and perhaps the commonest instrument in colour research today, is the colour CRT (cathode-ray tube). It offers the experimenter great freedom to manipulate stimuli in chromaticity, in space, and in time. It also offers many pitfalls for the unwary.

Colour-mixing is often performed by combining the beams of a multichannel Maxwellian-view system; but after many hours of constructing such a system, the experimenter may be disappointed to discover that the stimulus field is marred by striations or patches of different hue, or that there is a gradient of hue from one side to the other. Tweaking individual components will be of little help in this situation. The writer offers the following solutions:

1. Work up your courage and remove all the components except the source; it will take surprisingly little time to put them back, especially if you record their axial positions carefully (if you are using traditional optical benches, you can use spare saddles as stops to mark positions). Using a laser beam running from the position of the subject's pupil to the source, reintroduce each component in turn, ensuring that the laser beam remains centred on the source. The last fraction of a millimetre counts. For general advice on the alignment of Maxwellian-view systems, see Chapter 1 by W. Makous.
2. Use a ribbon-filament lamp as the source rather than, say, a tungsten–halogen projector lamp. The coiled filament of the latter provides a spatially inhomogeneous source and will contribute to inhomogeneity in the Maxwellian field.
3. (If you can afford some loss of output luminance) form a secondary source by imaging the real source on to a small aperture at which you have mounted a piece of diffusing material, and focus the secondary source on the subject's pupil.

4.3.6.1 The colour CRT

A schematic diagram of a typical colour monitor is shown in Fig. 4.6. Within a vacuum tube, streams of electrons are emitted from the heated cathodes of each of three guns and are drawn towards the front screen of the tube, which is coated with red, green, and blue phosphor dots or stripes. Just before the screen is a 'shadow mask', pierced with apertures, which ensure—more or less—that a given electron beam excites only its proper subset of phosphor elements. The electron beams are deflected electromagnetically from left to right, and, in successive passes, from the top to the bottom of the screen, so that the modulation of the electron flux in time draws out a detailed two-dimensional image.

In the Trinitron design, introduced by Sony, the electron beams are drawn from three horizontally aligned cathodes and the corresponding perforations are nearly continuous apertures between vertical wires. The vertically strung wires of a Trinitron display need one or two (depending on screen size) horizontal wires for support, and these are visible as dark lines. The experimenter will normally be able to work round them, once their nature is understood.

The 'pitch' of a colour monitor is the width of each triad of phosphor dots or lines: a typical value for a 19-inch (48-cm) graphics monitor is 0.28 mm.

4.3.6.2 The input signal

The CRT may be driven either from a 'framestore' that allows picture elements (pixels) to be addressed individually, or from a specialized waveform generator that is dedicated to producing repetitive stimuli with high precision (see Chapter 3). In both cases, a stream of numerical values is transformed by three digital-to-analog converters (DACs) into three corresponding voltage signals, which specify the

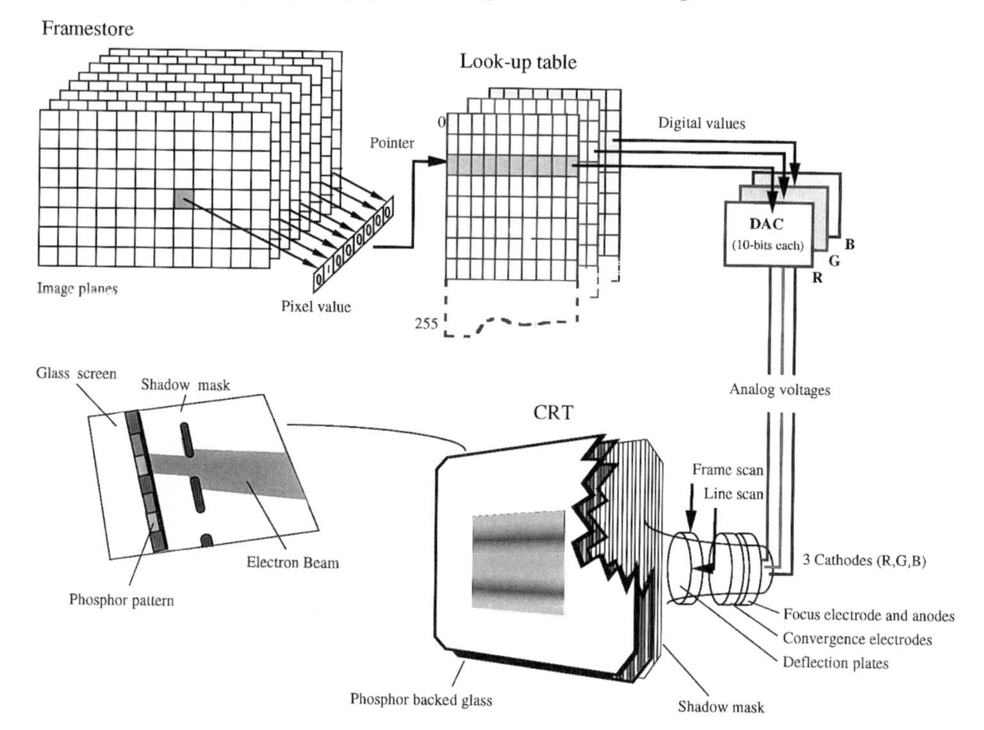

Fig. 4.6 Schema of a graphics board (above) and CRT monitor (below). In the framestore, the value stored for a particular pixel points to one entry in the look-up table. The look-up table depicted here has 256 entries (requiring 8 planes of framestore memory) and for each entry the gun voltage can be specified with a resolution of 10 bits. The monitor shown is based on the Sony Trinitron arrangement, in that the shadow mask or 'aperture grill' consists of vertical 'wires' and the phosphor pattern consists of vertical stripes.

required outputs of the red, green, and blue guns of the monitor. Any distortion of the temporal modulation of these input signals—introduced by the stimulus generator, the cables, the electronics of the monitor, or nearby equipment—will be faithfully translated into a spatial distortion on the screen. It is important to match the temporal properties of the monitor and of the graphics board, rather than buying them independently: more specifically, the 'pixel rate', the frequency with which the graphics board can send new voltages to the monitor, should be matched to the 'bandwidth' or 'video rate' of the monitor, the limiting frequency at which the electron flux of the guns can be modulated (for further details, see Chapter 3).

The upper left part of Fig. 4.6 schematically represents a framestore. The maximum number of individually addressable pixels in the image will be determined by the x, y dimensions of this store. The number of 'planes' of memory (i.e. the number of 'bits per pixel') will determine the number of colour/luminance combinations we can concurrently display. Many systems allow spatial resolution and palette size (number of colour/luminance combinations) to be traded off within a fixed memory array.

To sample colour space adequately we need to be able to control each of three guns with a resolution of 8 bits (giving us 256 different output levels) and if we wish to determine chromatic thresholds, we ideally need at least 10 bits per gun to avoid significant quantization in the measurements (Cowan, 1983). However, if you find yourself limited to 8 bits per gun there are four ways of securing the resolution needed to measure, say, chromatic thresholds on a white background:

1. Use 'dithering' or 'half-toning', gaining an extra bit by setting alternate raster lines to different values.
2. Reduce the contrast of the display by optically superposing it on a homogeneous background field (Cole *et al.*, 1993).
3. When setting up the monitor, turn down the contrast knob and turn up the brightness knob.
4. If the monitor has a high frame rate, then a temporal analog of (1) can be used, in that the same pixel can be set to alternate values on different frames (see Chapter 3.)

It is possible to have a framestore 24-bits deep (i.e. 24 bits per pixel) and thus directly to specify three gun values at each pixel with a precision of 8 bits per gun. But the more common arrangement in recent research has been that shown in the upper part of Fig. 4.6. What is stored for each pixel is not a set of three gun values but a pointer to an entry in a look-up table (Rodieck, 1983; see also Chapter 3). If there are 8 planes of framestore memory, then there will be only 256 possible entries in the look-up table, since this is the number of addresses that can be specified by 8 bits. However, each entry in the look-up table can specify with high precision (say, 10 or 12 bits or more) the signal to be sent to each gun from the digital-to-analog converters of the graphics board. Although the number of bits-per-pixel limits the number of colours concurrently displayed, this palette of colours can be rapidly changed between frames by changing just the entries in the look-up table. If a target area and its background field are linked to different positions in the look-up table, then the target can be made to appear and disappear according to whether the two entries in the look-up table contain the same set of gun values. Thus, although 24-bit graphics systems may recommend themselves for work with spatially complex stimuli (and would be necessary for work on natural scenes), many visual scientists will find it more useful to have a look-up table of limited size combined with DACs of high precision.

Notice that there is no automatic relationship between the pitch of a colour monitor (see previous section) and the number of pixels in a horizontal line: it is left to the user to determine what this relationship should be and to ensure that no aliasing occurs between the stimulus pattern and the pattern of phosphor triads.

4.3.6.3 Non-additivities and non-uniformities

In principle, if the experimenter knows the chromaticities of the three phosphors of a CRT and the gamma function for each gun (the relationship between the input voltage and the luminous output; see Chapter 3), then it is possible to calculate the gun values needed to generate any stimulus within the gamut of the monitor. However, this calculation depends on a number of assumptions, which need not hold.

A critical assumption is that of 'gun independence', the assumption that the luminous and chromatic output resulting from a given input to a given gun is independent of the signals sent to the other two guns. If we measure the X, Y, and Z outputs when given voltages are sent to each gun separately, we can determine whether the sum of the three sets of X, Y, and Z is obtained when the specified voltages are sent to the three guns simultaneously. This is a basic test to perform before paying for a new monitor. In the author's laboratory, good additivity has been found for Barco CD351, Mitsubishi HL20, and Sony GVM1400 monitors driven by a Cambridge Research Systems VSG board. However, failures of additivity may be observed in both cheap and expensive colour monitors, especially at high settings of contrast. In the case of cheap monitors, the non-additivity may arise from a low accelerating voltage present at the screen (Rodieck, 1983). In the case of modern expensive monitors, the culprit may be circuits deliberately introduced to ensure compliance with regulations on X-ray emission: these circuits place limits on total emission. Non-additivities of these kinds can often be eliminated or reduced by turning down the gain (the 'contrast') of the monitor (thus reducing the maximum luminance available from each gun). In principle, non-additivities may also arise in the graphics board.

An assumption that fails for all monitors known to the author is that of spatial uniformity of gun outputs across the screen: usually there is a falling off of luminous output near the edges of the screen and this may be as much as 20% (Mollon and Baker, 1995). However, Brainard (1989) found that most of the spatial variation could be described by a single scale factor applied to all gamma functions; so there is little change in chromaticity.

A third assumption is that the luminance and chromaticity from a given region of the screen are independent of the illumination of other regions. Mollon and Baker (1995) illustrate the complex ways in which this assumption can fail in research monitors. It is important for the experimenter to check thoroughly for artefacts of this kind, above all in experiments on colour constancy and simultaneous contrast, but I suspect that such checks are far too often neglected.

Another problem is that the time constants of the red, green, and blue phosphors are quite likely to differ. Vingrys and King-Smith (1986) have described how these differences may lead to detectable luminance artefacts when a temporal substitution is made between nominally equiluminant stimuli.

Finally, the nature of the raster scan means that different parts of a monitor screen are displayed at different times. It ought not to be necessary to remark on this. But I have refereed papers on electroretinogram (ERG) recording where the subject was placed 10 cm from a large monitor and a temporal red–green modulation was applied to the entire screen: under such conditions there will be a substantial phase shift between the signals generated in the inferior and superior retina.

4.3.7 Digital light processors (DLP)

This is the term that Texas Instruments have adopted for projection devices that incorporate their digital micromirror devices (see Section 4.3.4). In the single-chip version of the DLP, light from a high-intensity source passes through a rotating red–green–blue (RGB) colour filter and on to the mirror array. At any instant each pixel—each individual mirror—is reflecting light either into or out of the aperture of a lens that images the array on a screen. The input to the currently available system is the SVGA standard used for computer monitors, and the intensities of the separate RGB signals at each pixel are achieved by controlling the proportion of the time that the corresponding mirror is in the 'ON' state as the red, green, or blue filter is passing through the beam. As a result of this method of manipulating the intensity of each primary, the response of the device is linear. The images are bright and sharp. In principle, the individual pixels are independent of their neighbours and each primary is independent of the other primaries; so two of the types of interaction that characterize CRTs (Section 4.3.6) should be absent. And, in contrast to the case of a CRT, the chromaticities of the primaries could be modified to suit particular experimental purposes.

4.3.8 Munsell papers

Our colour vision evolved to discriminate coloured surfaces and to detect coloured targets in variegated backgrounds. It is surprising therefore that surface colours are rather little used in modern colour research, except in the study of colour deficiency and of colour naming. In the field of colour constancy, many experimenters simulate coloured surfaces on CRTs but seldom discuss the (very relevant) issue of how well they achieved the illusion.

An internationally recognized collection of coloured papers is that of the Munsell system, available from Macbeth (or in Europe from D. G. Colour). The papers can be bought both in atlas form and as individual sheets of varying size. Specially chosen subsets are available for particular purposes: thus there is a useful small atlas of skin, hair, and iris colours. General information about the Munsell system is available on the web at http://www.munsell.com. The papers are specified in terms of perceptual dimensions—Hue, Chroma (saturation), and Value (lightness)—but colorimetric specifications (CIE x, y values) are tabulated and plotted in Wyszecki and Stiles (1982, p. 840 ff). Notice that these chromaticity coordinates are valid only for the specified illuminant (CIE Illuminant C, an approximation to natural daylight of colour temperature 6774 K).

In general, since the spectral stimulus at the eye depends on the product of (a) the reflectance spectrum of a surface and (b) the spectral power distribution of the illuminant, it is essential when working with surface colours either to use a standard illuminant or to measure instrumentally the chromaticities (or spectral fluxes) actually presented to the eye by the coloured surfaces. For any exact work that relies on a standard illuminant, not just the chromaticity but also the spectral power distribution of the illuminant must be standard: otherwise the product of the spectral reflectance of the surface and the spectral power distribution of the illuminant may not give the intended spectral flux at the eye.

The Macbeth easel lamp, which consisted of a standard tungsten lamp and a blue glass filter, used to be a practical means of simulating Illuminant C. It has been replaced by the Macbeth 'Sol Source', an adjustable desk lamp with a tungsten–halogen source and blue filter. The 'Sol Source' is sold as a simulation of the standard Illuminant D65, which is nowadays preferred by the CIE as an approximation to natural daylight and which corresponds to a slightly lower colour temperature of 6500 K (Wyszecki and Stiles, 1982; p. 144). A cheap way of simulating Illuminant C, suitable for the routine administration of pseudoisochromatic plates and similar materials, is to illuminate the coloured materials with an ordinary tungsten source and to ask the subject to wear trial frames fitted with suitable colour filters: Pokorny *et al.* (1978) used a combination of Wratten filters nos 78B and 80B. Fluorescent lamps exist that approximate to D65 (e.g. Philips 'Colour 96' lamps): these would not serve for very exact work, since their spectral power distributions exhibit marked emission lines that are absent from D65, but they cost only a little more than ordinary fluorescent tubes and they could be useful where, say, experimenters wish to ensure that their subjects are in a known state of neutral adaptation.

4.3.9 Control of intensity when broad-band spectral stimuli are used

If broad-band colour stimuli, rather than strictly monochromatic beams, are used, then care is needed in the manipulation of intensity. 'Neutral' density filters will seldom attenuate all wavelengths by exactly the same factor; the extinction produced by crossed polaroids is very wavelength-dependent, typically being least at short wavelengths; varying the voltage or current available to a tungsten-filament lamp will change its colour temperature, making it redder the lower the current. A classical solution is physically to interrupt a known proportion of light, temporally with a rapidly rotating episcotister or spatially with a filter consisting of an array of small apertures; but often the best thing to do is to construct the required chromaticity by using monochromatic lights in the first place. Neutral-density filters can then be calibrated in the beam in which they are being used.

4.4 Measuring coloured stimuli

4.4.1 Spectroradiometry

If working in Maxwellian-view, the colour scientist can get a long way with a simple silicon photodiode that has been calibrated absolutely and spectrally. Any desired chromaticity can be constructed from individually measured monochromatic lights. A photodiode is also useful for measuring the gamma functions of the individual guns of a colour monitor.

A new generation of spectroradiometers are available in which the incoming light is dispersed on to an array of photodiodes and the energy in each spectral band is measured in parallel. Such devices especially recommend themselves where the stimuli are relatively broad-band. Instruments suitable for measuring surface reflectance are available for as little as $3000; the author has good reports of the Ocean Optics PS1000 instrument, which is small enough to be used in the field with a portable computer. Telespectroradiometers, suitable for measuring the spectral flux reaching an observer from a distant source, are more expensive. The author has had very satisfactory experience with a Photo Research 650 instrument, which has proved stable and accurate under the extreme conditions of a tropical rain forest.

When measuring CRT screens with radiometric devices it is important to remember that the stimulus is concentrated into brief, intense pulses, which may transiently saturate the detector and thus give you a distorted measurement. Some spectroradiometers have a special mode in which measurements are synchronized to the frame rate of a CRT (although I have found the Photo Research 650 unwilling to lock on to some monitors). Another factor to consider when calibrating CRTs is that the radiation emitted from the screen may be partly polarized and components in the detecting system may act as analysers.

To check the calibration of the spectroradiometer itself, the laboratory will need a standard light source and a very stable, constant-current power supply. A suitable system, after calibration at a standards laboratory, may cost $5000—more than some spectroradiometers.

4.4.2 Colorimeters

Electronic colorimeters directly estimate the CIE X, Y, and Z tristimulus values rather than first measuring the spectral power distribution of the stimulus. The colorimeter typically incorporates three detectors, which measure the light passed by three colour filters. The three detector/filter combinations have sensitivities that approximate to the CIE $\bar{x}$, $\bar{y}$, $\bar{z}$ functions, and such instruments will be accurate only to the degree to which they successfully simulate these functions. Some widely used instruments take a short-cut in simulating the $\bar{x}$ function: the detector/filter combination is unimodal in its spectral sensitivity (whereas $\bar{x}$ is bimodal) and the short-wave mode of $\bar{x}$ is simulated by adding the $\bar{z}$ signal, suitably scaled, to the long-wave lobe of $\bar{x}$.

The C1200 colorimeter made by Lichtmesstechnik was recommended in the survey of Berns *et al.* (1993). This instrument achieves high precision by the device of partial filtering: small pieces of secondary filter are placed in series with the primary filters to improve the approximation to the $\bar{x}$, $\bar{y}$, $\bar{z}$ functions. However, the price of such a colorimeter is comparable to that of a diode-array spectroradiometer, such as the Photo Research 650.

There is still a place for visual colorimeters, in which a match is made between the sample light and light drawn from a reference source and passed through calibrated glass filters. The Lovibond Tintometer (Tintometer Ltd) is an instrument of this type. Although such instruments are slow to use and the experimenter's vision may not be exactly that of the CIE standard observer, the calibrations are likely to be at least as valid as those made with a mid-priced electronic colorimeter, and it is certainly difficult to make gross errors with a visual colorimeter. For measurements in the short-wave corner of the CIE diagram, where colour matches become imprecise, Mollon and Baker (1995) recommend a method of triangulation using desaturated stimuli.

4.5 Screening tests for colour vision deficiencies

Some 8% of Caucasian males exhibit a hereditary anomaly or deficiency of colour vision. The experimenter will often wish to exclude such subjects from experiments in which colour is a variable. The most sensitive, widely available screening test is offered by the Ishihara plates, available from Keeler and from many other suppliers. These plates are not suitable for classifying colour-deficient observers, but will efficiently exclude all but a tiny minority of anomalous trichromats (the *minimalanomale Trichromaten* of Vierling). The plates must be presented under a standard illuminant. A computer-based test that quantifies the loss of colour discrimination is described by Regan *et al.* (1994) and is available to run on Cambridge Research Systems graphics boards. For a full treatment of tests for the diagnosis of colour deficiency, see Pokorny *et al.* (1979) or Birch (1993).

Acknowledgements

I am grateful to the editors and to C. R. Cavonius, G. Jordan, B. C. Regan, and Y. Tadmor for comments on the text.

References

Barlow, H. B. (1982). What causes trichromacy? A theoretical analysis using comb-filtered spectra. *Vision Research*, **22**, 635–43.

Baylor, D. A., Nunn, B. J., and Schnapf, J. L. (1987). Spectral sensitivity of cones of the monkey *Macaca fascicularis*. *Journal of Physiology*, **390**, 145–60.

Berns, R. S., Motta, R. J., and Gorzynski, M. E. (1993). CRT colorimetry. Part II. Metrology. *Color Research and Application*, **18**, 315–25.

Birch, J. (1993). *Diagnosis of defective colour vision*. Oxford: Oxford University Press.

Bonnardel, V., Bellemare, H., and Mollon, J. D. (1996). Measurements of human sensitivity to comb-filtered spectra. *Vision Research*, **36**, 2713–20.

Boynton, R. M. and Nagy, A. L. (1981). *The La Jolla analytic colorimeter: optics, calibrations, procedures and control experiments*, CHIP Report No. 109. Center for Human Information Processing, University of California, San Diego, CA.

Boynton, R. M. and Nagy, A. L. (1982). La Jolla analytic colorimeter. *Journal of the Optical Society of America*, **72**, 666–7.

Brainard, D. H. (1989). Calibration of a computer controlled color monitor. *Color Research and Application*, **14**, 23–34.

Cole, G. R., Hine, T., and McIlhagga, W. (1993). Detection mechanisms in L-, M- and S-cone contrast space. *Journal of the Optical Society of America*, **10**, 38–51.

Cowan, W. B. (1983). Discreteness artifacts in raster display systems. In *Colour vision: physiology and psychophysics* (ed. J. D. Mollon and L. T. Sharpe), pp. 145–53. Academic Press, London.

Derrington, A. M., Krauskopf, J., and Lennie, P. (1984). Chromatic mechanisms in lateral geniculate nucleus of macaque. *Journal of Physiology (London)*, **357**, 241–65.

Holtsmark, T. and Valberg, A. (1969). Colour discrimination and hue. *Nature*, **224**, 366–7.

Ives, H. E. (1915). A precision artificial eye. *Physical Review*, **6** (*New Series*), 334–44.

Jacobs, G. H., Neitz, J., and Deegan, J. F. (1991). Retinal receptors in rodents maximally sensitive to ultraviolet light. *Nature*, **353**, 655–6.

Kaye, G. W. C. and Laby, T. H. (1973). *Tables of physical and chemical constants* (14th edn). Longman, London.

MacLeod, D. I. A. and Boynton, R. M. (1979). Chromaticity diagram showing cone excitation by stimuli of equal luminance. *Journal of the Optical Society of America*, **69**, 1183–6.

Mansfield, R. J. W. (1985). Primate photopigments and cone mechanisms. In *The Visual System* (ed. A. Fein and J. S. Levine) pp. 89–106. Alan R, Liss, New York.

Mollon, J. D. and Baker, M. R. (1995). The use of CRT displays in research on colour vision. In *Colour vision deficiencies XII* (ed. B. Drum), pp. 423–44. Kluwer Academic, Dordrecht.

Mollon, J. D. and Cavonius, C. R. (1986). The discriminability of colours on c.r.t. displays. *Journal of the Institution of Electronic and Radio Engineers*, **56**, 107–10.

Mollon, J. D. and Cavonius, C. R. (1987). The chromatic antagonisms of opponent process theory are not the same as those revealed in studies of detection and discrimination. In *Colour vision deficiencies VIII* (ed. G. Verriest), pp. 473–83. Martinius Nijhoff/Dr W. Junk, Dordrecht.

Mollon, J. D. and Jordan, G. (1988). Eine evolutionäre Interpretation des menschlichen Farbensehens. *Die Farbe*, **35/36**, 139–70.

Palmer, D. A. and Whitlock, C. A. (1978). An improved monochromator using a single graded interference filter. *Journal of Physics*, *Section E: Scientific Instruments*, **11**, 996–7.

Pokorny, J., Smith, V. C., and Lund, D. (1978). Technical characteristics of 'Color-Test Glasses'. *Modern Problems in Ophthalmology*, **19**, 110–12.

Pokorny, J., Smith, V. C., Verriest, G., and Pinckers, A. J. L. G. (1979). *Congenital and acquired color vision defects*. Grune and Stratton, New York.

Regan, B. C. and Mollon, J. D. (1997). The relative salience of the cardinal axes of colour space in normal and anomalous trichromats. In *Colour vision deficiencies XIII* (ed. C. R. Cavonius and J. D. Moreland), pp. 261–70. Kluwer, Dordrecht.

Regan, B. C., Reffin, J. P., and Mollon, J. D. (1994). Luminance noise and the rapid determination of discrimination ellipses in colour deficiency. *Vision Research*, **34**, 1279–99.

Rodieck, R. W. (1983). Raster-based colour stimulators. In *Colour vision: physiology and psychophysics* (ed. J. D. Mollon and L. T. Sharpe), pp. 131–44. Academic Press, London.

Rushton, W. A. H. (1972). Pigments and signals in colour vision. *Journal of Physiology*, **220**, 1–31P.

Smith, V. C. and Pokorny, J. (1975). Spectral sensitivity of the foveal cone photopigments between 400 and 500 nm. *Vision Research*, **15**, 161–71.

Stiles, W. S. (1955). 18th Thomas Young Oration. The basic data of colour-matching. In *Physical Society Year Book*, pp. 44–65. *The Physical Society, London.*

Stromeyer, C. F., Cole, G. R., and Kronauer, R. E. (1985). Second-site adaptation in the red–green chromatic pathways. *Vision Research*, **25**, 219–37.

Turner, W. H. (1973). Photoluminescence of color filter glasses. *Applied Optics*, **12**, 480–6.

Vingrys, A. J. and King-Smith, E. (1986). Factors in using color video monitors for assessment of visual thresholds. *Color Research and Application*, **11**, 557–62.

Vos, J. J. (1978). Colorimetric and photometric properties of a 2 deg fundamental observer. *Color Research and Application*, **3**, 125–8.

Webster, M. A. and Mollon, J. D. (1995). Colour constancy influenced by contrast adaptation. *Nature*, **373**, 694–8.

Winderickx, J., Lindsay, D. T., Sanocki, E., Teller, D. Y., Motulsky, A. G., and Deeb, S. S. (1992). Polymorphism in red photopigment underlies variation in colour matching. *Nature*, **356**, 431–3.

Wyszecki, G. and Stiles, W. S. (1967). *Color science*. Wiley, New York.

Wyszecki, G. and Stiles, W. S. (1982). *Color science* (2nd edn). Wiley, New York.

Further reading

An indispensable piece of equipment in any colour laboratory is a copy of *Color science* by G. Wyszecki and W. S. Stiles (published by Wiley), which offers authoritative detail on the topics introduced above. For many purposes, the first (1967) edition is the better. Also very useful, and free, are the catalogues distributed by the larger suppliers of optical apparatus (for example, the *Monochromators and light sources* volume from Oriel): it has become the custom for these catalogues to contain detailed tutorials on several of the topics covered in the present chapter. Kaye and Laby's standard reference work *Tables of physical and chemical constants* (Longmans, London) contains useful sections on optics, radiation, and colorimetry.

Appendix: Addresses of suppliers

Beuth Verlag, Burggrafenstraße 4–10, Berlin, Germany.

Cambridge Research Systems, 80 Riverside Estate, Sir Thomas Langley Rd, Rochester, Kent ME2 4BH, UK (www.crsltd.com).

CRL, Dawley Rd, Hayes, Middlesex, UB3 1HH, UK.

D. G. Colour, 138 Greenwood Ave, Salisbury, Wilts SP1 1PE, UK.

Keeler, Clewer Hill Rd, Windsor, Berks SL4 4AA, UK.

Lichtmesstechnik, Helmholtzstr. 9, D-10586 Berlin, Germany.

Macbeth, Little Britain Rd, PO Box 230 Newburgh, NY 12550, USA; Macbeth House, Pacific Road, Altrincham, Cheshire WA14 5BJ, UK (www.munsell.com).

Melles Griot, 1770 Kettering St, Irvine, CA 92714, USA (www.mellesgriot.com).

Ocean Optics, 1237 Lady Marion Lane, Dunedin, FL 34698, USA (www.oceanoptics.com).

Oriel, PO Box 872, Stratford, CT 06497–0872, USA; 1 Mole Business Park, Leatherhead, Surrey K22 7BR, UK.

Photo Research, 9330 DeSoto Ave, PO Box 2192, Chatsworth, CA 91313–2192, USA (www.photoresearch.com).

Stage Electrics, Cofton Rd, Marsh Barton, Exeter EX2 8QW, UK.

Texas Instruments Inc., PO Box 655012, M/S 6, 13532 North Central Expressway, Dallas, Texas 75265, USA (www.ti.com).

Tintometer Ltd, Waterloo Rd, Salisbury, SP1 2JY, UK.

Notes

1. Although radiations outside these limits can normally be neglected with respect to their effect on human colour vision, the cautious experimenter has at least three reasons to exclude them from the stimulus: (a) ultraviolet (UV) radiation may be a health hazard, especially if an arc lamp is the source (see Chapter 1); (b) any extraspectral radiation may contaminate radiometric calibrations (see below); and (c) UV radiation may be re-emitted as visible fluorescence/luminescence by glass components such as colour filters and by the lens and other components of the eye. Turner (1973) gives examples of how glass colour filters may emit significant visible light when exposed to a xenon lamp emitting UV radiation.
2. In more formal treatments, X, Y, Z are the amounts of three imaginary reference lights that, when mixed together, produce the same sensation as the test colour.
3. In addition to the values x and y, we can also define $z = Z/(X + Y + Z)$.The latter does not appear directly in the diagram, but can be readily derived since the nature of the definitions mean that the sum of x, y, and z is always unity.

5

Psychophysical methods, or how to measure a threshold, and why

BART FARELL and DENIS G. PELLI

5.1 Introduction

This chapter explains how to measure visual effects. Psychophysical methods are usually described in a historical context, starting with Weber, Fechner, and Wundt in the 1800s and the development of the theoretical foundations; here we take a practical approach, focusing on what is most useful to know. Drawing conclusions about visual perception is difficult—not all questions are answerable. Psychophysics only considers questions that can be answered by measuring an observer's performance of a visual task. The art of psychophysical measurement is to channel one's curiosity into designing a question that retains the motivating interest and yet can be convincingly answered by measuring task performance. This chapter describes those tasks and measures that have proven to be most useful in vision research, and explains what kinds of question they answer.

Consider the complications in what might seem the simplest question, 'Do you see it?' One can simply present visual signals and put the question directly to the observer. But, on reflection, are we really interested in whether the observer says 'yes', or are we interested instead in whether the observer can prove that he or she has seen the signal, e.g. by correctly identifying it or locating it in time or space? In either case, when we collect the responses we find that the answer is probabilistic: in practice one measures the probability of each kind of allowed response to the signal. But then what does one do with these probabilities? Such complications need to be carefully considered on an experiment-by-experiment basis, but we will share with the reader the guidance offered by existing theory and practical experience about the most generally useful approaches to the most commonly encountered experimental problems.

5.2 Threshold

Probability measures of task performance, e.g. proportion correct, are usually much harder to interpret than the physical parameters of the stimulus. For example, theories of visual acuity directly relate known optical and anatomical properties of the eye—physical parameters—to the size of identifiable letters of an eye chart—another physical parameter—but would require speculative ancillary assumptions in order to predict the proportion correct. Consequently, the experimenter will almost always want to measure a 'threshold'. *Threshold* is the strength of the signal, as controlled by a particular stimulus dimension, that is required to attain a given level of task performance.

Fundamentally, there are two kinds of task that are used to obtain thresholds: adjustment and classification (Pelli and Farell, 1995). In *adjustment tasks* the observer is asked to adjust a knob controlling the stimulus to achieve some verbally described criterion, e.g. 'so you can just barely see it' or 'so it matches the standard stimulus'. Here, the observer directly sets the physical parameter of the stimulus. In *classification tasks* the observer is merely asked to identify the signal by placing it in one of a number of predetermined categories, e.g. 'is the screen displaying a pattern or a blank?' or 'is the test stimulus larger or smaller than the standard?' Repeated testing of the classification of a set of stimuli varying in signal strength measures the proportion of the identification responses to each stimulus that were correct, but it is usually most useful to find the threshold value of the stimulus parameter that would yield a certain proportion correct. So, in practice, both adjustment and classification tasks are used to estimate the threshold value of the signal parameter, i.e. the value that achieves a specified criterion, subjective in the case of adjustment, objective in the case of classification.

Thus, while one can imagine a wide variety of questions that might reasonably be asked to obtain a measure of psychophysical performance, the most useful methods that current vision science has to offer, and the most widely practised, are those that measure threshold.

5.3 Adjustment

In the days when vision research labs used analog function generators to synthesize their stimuli, it was very easy to continuously display a stimulus while adjusting it. Today, digital computers synthesize a vastly greater range of images to be used as stimuli, but, unfortunately, it takes some effort to get a computer dynamically to recompute the stimulus in response to the observer's adjustments. Nevertheless, computers are getting faster, and there are shortcuts to synthesizing certain stimuli. When feasible, adjustment tasks offer a quick direct measurement of a subjective match between the variable stimulus and a standard. Because of the subjective nature of the adjustment settings, this method is ideally suited for experimenters using themselves as observers, allowing them to quickly experience the full range of effects of a stimulus parameter on appearance.

When testing others, the instructions given to the observer are crucial. They are so crucial that if a published experiment relies on the method of adjustment, then the discussion section should convince the reader that the instructions used indeed bear on the aspect of perception that is the nominal topic of the paper. *Matching* instructions are particularly easy for observers to understand and are the most commonly used. In the matching paradigm, there are two objects, a standard and a test, and the observer is asked to adjust the test object to 'match' the standard. The criterion for matching is all important—what one asks the observer to do and what the observer actually does are not at all the same thing—and should not only be conceptually clear, but also, if at all possible, perceptually salient. For this reason, a particularly effective matching instruction is 'nulling'. This applies to cases where one presumes that the observer understands what the stimulus 'ought' to look like when it is 'undistorted' or 'neutral' and adjusts it to achieve that appearance. For example, adjusting a patch to eliminate any colour or motion or pattern; or adjusting a line to be straight.

Consider the influence of form on brightness and how it might be quantified by brightness matching. Benary (1924) and Adelson (1993) showed that the brightness of a surface depends on the perceived object structure. They presented two fairly similar images made up of contiguous uniform patches arranged to produce different three-dimensional interpretations: one patch in one image was adjustable, and the observer was asked to adjust its luminance to match the brightness of a particular patch in the other image. To explain brightness, as opposed to lightness, Adelson asked his observers to 'judge the shade of ink on the page' rather than make any inference about the surfaces of the objects portrayed.

In an effective use of the nulling technique, Cavanagh and Anstis (1991) employed motion nulling to measure the contribution of colour to the perception of motion. They showed observers a rightward-moving luminance grating superimposed on a leftward-moving chromatic grating. The observers adjusted the contrast of a second luminance grating, moving leftward in phase with the chromatic grating, to null the motion of the entire pattern. The difference between the contrasts of the leftward and rightward luminance gratings is then a measure of the contribution of colour to perceived motion.

5.4 Classification

At present, three kinds of classification are widely used: yes/no, two-alternative forced choice (2afc), and identification. Each task asks the observer to reply to a query: 'Did you see it?' (yes/no); 'Was the signal in the first or in the second interval?' (2afc); or 'Which signal was it?' (identification). All three call for the observer to classify stimuli (or their subjective responses). Those 2afc tasks that present a signal and a blank on each trial are said to be 'detection' tasks. In a 'discrimination' task, the signal is added to a constant background stimulus that appears in both intervals. Yes/no, 2afc, and identification all have their special niches, and all three tasks have been used to measure thresholds and convincingly establish important scientific conclusions. All other things being equal, however, readers who value their time will use identification if possible, otherwise 2afc, and yes/no only as a last resort.

5.4.1 Yes/no

If one must, then with some effort a *frequency of seeing curve* can be measured, which plots the probability of saying 'yes' to the question—'Did you see it?'—as a function of a stimulus parameter (e.g. contrast). Unfortunately, the observer in a yes/no experiment can't avoid introducing an internal subjective criterion in deciding whether each faint ambiguous percept deserves a 'yes' or a 'no'. The observer's personality, the instructions, and other experimental details may all affect the internal criterion, and thereby threshold. This thorny problem has been thoroughly analysed and various remedies have been devised, of which we endorse only one, though it is by far the most time-consuming. Theory of signal detectability shows that the frequency of seeing or 'hit rate' is uninformative unless one also measures the *false-alarm rate*, i.e. the probability of saying 'yes' when a blank is presented. By systematically changing the instructions one can push the observer's criterion up or down, and measure both the hit and false-alarm rates at each criterion level. This yields an ROC (receiver-operating characteristic) graph of hit vs. false-alarm rate, parameterized by the (unknown) internal criterion. The area under the ROC curve is an excellent measure of the visibility of the signal (Swets and Pickett, 1982). The ROC curve can be obtained more quickly by asking the observer to give a 1-to-5 confidence rating instead of merely saying yes or no, but this still entails a substantial effort on the part of the experimenter to collect and analyse the results to obtain the ROC area.

5.4.2 Two-alternative forced choice

The 2afc task gives the observer one of two stimulus arrangements and asks the observer to identify which it is. The advantage of the 2afc task is that it can be designed to avoid criterion effects by presenting a symmetric unbiased choice. Typically, this is achieved by having two stimulus arrangements both containing, say, a signal and a blank, which differ solely by the interchange of signal and blank. One might present the signal in a first interval and the blank in a second interval, or vice versa (randomly), and ask the observer in which interval the signal was presented ('two-interval forced choice'). Or one might present them simultaneously, side by side, and ask the observer on which side the signal is. Under mild theoretical assumptions, the measured proportion correct will equal the ROC area described in the previous section (Green and Swets, 1974; Nachmias, 1981), thus obtaining a similar result with much less effort.

This is an excellent technique. Its only drawback is that, because the observer has only two alternatives and thus will be right half the time even if the signal is invisible, a relatively large number of trials (about 60) is required to obtain a good threshold estimate. For greatest efficiency in 2afc tasks one should use a sequential estimation procedure (described below) to adjust the signal strength systematically and estimate threshold directly. Or, with somewhat more effort, one can measure the proportion correct as a function of signal strength (the 'psychometric' function). In each trial, one strength value is presented from a fairly small number (usually 5 or 7) that span the performance range (50–100%). Then the threshold for any level of performance can be read off the psychometric function. The shape of this function, like the shape of the ROC curve, can also be used to infer the distributions of internal stimulus representations on which decision processes operate (see Graham, 1989). However, usually you will just want to know threshold.

5.4.3 Identification

An even more efficient method is to present one of many signals and ask the observer to identify it. How many? Simulations show that four (or more) alternatives suffice to achieve a high efficiency accruing from minimizing the chance of blind guessing (Pelli *et al.*, 1988). Theoretical consideration of the ideal observer suggests that, in order to obtain a steep psychometric function (to estimate threshold quickly), the signals should all have approximately equal contrast energy and similar, pairwise cross-correlation (van Trees, 1968). Of course, observers must learn to identify the signals. Experiments involving the identification of foreign and novel alphabets show that observers learn to identify new symbols quickly, requiring only 2000 trials to attain the same threshold for letter identification as fluent readers of the alphabet (Pelli *et al.*, 1998). The observer's responses implicitly divide the high-dimensional stimulus space into many regions (one per kind of response) separated by category boundaries. In principle, these category boundaries are subjective and movable, like the observer's internal criterion in the yes/no task, but, on the one hand, theoretical understanding of the high-dimensional case is still wanting (Ashby, 1992), so there is nothing one can do about it, while, on the other hand, the problem is less worrisome, because high probabilities of correct identification are not attainable by blind guessing.

5.4.4 Sequential estimation: QUEST

Given that the experimenter is willing to run a reasonable number of trials (e.g. 40), and has some prior knowledge of the psychometric function and its parameters, one would like an efficient procedure for threshold estimation—a procedure for running each trial at whatever signal strength would contribute most to minimizing the variance of the final threshold estimate. Such a procedure combines the experimenter's prior knowledge and the observer's responses on past trials in choosing the signal strength for the next trial, and, at the end, estimating threshold. The best current procedure is called ZEST (King-Smith *et al.*, 1994), but QUEST (Watson and Pelli, 1983), which is nearly as efficient, can be implemented by a tiny C program, which we present below.

The only unknown is threshold, which is treated as a random variable, X, to be estimated. The experimenter supplies an initial guess, by specifying the mean and SD of a Gaussian probability density function. For the reader's convenience, we supply a one-line simulation of an observer with threshold *tActual*, so the program can be run on its own. To run a real experiment, that line must be replaced by code that presents a stimulus (at intensity x) and collects the observer's response (1 if right, 0 if wrong). After each response, the probability density function, q, is updated by Bayes's rule. Each trial is placed at x, the current maximum-probability estimate of threshold, i.e. the mode. The final threshold estimate is also the mode.

```
#include <math.h>
#include <stdio.h>
#include <stdlib.h>
#define DIM 400
#define DIM2 (2*DIM)
#define GRAIN 0.01
#define xi(i) (((i)-DIM/2)*GRAIN)
#define xii(ii) (((ii)-DIM2/2)*GRAIN)
#define iix(x) (int) (0.5+DIM2/2+(x)/GRAIN)
void main(void)
```

```
{
  float p[DIM2+1],s[2][DIM2+1],q[DIM+1];
  int i,ii,trialsDesired=40,k,imode,right;
  double beta=3.5,delta=0.01,gamma=0.5; /* parameters of psychometric
   function */
  double x,tGuess=-2.0,tGuessSD=4.0,tActual;
  char wrongRight[2][]={"wrong","right"},string[64];

  for(ii=0;ii<=DIM2;ii++){
    p[ii]=delta*gamma+(1-delta)* (1-(1-gamma)*exp(-pow(10,beta*xii(ii))));
    s[0][DIM2-ii]=log(1-p[ii]);
    s[1][DIM2-ii]=log(p[ii]);
  }
  for(i=0;i<=DIM;i++){
    x=xi(i)/tGuessSD;
    q[i]=-0.5*x*x;
  }
  printf("Estimate threshold:");
  gets(string);
  sscanf(string,"%lf",&tGuess);
  printf("Specify true threshold of simulated observer:");
  gets(string);
  sscanf(string,"%lf",&tActual);
  for(k=1;k<=trialsDesired;k++){
    for(imode=0,i=0;i<=DIM;i++)if(q[i]>q[imode])imode=i;
    x=xi(imode)+tGuess;
    /* to test a real observer, */
    /* replace the next line with your experimental task */
    right=p[iix(x-tActual)] > rand()/(RAND_MAX+1.0);
    printf("Trial %3d at %4.1f is %s\n",k,x,wrongRight[right]);
    for(i=0;i<=DIM;i++)q[i]+=s[right][i-imode+DIM2/2];
  }
  for(imode=0,i=0;i<=DIM;i++)if(q[i]>q[imode])imode=i;
  x=xi(imode)+tGuess;
  printf("Final (mode) threshold estimate is %4.1f\n",x);
}
```

5.5 Reaction time

Threshold is the stimulus strength required for a specified probability of correct decision. We typically assume that the observer is making a simple decision about a single elementary stimulus. In many practical situations, however, people do not respond on the basis of a single elementary decision, but only after making multiple decisions about the many stimuli present in a complex display. Searching for a particular face in a crowd is a familiar example. Thresholds could be found for tasks like this, but often the interest is in different types of questions than threshold measures answer. A researcher might be interested in measuring how long it takes to perform a task, or in analysing the theoretically more challenging question of how the component decisions leading up to a task response are distributed in time (Sternberg, 1969). Response times can be measured in any task, but one must not forget to measure response accuracy at the same time, because of trade-offs between the two (e.g. McElree and Dosher, 1989).

5.6 Devilish details

Having decided on a task, implementing it can bring a flood of new questions. Some are easy, e.g.: 'How many threshold estimates?' Enough to make the standard error small. Many other questions are harder, with answers that depend on details of your theory and experiment. The issues include practice, who triggers the trial (experimenter or observer), cueing (to warn of impending stimulus), manner of response (button, speech), allowing the observer to not respond ('I blinked and missed it'), feedback ('right'), and frequency of rest breaks. In general, you should look for the easiest way to obtain a convincing answer to your experimental question. You'll want to be very sure that the task is obvious to the observers. Counting on the observer's intelligence to figure out what the task really is invites huge individual differences in the results that are probably unrelated to the perceptual questions you are really interested in.

References

Adelson, E. H. (1993). Perceptual organization and the judgment of brightness. *Science*, **262**, 2042–4.

Ashby, F. G. (1992). *Multidimensional models of perception and cognition*. Lawrence Erlbaum, Hillsdale, NJ.

Benary, W. (1924). [The influence of form on brightness contrast, translated in Ellis, 1938] Beobachtungen zu einem Experiment über Helligkeitskontrast. *Psychologische Forschung*, **5**, 131–42.

Cavanagh, P. and Anstis, S. (1991). The contribution of color to motion in normal and color-deficient observers. *Vision Research*, **31**, 2109–48.

Ellis, W. D. (1938). *A source book of gestalt psychology*. Harcourt, Brace, and Co., New York.

Graham, N. V. S. (1989). *Visual pattern analysers*. Oxford: Oxford University Press.

Green, D. M. and Swets, J. A. (1974). *Signal detection theory and psychophysics*. Krieger, Huntington, NY.

King-Smith, P. E., Grigsby, S. S., Vingrys, A. J., Benes, S. C., and Supowit, A. (1994). Efficient and unbiased modifications of the QUEST threshold method: theory, simulations, experimental evaluation and practical implementation. *Vision Research*, **34**, 885–912.

McElree, B. and Dosher, B. A. (1989). Serial position and set size in short-term memory: Time course of recognition. *Journal of Experimental Psychology: General*, **118**, 346–73.

Nachmias, J. (1981). On the psychometric function for contrast detection. *Vision Research*, **21**, 215–23.

Pelli, D. G. and Farell, B. (1995). Psychophysical methods. In *Handbook of optics* (2nd edn), Vol. I (ed. M. Bass, E. W. Van Stryland, D. R. Williams, and W. L. Wolfe), pp. 29.1–29.13. McGraw-Hill, New York.

Pelli, D. G., Robson, J. G., and Wilkins, A. J. (1988). The design of a new letter chart for measuring contrast sensitivity. *Clinical Vision Sciences*, **2**, 187–99.

Pelli, D. G., Burns, C. W., Farell, B., and Moore, D. C. (1998). Identifying letters. *Vision Research* (In press.)

Sternberg, S. (1969). The discovery of processing stages: extensions of Donder's method. In *Attention and performance* II (ed. W. G. Koster), pp. 276–315. North-Holland, Amsterdam.

Swets, J. A. and Pickett, R. M. (1982). *Evaluation of diagnostic systems: methods from signal detection theory*. Academic Press, New York.

Van Trees, H. L. (1968). *Detection, estimation, and modulation theory*. Wiley, New York.

Watson, A. B. and Pelli, D. G. (1983). QUEST: a Bayesian adaptive psychometric method. Percept *Psychophys*, **33**, 113–20.

6

The behavioural analysis of animal vision

RANDOLPH BLAKE

6.1 Introduction

This chapter surveys techniques available for the behavioural study of animal vision, the discipline technically known as comparative visual psychophysics; it does not pretend to review the huge literature on behaviourally measured visual function in animals. As the reader will discover from the examples given within this chapter, the array of available techniques includes ones applicable to a very wide range of species, spanning mammals, birds, fish, and insects.

As a prelude to this survey of techniques, an obvious question begs answering: 'Why, in the first place, go to the trouble of studying vision in members of species other than our own, particularly given the technical challenges involved?' The answers are varied. For some, simple intellectual curiosity leads to the question: 'What does the world look like through the eyes of another species?' Naive realism can lull us into believing that all creatures experience the same visual world, but examination of the literature on comparative visual psychophysics forcefully undermines this belief—different species live in different visual worlds (see, for example, Fig. 6.1, which plots contrast sensitivity functions for a variety of species). For others, the study of animal vision is motivated by an appreciation of the great optical and neuroanatomical diversity found in visual systems of different species. As Gordon Walls (1967) so wonderfully documented, this structural diversity implies the existence of profound differences in visual capacities across species—documenting those differences is another reason for behavioural studies of animal vision. In a related vein, exciting discoveries in visual neurophysiology have sharpened our understanding of the neural events involved in the early stages of vision. Given this wealth of neurophysiological data, it is natural—indeed irresistible—to wonder to what extent aspects of visual perception can be related to neural events. This line of reasoning has culminated in the development and refinement of techniques for monitoring neural activity in alert, behaving animals engaged in visual tasks (Newsome, 1995). Finally, the study of animal vision affords the opportunity to assess visual consequences associated with disorders within the visual nervous system, either artificially produced (e.g. reduced cortical binocularity caused by early visual deprivation) or naturally occurring (e.g. disordered retinotopic maps caused by genetically based misrouting of optic fibres). These disorders can be quite selective and localized within the brain, in turn producing highly selective losses in vision (Daw, 1995).

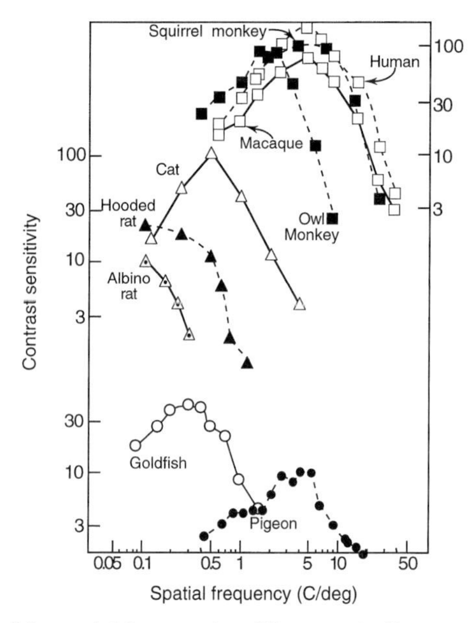

Fig. 6.1 The range of visible spatial frequencies differs markedly among species, as indicated by the differences in their contrast sensitivity functions (1/threshold contrast as the function of spatial frequency). Redrawn from Uhlrich *et al.* (1981). Cross-species correspondence of spatial contrast sensitivity functions. *Behavioural Brain Research*, **2**, 291–9. Copyright 1981, with kind permission from Elsevier Science—NL, Sara Burgerhartstraat 25, 1055 KV Amsterdam, The Netherlands.

Here, too, the study of animal vision can shed light on the neurobiological bases of visual perception (Blake, 1981).

6.2 Challenges and obligations

Comparative studies of animal vision are *de facto* behavioural studies: they concern themselves with behavioural indices of visual detection or visual discrimination. Accordingly, the bulk of this chapter focuses on techniques available for the establishment and maintenance of behavioural control in non-humans. The deployment of these techniques inevitably raises ethical issues, and these must be acknowledged and addressed explicitly from the outset of a project involving animal testing. The following section merely highlights some of these ethical issues; it is imperative that investigators consult published guidelines specifying the procedures to be followed when working with animals (e.g. *Guide for the care and use of laboratory animals*, 1985; *Handbook for the use of animals*, 1993).

6.2.1 Behavioural control during an experiment

Because spoken instructions are useless in comparative psychophysics, animals must be coaxed non-verbally to perform in a behavioural task. This coaxing typically capitalizes on one of two broad forms of behavioural control: appetitive or aversive.

As discussed in greater detail later in this chapter, appetitive control relies on rewarding a hungry or thirsty animal for correct performance, and aversive control entails presentation of some undesirable stimulus (e.g. a brief, loud noise) in association with a visual event or in consequence of an incorrect behaviour.

Effective appetitive control of an animal's behaviour is predicated on an appropriate motivational state (i.e. hunger or thirst) and, therefore, must include a strict regime of food or water deprivation. When employing a behavioural technique that entails appetitive control, it is crucial to monitor regularly the physiological and psychological status of the deprived animal, to ensure normal health and temperament as well as optimum performance of the task at hand. Towards this end, one must consult with veterinarian professionals in the design of any deprivation schedule and must rely on those individuals for assistance in daily monitoring of the health of the animals. As a rule, animals on a deprivation schedule should be tested at the same time each day, both for reasons of health and for optimum behavioural control. While 'Methods' sections of published papers do not always include this information, investigators often temporarily remove animals from strict deprivation periodically during the course of an experiment, in the interests of health and variety. Thoughtful selection of an appetitive reward must take into account the natural dietary habits of the species under test (e.g. cats love pureed liver but pigeons are indifferent to it). Fortunately, published literature provides the beginning investigator with fruitful pointers. Also, procedures employing appetitive control typically require restraining the animal in a small chamber or a specially constructed chair. While posing no physical threats, restraint for an extended period of time induces stress and, for this reason, investigators must take great care to optimize comfort when designing restraint devices. Moreover, it is essential to include frequent breaks during an extended behavioural testing involving restraint, as well as to provide psychological reassurance such as periodic physical contact (e.g. gentle petting).

Aversive control raises more serious ethical concerns. In the field of comparative visual psychophysics, no one accepts procedures that evoke pain. There are, however, techniques utilized that presumably induce in the test animal some mild degree of discomfort, albeit very briefly. For example, a technique used by some in the study of fish vision entails the delivery of a brief electric shock to the animal's body, with the intensity of the shock adjusted to a level just sufficient to produce a measurable change in the fish's respiration pattern. Similarly, investigators have tested vision in kittens and in rats utilizing a jumping-stand apparatus, whereby an animal forced to drop several centimetres on to one of two platforms loses its footing if the wrong platform is selected. Neither of these procedures involves bodily harm, but evidently both are unpleasant judging by animals' reactions to these kinds of events. In general, investigators should explore all means possible to avoid test procedures involving aversive stimulation, even if this requires more complex, time-consuming training regimes based on appetitive control. When elements of discomfort cannot be avoided, every effort must be made to ensure that discomfort is minimal and, by all means possible, that discomfort poses no threat to the animal's health or safety. In every case, it is imperative that all aspects of a behavioural testing procedure be evaluated and approved by the appropriate institutional review board.

6.2.2 Animal maintenance

While not engaged in actual testing, animals must be maintained in safe, sanitary facilities staffed by certified professionals. In many countries, legal requirements specify the particulars of those facilities, including cage sizes, quarantine periods, cleaning schedules, and conditions of social stimulation. Site visits of animal care facilities are routine, and failure to comply with regulations jeopardizes all projects supported by the facility.

For reasons of health and safety, it is also imperative that animal care facilities remain secure. Animals are susceptible to infection from uncontrolled contact with humans (and vice versa). Moreover, given the intense hostility of some to the use of animals in research (Singer, 1975), the lives of animals (and humans) can be threatened without proper security of an animal facility. Access to animal colony areas should be restricted to authorized care-takers and investigators responsible for the associated research programmes.

Finally, best practice, often legally enforced, is to ensure that experimental animals be acquired from registered suppliers only (never from individual sources or from the street), and newly acquired animals be quarantined before placement in the larger colony area; the duration of the quarantine period varies with species.

In brief, investigators should treat animal participants with the same degree of dignity, respect, and care afforded human subjects. Such treatment will be rewarded by sensitive, reliable visual performance.

6.3 Behavioural techniques for assessing vision

Regardless of the visual capacity under study, all testing procedures are grounded on two closely related principles of indistinguishability:

(1) from the viewpoint of the animal, a visual stimulus below threshold is indistinguishable from the absence of that stimulus; and

(2) from the viewpoint of an animal, two (or more) physically different stimuli can be rendered indistinguishable from one another by manipulation of the magnitude of their physical differences.

The first principle pertains to the measurement of so-called absolute, or detection, thresholds, and the second to the measurement of difference thresholds. Both types of threshold measurements, incidentally, represent Class A experiments within the scheme of Brindley (1970), for in both cases distinguishing stimulus information is lost. Capitalizing on these principles, the general strategy underlying animal psychophysics is straightforward: some component of the animal's behaviour is placed under visual stimulus control and, then, an attribute of that stimulus is varied to assess how the animal's behaviour is maintained consequent to that variation. The specifics of visual stimulation depend, of course, on the visual capacity under examination. In general, stimulus generation techniques described elsewhere in this volume are applicable to studies of animal vision.

The following sections describe behavioural techniques available for establishing stimulus control, and therefore assessing visual capability, along with some of the advantages and disadvantages associated with each.

6.3.1 Reflex responses

This category of behavioural responses has the attractive feature of requiring no training, because the animal's behavioural reaction occurs automatically, i.e. reflexively, in response to visual stimulation. The following are some of the more popular reflex responses used in comparative visual psychophysics.

6.3.1.1 Optokinetic nystagmus

Upon viewing a repetitive pattern of stripes moving continuously through the visual field, the eyes will involuntarily follow the motion, yielding a series of smooth movements punctuated periodically by saccadic returns. Termed optokinetic nystagmus (OKN), this visuomotor reflex can readily be induced and measured in a host of animals using essentially the same procedures as described in Chapter 9. (In freely moving animals with lateral eyes, this visuomotor reflex may include turning the entire body in the direction of motion.) It is possible, then, to estimate visual acuity by varying the spatial frequency of the inducing contours, thereby determining the highest value capable of driving OKN. Alternatively, one can determine the minimum grating contrast capable of generating OKN for different spatial frequencies, thereby deriving a contrast sensitivity function. In the domain of colour vision, it is possible to estimate the isoluminance value for chromatic gratings, by varying the relative intensities of the bars until OKN is maximally attenuated (Anstis and Cavanagh, 1983). Several laboratories (Logothetis and Charles, 1990; Dobkins and Albright, 1995) have utilized this technique to estimate isoluminance values in monkeys.

There are at least two limitations to this technique: one is the tendency for OKN to habituate with repeated stimulation, and the other is the necessity of using moving contours. This last limitation may be particularly serious when studying species whose visual systems include pathways populated by neurones relatively insensitive to motion (e.g. X cells and/or neurones of the parvocellular pathway). In these animals, OKN at best provides an incomplete picture of visual capability. In a related vein, the neural circuitry putatively involved in the production of the OKN (e.g. Collewijn, 1992) is distinct from those phylogenetically newer portions of the visual nervous system thought to mediate more complex visually guided behaviours. Other, more refined psychophysical techniques are therefore required for elucidating the visual capabilities mediated by those other portions of the nervous system.

6.3.1.2 Looming

A rapidly approaching (i.e. looming) object triggers reflexive head and body movements executed to avoid impending collision. This reflexive avoidance in response to visual looming can be employed as a crude gauge of visual function in animals (e.g. Blakemore and Cooper, 1970; Blake and Di Gianfilippo, 1980). Like OKN, however, this reflex provides very limited information about visual capacity. Indeed, animals with significant damage to the central retinae of both eyes may exhibit a normal startle reflex to a looming stimulus (Blake and Bellhorn, 1978), attesting to the poor sensitivity of this measure of visuomotor behaviour.

6.3.1.3 Visual cliff

One of the more famous tests of visual function is the visual cliff, an apparatus pictured in most introductory psychology texts. Devised by Gibson and Walk (1960), the cliff consists of a platform flanked by one or more lower steps. Capitalizing on an animal's natural tendency to avoid sharp drop-offs, one can place an animal on the central platform to see whether it will descend to a lower step. By varying the distance from platform to step, one can estimate depth perception based on the animal's reluctance to leave the platform. The visual composition of the lower step can also be manipulated (e.g. the size of texture elements on the surface of the step) to assess possible cues for depth. Experience indicates that the utility of this apparatus is limited primarily to smaller, younger animals.

6.3.1.4 Preferential looking

As described by Atkinson and Braddick in Chapter 7, human infants exhibit a natural preference for looking at the more complex or interesting of two visual objects. This same preference behaviour has been observed in young monkeys and kittens, thus introducing preferential looking as a procedure for evaluating vision in these species. As implemented by Kiorpes (1992), an infant monkey is held by an experimenter so that it directly faces two stimulus fields located side-by-side. A grating is displayed in one field and a homogeneous field of the same mean luminance appears in the other field. The experimenter cannot see these displays but can visualize the monkey's face by way of a video camera and monitor. By observing the animal's oculomotor behaviour, the experimenter judges on which side—left vs. right—the grating was presented, receiving error feedback following each trial. In a refined version of this procedure (Kiorpes and Kiper, 1996), the young monkey is taught to place its head within a face mask that allows it to view a large-screen video monitor; the monkey's eye position is monitored using a commercially available eye-tracker (see Chapter 9). When the monkey is looking at a central fixation point, a visual test stimulus (e.g. a patch of sinusoidal grating) is briefly presented either to the left or to the right of the fixation point. In some versions of the task, a correct judgement yields a reward (e.g. a squirt of milk) for the monkey. This particular implementation of preferential looking has the advantage of allowing rather precise delivery of test stimuli to given regions of the retina. For example, it has been possible to measure variations of contrast sensitivity with eccentricity in animals only a few months old (Kiorpes and Kiper, 1996). Once animals reach 4 months of age, it becomes feasible to replace preferential looking with an instrumental conditioning procedure of the sort described in Section 6.3.3. For a description of the application of preferential looking in the study of kitten vision, see either Wilkinson (1995) or Sireteanu (1985). According to Wilkinson (personal communication), the technique works best in 2–5-weeks-old kittens; for older kittens, one of the instrumental conditioning procedures described in Section 6.3.3 is preferable.

6.3.1.5 Orientation

There are other reflexive, species-specific behaviours that can be exploited for the study of vision. Some creatures, for example, are naturally attracted to light (phototaxis) while others are repelled by light (photophobia)—one can imagine manipulating light intensity to determine a behavioural threshold for seeing, indexed by the incidence of one of these orienting reflexes. Similarly, fish will attempt to orient their bodies in water so as to equalize light stimulation to both laterally placed eyes; this reflex, too, can be exploited for a behavioural analysis of vision. The biology literature is rich with examples of species-specific behaviours that afford 'windows' into the visual world of other species, as exemplified by Reichardt's seminal work on motion detection in the fly (Reichardt, 1957) and by von Frisch's studies of visually guided navigation in bees (von Frisch, 1967). Students of animal vision will profit from reading these classics.

6.3.2 Classical conditioning

Behavioural testing procedures based on principles of classical conditioning have proven particularly popular with students of comparative visual psychophysics, for the obvious reason that initial training is simple and relatively rapid. With this class of procedures, a visual stimulus (the so-called conditioned stimulus) is repeatedly paired with an unconditioned stimulus that itself reliably elicits a reflexive (i.e. unconditioned) response. Following relatively few such pairings, the visual stimulus on its own comes to elicit the reflex (termed the conditioned reflex), in anticipation of the unconditioned stimulus. The conditions that optimize training and testing are well documented (e.g. Kling, 1971), and only the highlights are mentioned here.

Of paramount importance is the time interval between the conditioned stimulus and the unconditioned stimulus—for optimal conditioning, the former must precede the latter by a duration dependent on the nature of the unconditioned response. As a rule of thumb, this interval must be sufficiently long for the conditioned response to occur prior to the presentation of the unconditioned stimulus, for otherwise it becomes difficult to distinguish the conditioned from the unconditioned responses. It is possible to omit the unconditioned stimulus entirely in a fraction of trials, which can be advantageous when one is unsure whether the conditioned stimulus is visible to the animal. (Administration of the unconditioned stimulus in trials where the conditioned stimulus is below threshold, constitutes, from the animal's perspective, an unsignalled unconditioned stimulus; these trials will increase the animal's level of anxiety and weaken the associative strength between the conditioned and unconditioned stimuli.)

Selection of the unconditioned stimulus and, by implication, the response it evokes, depends on the species under test. In principle, any stimulus reliably eliciting a measurable, reflexive response will suffice, and a survey of the classical conditioning literature reveals a broad range of stimuli including: brief, loud noise; air-puffs delivered to the eye; and food delivered to the mouth. Probably the most widely used unconditioned stimulus in comparative visual psychophysics is brief, mild electric shock delivered to an animal's feet (e.g. Shumake *et al.*, 1968) or, in the case of fish, to its body (e.g. Northmore and Dvorak, 1979). Shock itself typically produces a startle reaction that interrupts the animal's ongoing behaviour, including normal respiration. Thus anticipation of shock in reaction to visualization of the conditioned stimulus is occasioned by these same reactions—the incidence and strength of these responses provide the psychometric gauge of visual performance.

The conditioned stimulus will depend, of course, on the aspect of vision under test (e.g. assessment of acuity might dictate the use of a grating). It is important to realize that the visual stimulus itself may elicit an orienting response (e.g. visual fixation), so one must be cautious that this orienting response does not contaminate or mask the conditioned response under study. For example, both pupillary constriction and eye-blink have been utilized as conditioned responses in classical conditioning studies; being visually triggered responses, both, in principle, could occur independent of a visual stimulus. To avoid this problem, it is preferable to select an unconditioned stimulus whose associated response itself has no visually triggered component. Brief, mild electric shock, with its associated change in heart rate, represents one possibility.

6.3.2.1 Phases of training with classical conditioning

The rationale underlying the measurement of visual thresholds using classical conditioning is straightforward. Initially, an animal is conditioned using a stimulus that is easily seen. During this training phase, it is conventional to present this stimulus paired with the unconditioned stimulus on a somewhat irregular basis, so that elapsed time *per se* does not serve as a cue for the impending occurrence of the conditioned stimulus. Once the conditioned stimulus reliably elicits the conditioned response, one can systematically vary some aspect of this stimulus to determine the effect of that variation on the response. Ideally, the conditioned response would remain invariant in strength up to the point where the conditioned stimulus was no longer visible, thereby providing a clear index of the transition from stimulus visibility to stimulus invisibility. This ideal outcome, of course, never happens. Even in reaction to a conditioned stimulus of fixed intensity, the conditioned response will vary in strength from trial to trial. Thus the experimenter must evaluate the detectability of the conditioned stimulus against this background of response variability. One strategy is simply to plot response magnitude against stimulus intensity and to define 'threshold' as the stimulus value associated with a given reduction in response strength. Alternatively, one can use statistical procedures to decide on each trial whether or not the response exceeded some criterion value, thereby categorizing the detection response on each trial as 'yes' or 'no'. The resulting data are then plotted as a conventional method-of-constant-stimuli psychometric function (see Chapter 5) whose threshold is the stimulus value associated with 50% detection.

6.3.2.2 An example of classical conditioning: visual resolution in the goldfish

To illustrate the procedure at work, consider the following application of this technique for the measurement of visual contrast sensitivity in goldfish. With the fish restrained within an underwater Plexiglas chamber, Northmore and Dvorak (1979) first paired the 5-s presentation of a vertical, sinusoidal grating on a video monitor with a 0.2-s electric shock delivered through a pair of electrodes arranged on each side of the fish's body; this shock occurred coincident with the offset of the grating.

Within approximately 120 conditioning trials, the onset of the grating reliably elicited temporary interruption of breathing, which was monitored by a thermistor placed immediately in front of the fish's mouth (the thermistor sensed the change in water temperature associated with the transient cessation of respiration). To distinguish visually triggered changes in respiration from those occurring spontaneously, a computer sampled the respiration in four successive, 5-s intervals, selecting the interval in which the longest interbreath interval occurred. The grating was presented randomly during one of the four intervals, and a 'hit' was tallied for each 4-interval trial on which the computer correctly judged the interval containing the grating. An up/down staircase procedure (see Chapter 5) varied grating contrast from trial to trial, to determine the contrast associated with 50% correct performance. One technical detail is particularly worth noting: the shock was administered only on trials following a 'miss' on the previous trial. This regime, by reducing the incidence of shock presentations, presumably minimized stress on the fish and, moreover, avoided the need always to pair shock with subthreshold stimuli. The results from this study, incidentally, are summarized by the curve marked 'Goldfish' in Fig. 6.1.

6.3.2.3 Conditioned suppression

There is a version of classical conditioning, called conditioned suppression, in which an animal is first trained to emit a repetitive response in order to obtain a reward (this type of conditioning, called operant or instrumental learning, is covered in the next section). Superimposed on this baseline of steady responding is a classical conditioning procedure, whereby a stimulus (e.g. presentation of a grating) signals a forthcoming unconditioned stimulus (e.g. brief, electric shock to the foot). Upon detection of the warning stimulus an animal will cease responding until termination of the unconditioned stimulus, and it is this change in the rate of responding that indexes detectability of the warning stimulus. For a number of years this procedure proved popular with students of cat vision (Fox and Blake, 1971; Loop and Berkley, 1972; Lehmkuhle *et al.*, 1982), although it has now largely been replaced by the two-choice procedure described in the next section.

6.3.2.4 Limitations to classical conditioning

There are several disadvantages to classical conditioning and conditioned suppression as means for assessing vision behaviourally. First, the repeated presentations of an unconditioned stimulus can instil anxiety in animals, thereby raising their false alarm rates (i.e. the tendency for responses to occur spontaneously). Second, experimenters must adopt some rule for administering the unconditioned stimulus in trials where the animal's behaviour indicates failure to detect the conditioned stimulus. It is unwise to administer the unconditioned stimulus in all such non-detection trials, for this will weaken the association between that stimulus and the conditioned stimulus. But withholding the unconditioned stimulus in all trials where the conditioned response is weak or absent can unwittingly teach the animal to control the magnitude of the conditioned response, thereby undermining the sensitivity of the procedure. Third, classical conditioning does not lend itself to the measurement of discrimination thresholds, i.e. test conditions requiring the animal to respond to one visible stimulus but not to another. This is because, with classical conditioning, the response readily generalizes to stimuli other than the conditioned stimulus. For this reason, the technique works primarily for the measurement of detection thresholds, i.e. test conditions where the animal simply detects the presentation of a visual stimulus. Finally, the technique often, although not inevitably, involves the use of unpleasant stimulation, which raises ethical questions about the welfare of the animal.

6.3.2.5 Aversive conditioning

Before proceeding to the next category of behavioural technique, it is worth mentioning another procedure besides classical conditioning that also uses aversive stimulation. With this procedure—termed conditioned avoidance—an animal receives repeated pairings of a conditioned stimulus and an unconditioned stimulus. However, the animal can terminate or even prevent the unconditioned stimulus by executing some behaviour when the conditioned stimulus is detected. For example, rats will learn to move from one test chamber to another upon detection of a change in ambient illumination which signals that a forthcoming electric shock will be delivered through the floor of the initially occupied chamber. This method has been utilized infrequently in comparative visual psychophysics, in part because of the strong tendency for animals to produce high false alarm rates.

6.3.3 Instrumental conditioning

The term 'instrumental conditioning' is applied to procedures where the behaviour of the animal is instrumental in determining the consequences of that behaviour; this category of procedures stands in contrast to classical conditioning, where the animal's behaviour has no influence on the sequential presentation of the conditioned and unconditioned stimuli. (In some circles the term 'operant' conditioning is preferred over instrumental, because the animal's behaviour 'operates' to control certain consequences of that behaviour.) This broad category of behavioural techniques relies on the willingness of a motivated animal to emit a learned response in order to obtain a reward (or to terminate a negative event).[1] There are versions of instrumental learning where the animal is free to respond as often as it wants, in the interests of producing a reward—a rat trained to press a bar repetitively to obtain an occasional food reward exemplifies this kind of free-running instrumental behaviour. For comparative visual psychophysics, however, the instrumental behaviour is brought under the control of a visual stimulus. Specifically, animals learn that a reward may be obtained in the presence of one stimulus (or one class of stimuli) but not another; these discriminative stimuli, as they are called, cue the animal about which response produces reward.

The origins of instrumental learning as a technique for testing animal vision can be traced back to the seminal work of Yerkes and Watson (1911), who devised a two-choice test chamber in which hungry rats were trained to select one of two alleyways leading to a food reward; on each trial the correct alley was signalled by a given visual stimulus whose position varied randomly—left vs. right—from trial-to-trial. Over the years interesting modifications of this apparatus have been developed (see Fig. 6.2), including the jumping stand (Lashley, 1930), the Skinner box (Skinner, 1938), the Wisconsin General Testing Apparatus (Harlow, 1950) and the Berkley box (Berkley, 1970).

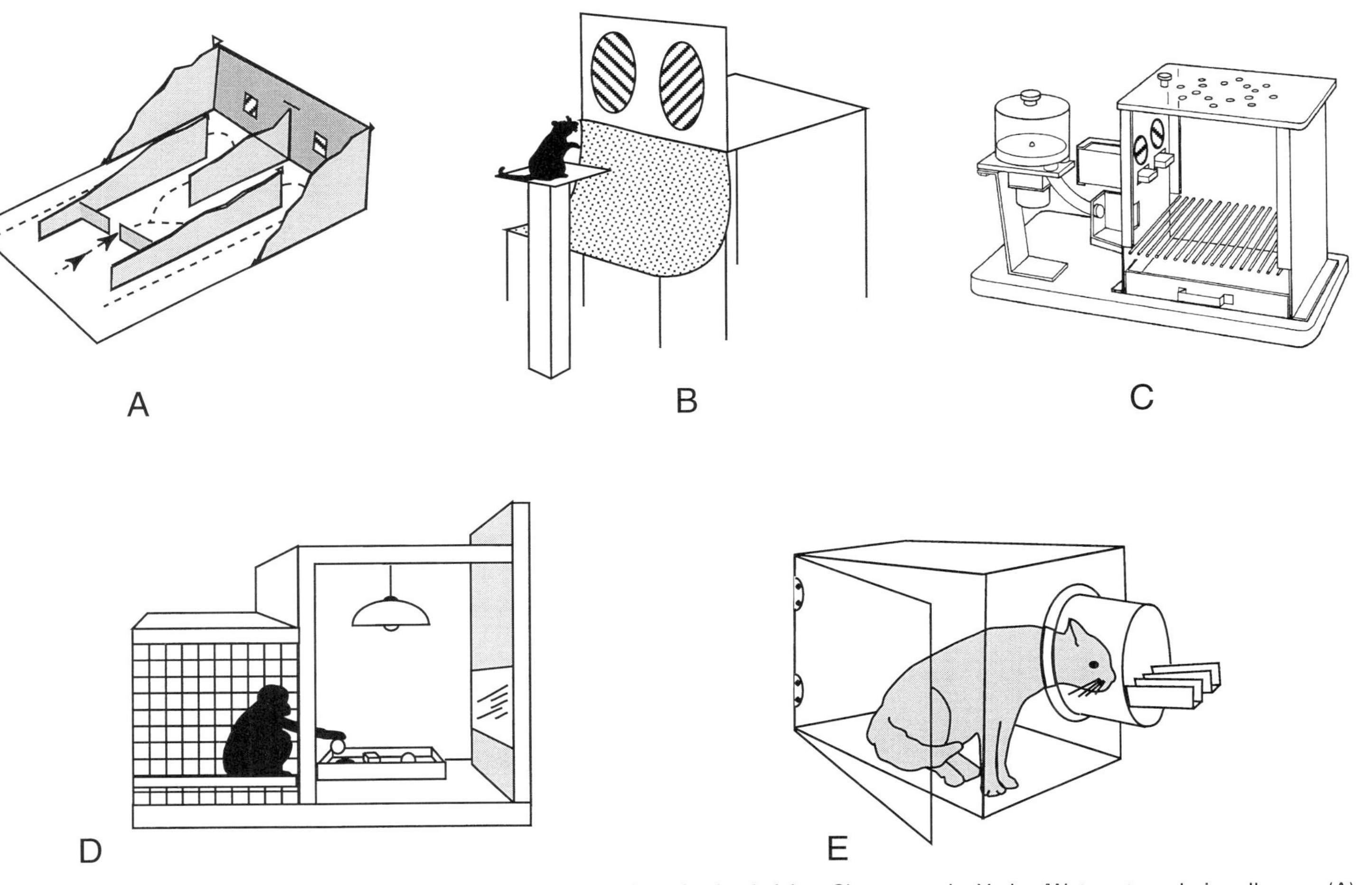

Fig. 6.2 Drawings of commonly used apparatus in behavioural testing of animal vision. Shown are the Yerkes/Watson two-choice alleyway (A), the Lashley jumping stand (B), the Skinner box (C), the Wisconsin General Testing Apparatus (D), and the Berkley box (E).

Regardless of details, the general procedure entails training an animal to make a given response to one of two (or more) visual stimuli presented simultaneously. The animal must select, in other words, which one of several visual stimuli is reliably associated with reward delivery. Once the animal's training performance reaches a plateau and is, ideally, nearly perfect, some dimension of the stimulus (e.g. grating contrast) is systematically varied over a range of values including those associated with chance performance (e.g. 50% when the task involves two equally likely alternatives). For ease of example, we shall assume that the task involves the presentation of two stimuli, although the technique works effectively with more than two (see, for example, DeValois *et al.*, 1974). For the measurement of detection thresholds, one of the visual stimuli is always sufficiently weak to be unequivocally invisible (e.g. zero contrast grating). The strength of the initially visible test stimulus is then varied systematically over trials, to find the value at which it becomes barely distinguishable from the genuinely blank one. For the measurement of discrimination thresholds, both stimuli are visible (i.e. above detection threshold) but differ noticeably along some dimension (e.g. grating spatial frequency). The animal is initially trained to respond to one of the two easily discriminable stimuli, and then the magnitude of this stimulus difference is varied to find the value at which the two are barely discriminable. (The terms 'barely detectable' and 'barely discriminable' are defined in terms of some arbitrary level of performance intermediate between chance and 100% correct, exactly as one does in human psychophysics—see Chapter 5) The actual trial-to-trial variation in stimulus value can obey a method of constant stimuli or can follow a staircase rule that adaptively homes on to the stimulus value associated with a given level of performance.

6.3.3.1 Response and reward selection

Instrumental conditioning can involve relatively lengthy training periods. Animals must be systematically 'shaped' to emit the desired response and then taught to associate that response with a given visual stimulus. Shaping involves rewarding behaviours that more and more closely approximate the final, desired response. (Anyone who has trained a pet to perform tricks intuitively understands the idea of shaping.) Fortunately, the problem of shaping can be simplified by the selection of a response that is part of the animal's normal behavioural repertoire and by utilization of a reward selected from the animal's natural diet. To give some examples, a thirsty monkey readily learns to manipulate a lever to the left or the right based on visual cues, in order to obtain a squirt of fruit juice (e.g. Crawford, 1976). Pigeons quickly learn to peck at one of two discs situated in line with one of two visual stimuli, in order to receive a pellet of grain (e.g. Wright, 1972). Bees willingly fly down one of two visually cued tubes to gain access to sugar-flavoured liquid (e.g. Horridge *et al.*, 1992). Fish can be taught to swim towards one of several illuminated test plates, one of which affords access to food (e.g. Neumeyer, 1986). Kittens quickly learn to jump on to one of two textured platforms to obtain a small piece of liver (e.g. Mitchell *et al.*, 1977). Adult cats easily master the task of running from a start box to one of two visually marked stations where food can be obtained (e.g. Berlucchi and Marzi, 1970). And predatory birds will fly from a perch to one of two stimulus displays, for the opportunity of snacking on a morsel of beef-heart (Fox *et al.*, 1976).

Another problem with instrumental procedures based on reward is that the length of the test session is determined by the animal's appetite or level of thirst. Performance becomes more erratic as the animal approaches satiation, sometimes forcing the experimenter to discard trials towards the end of a test session. To minimize the consequences of this tendency, experimenters often use adaptive staircase techniques that concentrate test trials on stimulus values in the threshold region (see Chapter 5). An efficiently devised staircase procedure can derive an estimate of threshold within several dozen test trials, while the animal is still highly motivated. Still, though, visual psychophysics typically requires multiple estimates of thresholds measured at each of many different stimulus values. Thus to measure, say, a spatial frequency discrimination function in the cat may entail several month's work, as the author has laboriously learned (Blake *et al.*, 1986).

A third problem associated with instrumental conditioning is the possible development of a position habit by an animal. Rather than relying on visual cues for response selection, an animal may simply select the same response (e.g. peck the left-hand response key) trial after trial. This strategy is not as perverse as it may seem, for such a dogged animal will be rewarded on approximately half the test trials. Fortunately, positional habits can be quickly extinguished by restricting the rewarded stimulus to the non-preferred response until the animal is again brought under stimulus control. With computer-controlled testing regimes, it is simple to detect positional biases and implement correction procedures. It is also a good idea throughout a test

A few additional hints that may simplify testing and maintain the animal under more stringent stimulus control are shown below.

1. When using a two-choice procedure, it can be helpful to signal the presentation of the visual display by simultaneously broadcasting a tone, thereby notifying the animal that the next trial is at hand.
2. To discourage animals from responding during the time-out periods between trials, it is customary to cause each response during these periods to retard the initiation of the next trial by a couple of seconds—animals readily learn not to respond unless the events defining the trial onset are present.
3. During the initial phase of discrimination training, start with easily discriminable visual targets and present the positive target (i.e. the one associated with reward) a second or two before the negative one, and extinguish the negative one after only a few seconds; this differential exposure method highlights the correct stimulus and hastens discrimination learning (De Weerd *et al.*, 1990*a*).
4. Occasionally insert blank trials in which all trial-related events occur except for the presentation of the visual targets. Animals should perform at chance levels on these trials, and evidence to the contrary implies the existence of extraneous cues that must be isolated and eliminated. Animals are adept at picking up subtle artefacts such as the click of switch closures and the hum of transformers; if those artefacts are correlated with the positive or with the negative stimulus, the animal's level of performance may be telling you more about its ears than its eyes.
5. Ensure that the visual stimulus (e.g. a grating) is varying only along the intended dimension (e.g. contrast, not average luminance), which may entail repeated, careful calibration and regular visual inspection by the human experimenter).

session to intersperse trials in which the stimulus associated with reward is easily detected or discriminated; occasional 'easy' trials minimize frustration and reinforce the association between a given stimulus and reward. A third trick to encourage animals to stay on task is to impose a longer intertrial interval following incorrect responses, a manoeuvre that forces the animal to wait longer before the next opportunity to obtain food. As a rule, intervals between trials following correct responses should be just long enough to allow the animal to consume the reward; intervals following incorrect responses are typically twice this duration.

In closing this section, it is worth repeating that one must carefully monitor the animal's body weight, temperament, and general health when using procedures involving food or water deprivation, and one should test the animal at the same time each day to maximize the effectiveness of a reward.

6.3.3.2 Visual reaction time

The instrumental learning procedure just outlined can also be adapted to the study of visually triggered reaction time (RT), as exemplified by the work of Boltz *et al.* (1979). They trained monkeys to depress and hold a lever upon presentation of an auditory cue. Following a variable period of time, a sinusoidal grating appeared on a CRT located directly in front of the monkey's eyes. If the monkey released the lever within 0.70 s after grating onset, it received a liquid reward. If the monkey released the lever prior to grating onset, it was punished with an exceptionally long intertrial interval. Using this procedure, Boltz *et al.* were able to manipulate grating contrast from trial to trial, to estimate the contrast at which RTs achieved a criterion level.

6.4 Additional testing challenges

Besides the considerations discussed above, instrumental procedures present other methodological challenges to the investigator of animal vision. Several of those challenges, along with possible ways to meet them, are outlined here.

6.4.1 Viewing distance

In vision studies it is often imperative that the viewing distance be carefully controlled, for otherwise it is impossible to specify the retinal angular subtense or the spatial frequency content of visual targets. The lack of control over viewing distance in fact represents a major drawback to devices such as the jumping stand and the visual cliff, both of which allow an animal too much leeway in body movement. A common strategy for controlling viewing distance is to restrict the animal's body and, sometimes, head position. How this is accomplished depends on the animal under study. Monkeys can be placed in chairs that force them to maintain proper body orientation, and they can be trained to position their heads within a face mask containing holes for the eyes (e.g. Boothe *et al.*, 1988) or to position one eye so that the natural pupil coincides with the exit pupil of an optical system (e.g. Harwerth and Sperling, 1975). Cats do not mind being confined within a small test box, and they will readily extend their heads through a porthole at one end of the box in order to view a visual display and to gain access to a food-delivery tube (Blake *et al.*, 1974). Goldfish can be placed in a specially constructed, underwater 'straight jacket' that comfortably maintains the body in a fixed orientation relative to a display monitor (Billota and Powers, 1991). Imagination and a talented workshop technician are two primary ingredients in developing an effective, safe means for restraining animals and, thereby, controlling viewing distance.

6.4.2 Fixation

In some studies one must know where an animal is looking in order to image visual targets on given regions of the retina. This challenging requirement has led to the development of several strategies for monitoring eye position and thereby controlling fixation; these strategies have been successfully implemented by a number of laboratories studying primate vision (referenced below) and by at least one laboratory studying cat vision (Pasternak and Horn, 1991). How is this feat of monitoring and controlling fixation achieved?

Today's most widely used procedure involves the sterile surgical implantation of a wire coil just under the surface of the sclera at the margin of the conjunctiva of one or, in some applications, both eyes (Robinson, 1963). With the animal's head placed within an alternating magnetic field, the current evoked in the coil provides the index of eye position, typically within an accuracy of several minutes of arc. There are two less popular techniques for monitoring eye position: these rely on infrared reflection from the cornea (e.g. Motter and Poggio, 1984) or electro-oculographic signals (Mohler and Wurtz, 1977) (see Chapter 9 for a description of these techniques). With any of these procedures, the animal is trained to fixate a small target appearing somewhere within the boundaries of a video monitor or a tangent screen. Training is accomplished by conditioning the animal to shift its gaze to this small target and to maintain fixation for a period of time in order to obtain a reward. Deviations in eye position sufficient to move the target outside the virtual window defining central fixation cause extinction of the fixation target and termination of the trial with no reward. It is worth noting that these techniques for monitoring eye position, besides being used in behavioural studies of visual capacity, are now also widely used in neurophysiological experiments where controlled fixation is needed to ensure visual stimulation within the receptive field of a visually activated neurone (e.g. Snowden *et al.*, 1991). In addition, they are also routinely employed in studies of oculomotor control (e.g. Sparks, 1975; Hanes and Schall, 1995; Munoz and Wurtz, 1995).

6.4.3 Fixation combined with instrumental conditioning

From laboratory to laboratory, there are interesting variations involving combinations of eye-movement monitoring and instrumental conditioning. Some vision laboratories (e.g. Mohler and Wurtz, 1977) have trained monkeys to depress a bar to turn on the fixation light which, in some trials, may be dimmed a few seconds after presentation; the monkey is rewarded if it releases the bar within a specified time following dimming of this light. In other trials, the intensity of the fixation light remains undimmed and, instead, another visual target appears briefly at another position on the screen. During these trials, the monkey is rewarded for releasing the bar within a specified time. Finally, in a small fraction of trials the fixation spot remains steady and no peripheral target is presented; here behaviour provides a point of comparison for the animal's behaviour in trials where the peripheral stimulus was presented but at a subthreshold intensity value.

An interesting variant of this procedure—called 'delayed match to sample'—offers a behavioural technique for measuring visual memory (e.g. Chelazzi *et al.*, 1993). While fixating a central spot and depressing a lever, a monkey views a briefly presented cue, or sample stimulus, which then, after some delay, is followed by a sequence of visual stimuli—one of which matches the sample. The monkey's task is to release the lever quickly upon recognition of this matching test. This delayed match to sample procedure has been employed, for example, by Ferrera *et al.* (1994) to examine the extent to which short-term visual memory modulates the activity of neurones in parietal and temporal visual pathways.

Another variation of this procedure has been used to control shifts in visual attention. For instance, Motter (1994) used instrumental conditioning to teach monkeys to attend to one particular coloured bar among an array of different coloured bars that also varied in orientation—the bar to be attended to was signalled by the colour of the fixation spot. The animals then indicated, by way of a manual response, the orientation of the attended bar—leftward vs. rightward—all the while maintaining gaze on the central fixation spot. This procedure effectively manipulated attention without changing the physical layout of the array of targets—it was the behavioural relevance of a given stimulus that changed from trial to trial. Motter was able to administer over a thousand trials per day, with the monkeys' performance stabilized around 85% correct on this two-alternative, forced-choice task.

Other investigators have successfully used voluntary saccadic eye movements, not manual responses with a bar, as the behavioural index of visual discrimination. Monkeys are trained to maintain fixation on a small, central target until a positive stimulus appears elsewhere in the visual field. If the monkey executes a saccadic eye movement to that peripheral target (as documented by the eye position signal from the monitoring system), the animal receives a reward; failure to execute a saccade, or execution of a saccade to an incorrect position, produces a prolonged time-out. For the measurement of detection thresholds, a single positive target is presented at one of multiple possible locations within the visual field, with its strength (e.g. contrast) varied over trials; for discrimination thresholds, the positive target is presented amongst an array of negative stimuli, with the difference between positive and negative stimuli varied over trials. Schiller and colleagues have fruitfully employed this technique for the study of 'ON' and 'OFF' channels in primate vision (Schiller *et al.*, 1986), colour-opponent and broad-band channels in primate vision (Schiller and Logothetis, 1990), and the contribution of cortical area V4 to visual object recognition in primates (Schiller, 1995).

The saccadic technique can be modified to create more elaborate task requirements that address questions other than simply the detection and discrimination of visual targets. Using this saccadic eye movement technique, Schall (1991) was able to train monkeys to withhold the saccadic response to a peripherally presented target until the central fixation spot changed colour, which might occur several seconds after presentation of the peripheral target. Moreover, one colour change (e.g. from initially yellow to green) signalled the animal to execute the saccade ('go trials'), while another colour change (e.g. yellow to red) warned the animal to maintain fixation ('no-go trials'). By recording neural activity from visual areas in monkeys engaged in this go/no-go task, Schall was able to dissociate visual evoked activity associated with identification of the target from activity involved in the planning and execution of voluntary saccadic eye movements.

Among the most ambitious research programmes employing the saccadic eye movement technique has been that of Newsome and colleagues who have recorded neural activity from visual area MT in monkeys concurrently engaged in visual psychophysical tasks involving motion perception. For this combined behavioural/physiological task, monkeys fitted with scleral search coils fixate a spot of light, while at the same time a circular patch of dynamic random dots is positioned on the video monitor so as to fall within the receptive field of the neurone(s) whose activity is being monitored. Following each brief presentation of motion, the monkey moves its eyes in a direction mirroring the perceived direction of motion portrayed in the motion sequence. By varying the percentage of dots moving in a given direction, it is possible to vary the perceived coherence of motion and the level of neural activity within MT neurones. In an elegant series of experiments, Newsome and colleagues showed that: (a) the psychophysical threshold for the detection of coherent motion corresponds closely to the minimum motion signal necessary to evoke reliable activity in MT (Britten *et al.*, 1992); and (b) the perceived direction of motion can be biased by electrical stimulation of MT neurones coincident in time with the presentation of the visual display (Salzman *et al.*, 1990). These findings underscore the power of animal psychophysics paired with neurophysiological recording.

6.4.4 Measuring subjective visual phenomena in animals

Some of the most intriguing phenomena in vision entail perception of objects or events that have no explicit embodiment in the physical stimulus. Some well-known examples include: illusory motion of a stationary object following prolonged inspection of real motion (the motion after-effect), perception of bounded objects in the absence of physically defined edges (illusory figures), physically identical lines whose lengths are perceived as different (Müller–Lyer illusion), and the appearance of three-dimensional objects from flat, two-dimensional images devoid of object contours (random-dot stereopsis). These kinds of subjective phenomena, because they presumably reveal important properties of normal visual information processing, are widely used as inferential psychophysical tools. But given their inherently subjective nature, how are these phenomena to be studied in non-verbal species? People can easily reach a consensus on the appearance of the visual targets presented in Fig. 6.3 simply by describing what they see (e.g. 'a black square resting on top of some white discs'). But imagine wanting to learn, say, whether a cat perceives illusory contours and figures (Bravo *et al.*, 1988; De Weerd *et al.*, 1990*b*) or whether a pigeon sees complementary after-images (Williams, 1974). How can these methodologically challenging questions be tackled by students of comparative visual psychophysics?

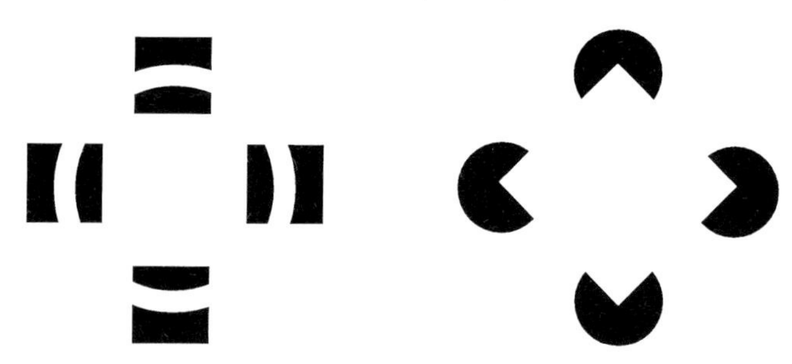

Fig. 6.3 Illusory figures defined by subjective contours. Can animals perceive these subjective phenomena?

There are two complementary strategies available. The first entails initially training an animal using a positive, rewarded stimulus that physically (i.e. explicitly) embodies the stimulus characteristic implied by the subjective phenomenon of interest; the non-rewarded foil stimulus contains the same local features as the positive stimulus but not the same global visual organization. Once the animal reliably discriminates this positive stimulus from the foil, the positive stimulus is modified over successive trials so as to remove the physical representation of the object, leaving only the stimulus conditions sufficient to generate the subjective phenomenon. At this point, the positive stimulus and the foil differ only in terms of their global perceptual organization, and continued successful discrimination implies that the animal can visually register this difference. (It is crucial to eliminate all possible extraneous cues, to be certain that the positive and foil targets are being discriminated based on the relevant dimension.) To give an example of this strategy, one could initially train animals to discriminate a field of random dots containing a square subset of lighter dots (i.e. a luminance-defined square) from a field of random dots all equal in lightness. Then the animals could be transferred to conditions where the square within the field of dots was defined exclusively by horizontal retinal disparity. Successful transfer would constitute evidence that this species possessed global stereopsis (e.g. Fox *et al.*, 1977).

A complementary strategy involves starting with a positive stimulus that generates the subjective phenomenon of interest, and then over blocks of trials altering the foil stimulus until it comprises all the local features of the positive stimulus except for the global organization instantiating the subjective phenomenon. To see this strategy in action, consider the author's investigation of whether cats can see biological motion in point-light animation sequences (Blake, 1993). A spatial two-choice discrimination procedure was used, wherein, in each trial, the cat viewed two spatially adjacent animation sequences presented on a pair of monochrome video monitors. Both animation sequences consisted of an array of approximately 14 dots whose positions changed from frame to frame of the sequences (see Fig. 6.4). In each trial the dots in one sequence (the positive stimulus) defined strategic positions on the body and limbs of a walking cat; the other, foil sequence, contained the same number of dots but whose positions and movement patterns differed from those defining a walking cat. Initially, the cats were trained to discriminate the sequence portraying biological motion from foil dots that were randomly scrambled in position and were free to

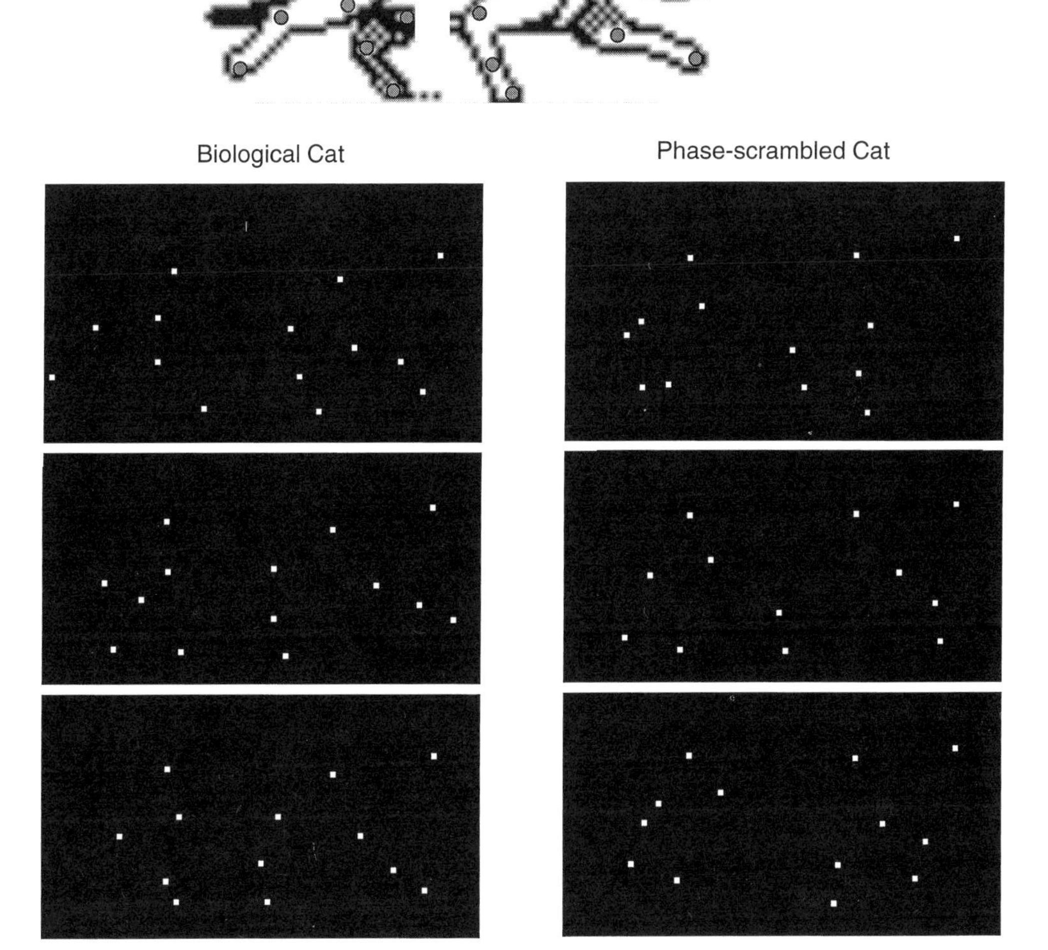

Fig. 6.4 Computer generation of a point-light animation sequence of a walking cat. For each individual frame of a film depicting a walking cat, small points of light are electronically 'attached' to strategic body parts of the animal, including the limbs and joints. When replayed so that the dots alone are visible, the animation sequence is readily perceived by human observers as a walking cat. Cats can discriminate a normal sequence portraying this biological motion (left-hand panel, which shows 3 frames from a 35-frame film) from one in which the spatiotemporal phases of the dot motions are scrambled (right-hand panel, also showing 3 frames from a 35-frame sequence); cats act, in other words, as if they too perceive biological motion. Reprinted from Blake, R. (1993). Cats perceive biological motion. *Psychological Science*, **4**, 54–7. Copyright 1993 with kind permission of Cambridge University Press, North American Branch, 40 West 20th Street, New York, NY 10011, USA.

move in any direction from frame to frame. After mastering this simple discrimination, the cats were graduated to a more difficult foil stimulus in which the dots moved in exactly the same manner as the individual dots in the biological sequence, except that their positions were scrambled. Despite portraying the same local motion cues, this foil stimulus, too, was easily discriminated from the biological motion sequence. Finally, a foil sequence was created in which the same dot motions appeared in the same positions as those defining the walking cat, except that the temporal phase relationship among dots was randomized. This manoeuvre rendered the positive and foil sequences identical in all respects except the global, spatiotemporal relationship of the dots. Still, cats could reliably discriminate the two types of sequences (as well as other sequences portraying walking humans and running cats), indicating that this species is sensitive to the kinematic invariants portrayed in these point-light displays. Cats, in other words, perceive biological motion, at least it would seem so.

There is a provocative philosophical issue raised by this last conclusion, that cats perceive biological motion. A sceptic could argue that one can never know what a cat (or any other animal) actually sees, where 'seeing' implies a subjective phenomenal experience. Strictly speaking, that sceptic would be correct, but the sceptic's argument would apply to other humans as well. Just because two people use the same words to describe a given visual stimulus does not prove that they see the same thing—for example, a trichromat and a dichromat may both label a colour 'orange' while actually having different visual experiences. It is easy to forget that verbal descriptions of what one sees represent the culmination of complex cognitive acts in which visual experiences are translated into symbolic descriptions. Psychophysical testing, in contrast, does not ask what something looks like, but rather simply asks whether one stimulus is discriminable from another. If a carefully designed psychophysical experiment leads to the answer 'no', then we can justifiably conclude that the two physically different stimuli look alike. If the answer is 'yes' then we know they do not look exactly the same. This reasoning based on discriminability applies with equal force to the study of human vision and to the study of animal vision. So, to conclude that cats perceive biological motion (or that monkeys experience the motion after-effect, or that falcons see figures portrayed by random-dot stereograms) is no less justified than concluding that humans (aside from one's self) perceive biological motion. The challenge in the case of animal and, for that matter, human psychophysics is to design visual conditions that rule out all possible alternative, extraneous cues that might be supporting successful discrimination. The sceptic is correct to search for alternative cues that animals might be employing when asked to discriminate subjective visual phenomena from foils. When arguing that one cannot know what an animal actually sees, however, the sceptic relies on an argument that generalizes to all other individuals—animal and human—except for the sceptic himself. While philosophically stimulating, the argument is self-defeating when it comes to understanding visual perception.

6.5 Concluding remarks

The behavioural study of animal vision has been significantly advanced by the development and refinement of powerful conditioning paradigms, some of which have been synergistically combined with eye position monitoring. As illustrated by the examples given in earlier sections, these conditioning regimes can be used to train animals to perform remarkably complex visual tasks involving detection, discrimination, visual memory, and selective attention. Moreover, by recording single cell activity from behaving animals, investigators can look for variations in neural responsiveness contingent on the task demands and on the animal's behavioural state (Maunsell, 1995). In effect, the evolution of these sophisticated behavioural techniques sets the stage for discovering the neural concomitants of visual perception and cognition.

Acknowledgements

This chapter is dedicated to the memory of the late Mark Berkley, whose groundbreaking behavioural work on cat vision inspired a generation of students of animal vision. The author is grateful to Jeffrey Schall, Yuede Yang, and Frances Wilkinson for helpful discussion during the preparation of this chapter. The author is supported by a grant from the National Institutes of Health (EY07760).

References

Anstis, S. M. and Cavanagh, P. (1983). A minimum motion technique for judging equiluminance. In *Colour vision: physiology and psychophysics* (ed. J. D. Mollon and L. T. Sharpe), pp. 156–66. Academic Press, London.

Berkley, M. A. (1970). Visual discriminations in the cat. In *Animal psychophysics: the design and conduct of sensory experiments* (ed. W. Stebbins), pp. 231–47. Appleton-Century-Crofts, New York.

Berlucchi, G. and Marzi, C. A. (1970). Veridical interocular transfer of lateral mirror-image discriminations in split chiasm cats. *Journal of Comparative Physiological Psychology*, **72**, 1–7.

Billota, J. and Powers, M. K. (1991). Spatial contrast sensitivity of goldfish: mean luminance, temporal frequency and a new psychophysical technique. *Vision Research*, **31**, 577–85.

Blake, R. (1981). Strategies for assessing visual deficits in animals with selective neural deficits. In *Development of perception: psychobiological perspectives* (ed. R. N. Aslin, J. R. Alberts, and M. R. Petersen), pp. 95–110. Academic Press, New York.

Blake, R. (1993). Cats perceive biological motion. *Psychological Science*, **4**, 54–7.

Blake, R. and Bellhorn, R. (1978). Visual acuity in cats with central retinal lesions. *Vision Research*, **18**, 15–18.

Blake, R. and Di Gianfilippo, A. (1980). Spatial vision in cats with selective neural deficits. *Journal of Neurophysiology*, **43**, 1197–205.

Blake, R., Cool, S., and Crawford, M. L. J. (1974). Visual resolution in the cat. *Vision Research*, **14**, 1211–17.

Blake, R., Holopigian, K., and Wilson, H. R. (1986). Spatial frequency discrimination in cats. *Journal of the Optical Society of America A*, **3**, 1443–9.

Blakemore, C. and Cooper, G. F. (1970). Development of the brain depends on the visual environment. *Nature*, **228**, 477–8.

Boltz, R. L., Harwerth, R. S., and Smith, E. L. (1979). Orientation anisotropy of visual stimuli in rhesus monkey: a behavioral study. *Science*, **205**, 511–13.

Boothe, R. G., Kiorpes, L., Williams, R. A., and Teller, D. Y. (1988). Operant measurements of spatial contrast sensitivity in infant macaque monkeys during normal development. *Vision Research*, **28**, 387–96.

Bravo, M., Blake, R., and Morrison, S. (1988). Cats see subjective contours. *Vision Research*, **28**, 861–5.

Brindley, G. S. (1970). *Physiology of the retina and visual pathways* (2nd edn). Edward Arnold, London.

Britten, K. H., Shadlen, M. N., Newsome, W. T., and Movshon, J. A. (1992). The analysis of visual motion: a comparison of neuronal and psychophysical performance. *Journal of Neuroscience*, **12**, 4745–65.

Chelazzi, L., Miller, E. K., Duncan, J., and Desimone, R. (1993). A neural basis for visual search in inferior temporal cortex. *Nature*, **363**, 345–7.

Collewijn, H. (1992). The optokinetic contribution. In *Eye movements* (ed. R. H. S. Carpenter), pp. 45–70. MacMillan, London.

Crawford, M. L. J. (1976). Central vision of man and macaque: cone and rod sensitivity. *Brain Research*, **119**, 345–56.

Daw, N. (1995). *Visual development*. Plenum, New York.

DeValois, R. L., Morgan, H., and Snodderly, D. M. (1974). Psychophysical studies of monkey vision—III. Spatial luminance contrast sensitivity tests of macaque and human observers. *Vision Research*, **14**, 75–81.

De Weerd, P., Vandenbussche, E., and Orban, G. A. (1990*a*). Speeding up visual discrimination learning in cats by differential exposure of positive and negative stimuli. *Behavioural Brain Research*, **36**, 1–12.

De Weerd, P., Vandenbussche, E., De Bruyn, B., and Orban, G. A. (1990*b*). Illusory contour orientation discrimination in the cat. *Behavioural Brain Research*, **39**, 1–17.

Dobkins, K. R. and Albright, T. D. (1995). Behavioral and neural effects of chromatic isoluminance in the primate visual motion system. *Visual Neuroscience*, **12**, 321–32.

Ferrera, V. P., Rudolph, K. K., and Maunsell, J. H. R. (1994). *Journal of Neuroscience*, **14**, 6171–86.

Fox, R. and Blake, R. (1971). Stereoscopic vision in the cat. *Nature*, **233**, 55–6.

Fox, R., Lehmkuhle, S., and Westendorf, D. H. (1976). Falcon visual acuity. *Science*, **192**, 263–5.

Fox, R., Lehmkuhle, S., and Bush, R. C. (1977). Stereopsis in the falcon. *Science*, **197**, 79–81.

Gibson, E. J. and Walk, R. D. (1960). The visual cliff. *Scientific American*, April, 64–71.

Guide for the care and use of laboratory animals (1985). Publication No. 86–23. National Institutes of Health, Bethesda, MD.

Handbook for the use of animals in biomedical research (2nd edn) (1983). Association for Research in Vision and Ophthalmology, Inc.

Hanes, D. P. and Schall, J. D. (1995). Countermanding saccades in macaque. *Visual Neuroscience*, **12**, 929–37.

Harlow, H. F. (1950). Analysis of discrimination learning by monkeys. *Journal of Experimental Psychology*, **40**, 26–39.

Harwerth, R. S. and Sperling, H. G. (1975). Effects of intense visible radiation on the increment-threshold spectral sensitivity of the rhesus monkey eye. *Vision Research*, **15**, 1193–204.

Horridge, G. A., Zhang, S-W., and O'Carroll, D. (1992). Insect perception of illusory contours. *Philosophical Transactions of the Royal Society, London B*, **337**, 59–64.

Kiorpes, L. (1992). Development of vernier acuity and grating acuity in normally reared monkeys. *Visual Neuroscience*, **9**, 243–51.

Kiorpes, L. and Kiper, D. C. (1996). Development of contrast sensitivity across the visual field in macaque monkeys (*Macaca nemestrina*). *Vision Research*, **36**, 239–47.

Kling, J. W. (1971). Learning: introductory survey. In *Experimental psychology* (3rd edn) (ed. J. W. Kling and L. A. Riggs), pp. 000. Holt, Rinehart, and Winston, New York.

Lashley, K. S. (1930). The mechanism of vision: I. A method for rapid analysis of pattern vision in the rat. *Journal of Genetic Psychology*, **37**, 453–60.

Lehmkuhle, S., Kratz, K. E., and Sherman, S. M. (1982). Spatial and temporal sensitivity of normal and amblyopic cats. *Journal of Neurophysiology*, **48**, 372–87.

Logothetis, N. K. and Charles, E. R. (1990). The minimum motion technique applied to determine isoluminance in psychophysical experiments with monkeys. *Vision Research*, **30**, 829–38.

Maunsell, J. H. R. (1995). The brain's visual world: representation of visual targets in cerebral cortex. *Science*, **270**, 764–9.

Mitchell, D. E., Giffin, F., and Timney, B. (1977). A behavioural technique for the rapid assessment of visual capabilities of kittens. *Perception*, **6**, 181–93.

Mohler, C. W. and Wurtz, R. H. (1977). Role of striate cortex and superior colliculus in visual guidance of saccadic eye movements in monkeys. *Journal of Neurophysiology*, **40**, 74–94.

Motter, B. C. (1994). Neural correlates of attentive selection for color or luminance in extrastriate area V4. *Journal of Neuroscience*, **14**, 2178–89.

Motter, B. C. and Poggio, G. F. (1984). Binocular fixation in the rhesus monkey: spatial and temporal characteristics. *Experimental Brain Research*, **54**, 304–14.

Munoz, D. P. and Wurtz, R. H. (1995). Saccade-related activity in monkey superior colliculus. I. Characteristics of burst and buildup cells. *Journal of Neurophysiology*, **73**, 2313–33.

Neumeyer, C. (1986). Wavelength discrimination in the goldfish. *Journal of Comparative Physiology*, **158**, 203–13.

Newsome, W. T. (1995). On neural codes and perception. *Journal of Cognitive Neuroscience*, **1**, 95–100.

Northmore, D. P. M. and Dvorak, C. A. (1979). Contrast sensitivity and acuity of the goldfish. *Vision Research*, **19**, 255–62.

Pasternak, T. and Horn, K. (1991). Spatial vision of the cat: variation with eccentricity. *Visual Neuroscience*, **6**, 151–8.

Reichardt, W. (1957). Autocorrelation, a principle for the evaluation of sensory information by the central nervous system. In *Sensory communication* (ed. W. A. Rosenblith), pp. 000. Wiley, New York.

Robinson, D. A. (1963). A method of measuring eye movement using a scleral search coil in a magnetic field. *IEEE Transactions in Biomedical Engineering*, **10**, 137–45.

Salzman, C. D., Britten, K. H., Newsome, W. T. (1990). Cortical microstimulation influences perceptual judgements of motion direction. *Nature*, **346**, 174–7.

Schall, J. D. (1991). Neuronal activity related to visually guided saccadic eye movements in the supplementary motor area of rhesus monkeys. *Journal of Neurophysiology*, **66**, 530–58.

Schiller, P. H. (1995). Effect of lesions in visual cortical area V4 on the recognition of transformed objects. *Nature*, **376**, 342–4.

Schiller, P. H. and Logothetis, N. K. (1990). The color-opponent and broad-band channels of the primate visual system. *Trends in Neuroscience*, **13**, 392–8.

Schiller, P. H., Sandell, J. H., and Maunsell, J. H. R. (1986). Functions of the ON and OFF channels of the visual system. *Nature*, **322**, 824–5.

Shumake, S. A., Smith, J. C., and Taylor, H. L. (1968). Critical fusion frequency in rhesus monkeys. *Psychological Record*, **18**, 537–42.

Singer, P. (1975). *Animal liberation*. New York Review of Books, New York.

Sireteanu, R. (1985). The development of visual acuity in very young kittens: a study with forced-choice preferential looking. *Vision Research*, **25**, 781–8.

Skinner, B. F. (1938). *The behavior of organisms*. Appleton-Century-Crofts, New York.

Snowden, R. J., Treue, S., Erickson, R. G., and Andersen, R. A. (1991). The response of area MT and V1 neurons to transparent motion. *Journal of Neuroscience*, **11**, 2768–85.

Sparks, D. L. (1975). Response properties of eye-movement related neurons in monkey superior colliculus. *Brain Research*, **90**, 147–52.

Uhlrich, D. J., Essock, E. A., and Lehmkuhle, S. (1981). Cross-species correspondence of spatial contrast sensitivity functions. *Behavioural Brain Research*, **2**, 291–9.

von Frisch, K. (1967). *The dance language and orientation of bees* (trans. L. Chadwick). Belknap Press, Cambridge.

Walls, G. L. (1967). *The vertebrate eye and its adaptive radiation*. Hafner, New York.

Wilkinson, F. (1995). Acuities for textures and gratings in kittens assessed by preferential looking. *Behavioural Brain Research*, **68**, 185–99.

Williams, J. L. (1974). Evidence of complementary afterimages in the pigeon. *Journal of the Experimental Analysis of Behavior*, **21**, 421–4.

Wright, A. A. (1972). Psychometric and psychophysical hue discrimination functions for the pigeon. *Vision Research*, **12**, 1447–64.

Yerkes, R. M and Watson, J. B. (1911). Methods of studying vision in animals. *Behavioral Monographs*, whole of No. 2.

Note

1. In the field of learning, psychologists refer to 'reinforcement' as any stimulus that increases the probability of a response. A positive reinforcement involves the presentation of a desirable stimulus (e.g. food to a hungry animal), while a negative reinforcement involves the removal of an unpleasant stimulus (e.g. termination of a loud noise). Both forms of reinforcement serve to increase the incidence of the trained behaviour. The remainder of this chapter will focus on situations involving positive reinforcement, and the term 'reward' will be used synonymously with positive reinforcement.

7

Research methods in infant vision

JANETTE ATKINSON and OLIVER BRADDICK

7.1 Introduction

There are several reasons why it may be important to investigate the visual performance of infants. We may wish to understand the process of development, as a scientific problem in itself or for light it may throw on the organization of the mature visual system. We need to know what information the infant receives through vision if we are to study cognitive and sensory-motor processes that use this information. The relationship between developing visual performance and the anatomical development of the infant's visual pathways may help us to understand the broader relationship between structure and function. Finally, measurements of visual function are important for the detection and diagnosis of pathological conditions and for monitoring the benefits of particular treatments in paediatric ophthalmology and paediatric neurology.

7.2 Constraints on vision testing with infants

7.2.1 Introduction

The investigator faces serious constraints in attempting to measure an infant's visual abilities. Unlike adult subjects, infants cannot follow verbal instructions and cannot deliver verbal responses. Compared with behavioural experiments with animals, it is difficult to teach infants stimulus–response relationships that can be used to indicate performance; usually there is no possibility of exposing infants to controlled stimulus contingencies over an extended period, and the motivational manipulations that help to establish learning are usually not ethically or practically possible. Research must use either non-invasive physical methods (such as evoked potentials or optical measurements) or the limited range of behaviour that is in the infant's normal repertoire. This behavioural repertoire varies widely with age, making the analysis of individual changes in vision, measured longitudinally, quite complex. Fluctuations in the infant's overall state of arousal limit testing to short time-windows (Section 7.2.4).

7.2.2 Motor development

The newborn (the first postnatal month) has some control over head and eye movements and so fixation behaviour can be used as a visual indicator. Changes in fixation are slow compared to later ages, partly because the current target of fixation exerts a strong 'capture' effect and partly because saccades to new targets are often inaccurate and tend to undershoot. There is some evidence that whole arm movements may be crudely directed towards targets of interest, but in general they are not sufficiently reliable to provide a useful response measure in vision testing on individual infants.

Between birth and 6 months, infants show an increasing readiness to shift fixation briskly and accurately between targets in their field of view; the fastest saccades are essentially adult-like although there is more variability than in the adult (Hainline *et al.*, 1984). Most behavioural studies of infant vision have depended on fixation behaviour, and have been most successful in this age range.

From about 4 months on, infants start to make visually directed arm movements; initially these reach a target only in a number of corrective steps, but over the period 4–12 months an increasing proportion of reaches occur in a single smooth ballistic action. From 6–12 months the behaviour of reaching for a small visually presented object can appear quite compulsive, although little is known about the visual properties that elicit this behaviour. Visually guided reaching has been used as a test behaviour principally in studies of distance perception (Yonas and Granrud, 1985), since it can be taken to indicate not simply that infants discriminate stimuli which differ in distance-related cues, but also that the infant specifically relates these cues to a cross-modal representation of space.

7.2.3 Posture

Infants require considerable postural support for any of these behaviours to be studied. Some work in the first 2 months of life has presented visual displays above a supine infant (e.g. Fantz, *et al.*, 1962; Aslin, 1981). This may make access for comforting or adjusting the infant's position difficult, and there is also a concern that the youngest infants are more likely to go into a drowsy state than when they are more upright. Most studies have tested infants in a seated posture. Special infant seats have been designed; e.g. infants of 1 month and under have been held by a support cushion, a large padded band around the midriff, and a head support, all adjustable on a near-vertical shaft. However, the most common and flexible method is for the infant to be held by a human being. Alternative arrangements are given below:

1. The infant can be seated on the lap of a holder who is her/himself seated on an adjustable chair. This can provide head support against the holder's chest with little restraint.
2. The infant can be supported by a standing observer, for whom the stimulus display is at chest level. One hand can support the infant's bottom or be placed between its legs, while the other is placed around the waist and may provide support under the chin.
3. The infant can be held over the shoulder with the observer facing away from a shoulder-height display.

Method 1 is the most widely used and the most comfortable for infants over 6 months of age. Methods 2 and 3 make it easier to arrange the layout so that the holder does not see the visual display—so for instance, in Teller's laboratory (Teller,

1983) a 'method 2' holder, viewing the infant's face on a video monitor, can act as a 'blind' observer in a preferential-looking experiment (Section 7.3.2). However, method 3 reduces one of the main advantages of a human holder, namely the ability to easily adjust the infant's position relative to the display.

If possible, the holder should be a member of the research team, familiar with the procedure and requirements of the test. However, from the age of about 8 months, infants start to become anxious with strangers and may only be content when held by a familiar care-giver. In this case it is important to brief this person about the required holding position, the sequence of the test, and the times at which it is acceptable to move the child away.

All procedures involve some element of restraint. As infants get older they accept this less readily and for shorter periods; this is one of the reasons why more reliable data is available about infant vision in the first 6 months of life than at later ages.

7.2.4 'State' variations and test duration.

Behavioural studies have categorized infants' overall states of arousal, ranging from deep sleep to awake and crying with vigorous limb movement (Prechtl, 1974). Visual testing with any method should only be undertaken when the infant is in an alert state with eyes open, not persistently crying; the limbs may be moving gently. Younger infants (e.g. at 2 months) will not usually be in this state for more than half an hour at a time and it cannot be produced on demand. The planned schedule of testing must allow testers and families to wait for 1–2 hours if necessary after the infant arrives in the unit, for an appropriate state. In the first 3 months, the best state is often achieved shortly before and after the baby feeds. This needs to be explained to families when arranging visits, and there must be facilities that allow comfortable arrangements for feeding and changing babies close to the test room.

Test procedures can be demanding for an infant and the amount of testing time that is practical on a single visit is usually limited to a maximum of about 20 minutes, with breaks (less for infants in the first month). Procedures which allow the test sequence to be interrupted at any time (e.g. preferential looking) have higher success rates than those (such as habituation) which depend on the completion of a specified sequence of trials within time constraints. Any preparation of the infant, such as cleaning the skin and attaching surface electrodes for VEP or EOG recording, uses time within the limited window in which the infant is in a cooperative state.

7.2.5 The interpersonal and ethical context of infant testing

Infant vision research depends on working with vulnerable subjects who cannot themselves give consent to test procedures. They are in the test unit with the consent of their parents who are concerned to protect their child from any risk or discomfort. Ethically it is essential, not only that there are no objective risks to the child, but also that parents are entirely confident that their baby is happy and comfortable, that they fully understand what is going on, and that they feel free to withdraw their child at any time.

In addition to ethical requirements, if the confidence of families is not maintained, the research programme will soon cease for lack of volunteers. The staff working on the programme must therefore be sensitive both to the state of the infant subjects and to the concerns, including unspoken concerns, of parents. This includes the concern that they and their child are giving up their time for a worthwhile and interesting purpose. The researchers need to explain in advance to the infant's parents just what will happen in the test procedures and how long it will take. They also need to explain their scientific objectives in a comprehensible but not patronizing manner, and to respond to questions about the individual child's performance with honesty but appropriate reassurance.

This consideration of interpersonal skills may be outside the scope of most areas of vision research, but for the success of research in infant vision it is as significant as skill in display calibration or data analysis.

7.3 Behavioural techniques: preferential looking

7.3.1 Introduction

Preferential looking is the most widely used method for testing infants' ability to make visual discriminations. It is based on the logic that an infant who shows a statistically reliable preference for stimulus A (the 'positive stimulus') over stimulus B (the 'negative stimulus' or 'foil') must be able to discriminate between A and B. In particular, infants prefer to look at a patterned field rather than a homogeneous field, and this preference can be used to assess acuity or contrast sensitivity.

7.3.2 The forced-choice preferential-looking procedure

The method was introduced by Fantz (e.g. Fantz *et al.*, 1962), but most modern work has used the version known as forced-choice preferential looking (FPL), introduced by Teller (Teller *et al.*, 1974; Teller, 1979, 1983). Stimuli are displayed at two locations, at the child's eye level, either side of the midline. For each trial, one side displays the positive (patterned) stimulus, and the other a uniform field matched in mean luminance. The observer views the child's face either directly through a peephole or via a video camera, from a midline location. The observer does not know which is the positive side in each trial (this is randomly selected either by a computer controlling the stimulus, or by a second operator) and has to make a 'blind' decision as to which side the infant preferred to fixate. The observation period is under the observer's control and does not usually exceed 10 s.

In a sense, the method combines forced-choice responding by both the infant and the observer. The observer may use information about the side of first fixation, or the number and duration of fixations on either side. The method has been shown to be more sensitive than using either of these measures directly (Atkinson *et al.*, 1977), presumably because the observer can optimally combine these with qualitative judgements of the child's attention; optimal performance is encouraged by giving the observer trial-by-trial feedback on which side the positive stimulus was presented. An infant, particularly in the first month, may show a marked bias towards fixating one or other side; with feedback, observers may be able to make some allowance for this in their judgements.

The method can be used for monocular testing, although sensitivity may be reduced by infants' less optimal state when wearing an eye patch (Atkinson *et al.*, 1982).

7.3.3 Alternative display layouts

In preferential-looking testing, three alternative layouts have been used:

1. The positive and negative stimuli are displayed on two separate video monitors (Atkinson *et al.*, 1979). Between these, a central set of flashing lights initially attract the infant's attention; the observer, when satisfied that the child is fixating centrally, presses a button which turns off the central lights and initiates the trial. The space-average luminance of the screen does not change at pattern onset. A similar display can be achieved by back-projecting stimuli on to two screens (Gwiazda *et al.*, 1978), although it is then necessary to have a large luminance change on both screens at the pattern onset.
2. A single large video monitor initially displays a central dynamic fixation pattern, followed by the positive display on one half and the negative on the other half of the screen (e.g. Atkinson and Braddick, 1992).
3. Patterns are mounted on a large disc which can be rotated so as to present them through apertures in a panel (Teller, 1979). The surface of this panel is made of identical material to the homogeneous negative stimulus.

In all cases the infant and the display have to be within an enclosure, screened so that the infant cannot see any part of the observer or other distracting external objects.

Method 3 has the feature that the positive target appears in a nearly uniform field. This is also readily achieved with method 2, but not with method 1 where it is difficult to match the surround to the illuminated CRT screen, so the negative stimulus provided by the blank screen still provides a conspicuous target. (Increasing the screen size may reduce the effects on performance of the screen's edge (Atkinson *et al.*, 1983).) Method 3 suffers, however, from inflexibility. In particular, the mechanical stimulus change requires the infant to be turned away; the procedure in Teller's laboratory has been then to turn the infant towards positive and negative locations alternately, ensuring that the infant samples both locations. With the video-based methods, since the infant is initially fixating centrally, a free sequence of fixations can be observed.

The main advantage of computer-controlled video displays is that trials can be run in rapid succession to exploit the limited time for which the infant is alert and co-operative. However, it is still necessary to turn the infant away from the display, every five trials or so, to maintain this state.

The methods potentially differ in the relative contribution of peripheral and central vision; in method 3, the infant normally views each location centrally before the observer's decision. In method 1, a decision based solely on the initial fixation will reflect visual sensitivity at the eccentricity of the screens from the central target, and this has been used to study infants' peripheral vision (Sireteanu *et al.* 1994). However, observers' judgements usually include behaviour after the target has been fixated, and the lack of variation with screen separation suggests that they reflect central vision (Atkinson *et al.*, 1983).

7.3.4 Trial sequences and staircases in FPL

With a 2–4-month-old infant it may be possible to complete 40–50 FPL trials in a single visit. For older or younger infants the number is usually less. To estimate a threshold such as acuity or contrast sensitivity from this limited data, the stimulus values (spatial frequency or contrast) must be chosen efficiently. Chapter 5 discusses staircases and other adaptive psychophysical methods which are designed to do this. A simple approach in infant testing is a version of the 'two-up/one-down' staircase:

A staircase procedure for infant acuity testing

1. For the first trial, choose a spatial frequency believed to be at least 2 octaves below the acuity limit for the subject's age.
2. For each correct judgement by the observer, increase spatial frequency by one step (two-thirds of an octave is a suitable step size). When an error occurs, go back one step and start the staircase proper.
3. If there are two successive correct judgements at one spatial frequency, go up one step. Whenever there is an incorrect judgement, go down one step.
4. Terminate the sequence after its direction has changed a selected number of times (reversals).

The average of these reversals estimates the 71% correct point (Weatherill and Levitt, 1965). Empirical investigation has shown that a short staircase terminating after three reversals estimates infant acuity as reproducibly as a staircase including over 40 trials (Atkinson *et al.*, 1986*b*), or as an alternative staircase rule basing reversal decisions on blocks of five trials (Atkinson *et al.*, 1982). The method used, of obtaining parallel estimates from two interleaved staircases, can be applied by experimenters who need to check the reliability of threshold estimates in their own procedures.

There has been a good deal of theoretical analysis of the variability of staircase estimates (McKee *et al.*, 1985). However, assumptions used in the analysis of adults' psychometric functions may not be appropriate for infants. First, infant FPL performance may not reach 100% for any stimulus, which tends to lead to underestimates of sensitivity compared to adults (Swanson and Birch, 1992). Second, variations in infants' state and attention mean that the detection process is almost certainly not stationary over time. Thus the inclusion of more trials, which on McKee *et al.*'s analysis is needed to increase precision, may end up reducing it. This is presumably the reason that Atkinson *et al.* (1986*b*) found short staircases to be as good as long ones.

In principle, it should be possible to gain more information in fewer trials by increasing the number of alternative locations, e.g. to a 4-choice version of FPL. No laboratory seems to have pursued this successfully.

7.3.5 The acuity card procedure

Acuity cards (McDonald *et al.*, 1985; versions commercially available as the Teller Acuity Cards or the Keeler Acuity Cards) are an adaptation of the preferential-looking procedure for the quick and simple estimation of infants' acuity in clinical settings. A grating is printed on one side of a card, with the rest a matching grey.

Each card is held up by the observer, who views the infant through a central peep-hole. The observer, who can choose to present the card either or both ways round, has to judge whether the infant can see the grating. Reliability of the acuity estimates within about 1 octave has been demonstrated (Mash *et al.*, 1995). However, this is not a forced-choice procedure, and is not necessarily carried out 'blind' as in many clinical settings one tester will arrange the order of the cards and know the side of grating on presentation.

7.3.6 FPL tests of performance other than pattern detection

The underlying logic of FPL can be applied in any case where infants are shown to have an intrinsic preference. An obvious extension is to the detection of temporal structure: Regal (1981) used FPL between a flickering and an unmodulated uniform field, to measure infants' critical flicker-fusion frequencies. A further extension is to test for a preference between two patterned fields, one of which contains some feature or large-scale organization lacking in the other. Table 7.1 summarizes some of the diverse, successful applications of this approach. However, in any new test of this kind, the existence of an infant preference must be empirically demonstrated. In at least some of the cases listed, the preference is clearly less marked than in acuity testing, presumably

Table 7.1 Some applications of forced-choice preferential looking other than simple pattern detection

Function studied	Positive stimulus	Negative stimulus	Reference, e.g.
Stereoacuity	Vertical lines with relative disparity	Similarly spaced lines in a single disparity plane	Held *et al.*, 1980
Stereo/ binocularity	Dynamic random-dot stereogram with alternating disparity- or correlation-defined pattern	Dynamic random-dot pattern with uniform disparity	Smith *et al.*, 1991
Texture segmentation	Oriented texture containing a region with differently oriented elements	Uniformly oriented texture	Atkinson and Braddick, 1992; Rieth and Sireteanu, 1994
Vernier acuity	Vertical lines containing vernier breaks which travel up and down the lines	Unbroken vertical lines	Shimojo *et al.*, 1984
Directional discrimination	Random-dot kinematogram in which alternate strips oscillate in opposite phase, providing motion-shear boundary	Random-dot kinematogram—whole pattern oscillates uniformly	Wattam-Bell, 1992

because even the 'negative stimulus' contains enough detail and contrast to be attractive to the child. Compared to acuity testing, the observer may well require to observe a longer sample of fixation behaviour before reaching a decision on each trial.

7.3.7 The forced choice method using other responses

FPL, as described above, depends on stimuli which elicit fixations to particular locations. Other responses are possible. For example, Fox *et al.* (1980) tested for stereopsis with a cyclopean bar which moved left or right from a central position; the observer had to distinguish the direction of pursuit eye movements. The direction or occurrence of optokinetic nystagmus is another possible indicator (Hainline *et al.*, 1986; Teller and Palmer, 1996). These methods have been less popular than FPL, mostly because they require longer individual trials.

7.3.8 Acuity measures beyond infancy

It is sometimes difficult to sustain interest long enough, even with a short staircase procedure, to get a reliable FPL acuity measure from a child aged over 12–18 months. Some results in older children have been obtained with an operant variant, where fixating the patterned stimulus is rewarded by the appearance of an animated toy (e.g. Mayer and Dobson, 1980). From about 3 years of age, children can usually understand the demands of a simple matching task, which can be used to test optotypes of graded sizes without the need for naming. Single optotypes, such as the STYCAR letters (Sheridan, 1976) are known to give higher estimates of acuity than arrangements such as the Snellen chart, which introduces 'crowding' interactions between the letters. While the Snellen chart is not practical with most 3–6-year-olds, 'crowded' acuity can be satisfactorily tested for example with the Cambridge Crowding Cards, which require the child to match only the central letter from a cross-shaped array of five (Atkinson *et al.* 1986*a*, 1988*a*). Such tests are particularly useful in the study of amblyopia.

7.4 Habituation methods

7.4.1 Introduction

A visual discrimination which the researcher wishes to test in infants will not necessarily be associated with any intrinsic preference, so in many cases the FPL method described in Section 7.3 is not useful. It may, however, be possible to establish a preference by the 'habituation-recovery' method, in which the infant becomes habituated by repeated exposure to a stimulus A. Preference for a novel test stimulus B then provides evidence that the infants can discriminate A and B. This approach has been used to study many different aspects of visual performance, ranging from colour categorization, through spatial phase discrimination, to shape constancy.

7.4.2 Infant control procedures

There have been many variants of the basic habituation procedure (Bornstein, 1985). However, to meet the requirement that habituation has occurred to a common level across individuals, the 'infant-control' procedure (Horowitz *et al.*, 1972) is desirable and forms of it have been widely adopted. The infant is tested with display arrangements similar to those described for FPL in Section 7.3.3. A sequence which can be used is shown below:

Habituation phase

1. The infant is turned towards the screen; when fixating centrally, the habituation pattern is turned on.
2. The observer presses the button while the infant is fixating the pattern. When the infant looks away for 2 s, the pattern display is terminated and the infant turned away. (1) and (2) constitute one habituation trial.
3. The looking time over each set of three consecutive trials is totalled. Habituation trials are repeated until this three-trial total has fallen to the 'habituation criterion' of 50% or less of its maximum value (implying a minimum of six habituation trials), at which point the test phase begins.

Test phase (sequential)

1. Four successive trials are presented in which the two stimuli ('novel' and 'habituated') are alternated.
2. Timing procedures are as in the habituation phase.
3. The choice of novel or habituated stimulus for first presentation is randomized or counterbalanced across successive runs (usually with different infants).

or

Test phase (simultaneous)

1. Two trials are presented, in each of which the two test stimuli are presented equally spaced either side of the midline.
2. The observer records, on two buttons, the duration of fixations to each stimulus; a trial is terminated when the total looking time for the two stimuli reaches 20 s.
3. Which stimulus is presented on the left is randomly selected for the first trial, and reversed for the second trial, so that the effects of any side preference cancel out.

These procedures have successfully yielded results with a number of different discriminations and ages. However, they contain a number of parameters (e.g. habituation criterion, test trial duration) which have been selected without exhaustive testing and so are not necessarily optimal.

The use of simultaneous presentation in the test phase has been found to be more sensitive than successive presentation with infants aged 0–1 months (Atkinson *et al.*, 1988*b*; Slater *et al.*, 1988); whether this is true for older infants is not yet known. Looking times and habituation rate show marked variation with age (Bornstein, 1985; Hood *et al.*, 1996); the use of the infant control procedure is intended to minimize the effect of these variables on visual discrimination measures, but cannot be considered to make the method age-independent.

The nature of habituation procedures means that relatively few test trials can be obtained from a fairly lengthy procedure. This means that, in contrast to FPL experiments where enough data can be gathered on an individual infant to yield a statistically meaningful measure of visual performance, any conclusions from a habituation experiment almost always have to rest on analysis of group data.

7.4.3 Generalization during habituation

In some experiments, it is not possible to confine differences between the habituated and novel stimuli to the variable that the experimenter wishes to investigate. For example, in investigating infants' shape constancy, the desired question is whether an infant, habituated to a rectangle at one slant, responds to the novelty of a trapezoid at another slant which has a similar perspective projection. Unfortunately, discrimination in this case could rest on the slant difference rather than on any sensitivity to real shape. This problem can be resolved by varying the habituation stimulus in the dimension in which the experimenter is not interested. In this instance, all habituation trials would use the same physical shape, but presented at different slant angles. The novel stimulus would then be a different physical shape, whose slant and perspective projection fell within the range to which the infant had been exposed (Caron *et al.*, 1978). Longer looking at the novel stimulus could then only be based on sensitivity to physical shape regardless of slant. An analogous example is the variation of absolute contrast during habituation to a compound grating stimulus (Braddick *et al.*, 1986). This ensured that infants' responses to the novelty of a stimulus with different spatial phase relationships could not be based on the difference in peak-to-trough contrast rather than in spatial configuration.

7.5 Refractive measurements with infants

7.5.1 Introduction

Measurements of where an infant's eyes are focused are important because:

- they determine the spatial quality of the image on which most aspects of vision depend;
- the development of accommodation and refraction are significant elements in the overall process of early visual development; and
- refractive errors and anomalies of accommodative behaviour in infants are of clinical concern. In particular, they are associated with the common problems of strabismus and amblyopia.

7.5.2 Refraction and accommodation

The eyes of a 1-month-old infant will frequently be found to be focused at a distance of around 50 cm, even when the visual stimulus is at a greater distance. However, if accommodation is relaxed by cycloplegia, the average infant eye is slightly hyperopic (1–2 dioptres). The distinction must therefore be kept clear between measurements of accommodative behaviour and those of the structurally determined refraction of the eyes. Cycloplegia is the only way to assure that structural refraction is being measured. However, errors of focus in older infants accommodating on a target are strong indicators of refractive error (Braddick *et al.* 1988; Anker *et al.* 1995), and the non-cycloplegic methods of 'near' or 'dark' retinoscopy (Section 7.5.5) have been found to yield measurements that are systematically related to refraction.

7.5.3 Retinoscopy

Retinoscopy (sometimes called skiascopy) is the standard clinical method for assessing refraction without the need for subjective report from the patient. The observer rotates the retinoscope to and fro so that its input beam moves across the pupil; the beam forms an image on the retina and the reflected light ('reflex') from this image is observed in the pupil. The speed and direction of movement of the reflex depends on the distance at which the subject's eye is focused. Plus or minus lenses are placed in front of the eye until a lens is found with which the reflex is seen to move neither with nor opposite to the movement of the input beam, but rather the pupil is immediately and uniformly flooded with light as the input beam enters it. This null point is where the retina is optically conjugate with the light source, i.e. the lens used represents the refraction relative to the retinoscopist's viewing distance. Chapter 10 provides further details.

Satisfactory retinoscopic refraction of infants is possible, but requires high levels of skill, and extensive practice specifically with infants. The need to find a lens that neutralizes the refractive error means that there are no limits on the range of measurement. Each eye must be measured separately, and since many infants show significant astigmatism, orthogonal axes must be tested in each eye. The need to conduct these tests with multiple lenses within the limited time tolerated by an infant makes the method very demanding. Most experienced retinoscopists consider 10–15 min as the maximum time available to carry out an examination on an infant or young child.

Lenses held close to the eye are often found aversive by infants, which increases the difficulty of the procedure. Infants will not tolerate the usual kind of trial frame. Some practitioners prefer to use a 'lens bar' (or 'retinoscopy rack') which allows the power to be varied more readily than interchanging separate lenses from a trial set.

7.5.4 Photorefraction and videorefraction

This is a family of techniques which uses a small flash source close to the camera to illuminate the eyes from some distance away (e.g. 1 m). If the eye is focused at the source distance then, in principle, the light will return from the point image on the retina, along its path, to the conjugate point at the source. If the eye is defocused, the returning rays form a diverging cone. The distribution of light returning from the pupils is recorded on film (photorefraction) or digitally (videorefraction). With either photo or video recording, there are alternative optical arrangements, described in Sections 7.5.4.1–7.5.4.3, for creating a light distribution which is dependent on the divergence of the returning cone and hence on the dioptric defocus of the eye.

Photo- and videorefraction require much less cooperation from the infant subject than does retinoscopy. Since the subject's attention is required only for the brief presentation of the flash, both eyes are imaged simultaneously, no lens or eyepiece is introduced close to the face, and head position is not closely constrained, these methods are well adapted to obtain refractive measures from infants.

7.5.4.1 Orthogonal photorefraction

This was the first photorefractive method to be introduced (Howland and Howland 1974) and used in infant research (Howland *et al.*, 1978; Braddick *et al.*, 1979). The flash is provided via a fibre-optic light guide centred in the camera lens. This is surrounded by two pairs of cylinder lens segments with orthogonal axes, which image the light returning from the eye as a cross. The length of the cross-arms increases with increasing dioptric defocus of the eye in the corresponding meridian. These images can be difficult to interpret; in particular they do not give a direct indication of the sign (myopic or hyperopic defocus). Isotropic and eccentric photorefraction have been more widely used in recent years.

7.5.4.2 Isotropic photorefraction

The light source is again a fibre-optic tip centred in the camera lens, but the image is defocused by adjusting the setting of the camera lens (Howland *et al.*, 1983; Atkinson *et al.*, 1984). The light returning from the eye is imaged as a blur circle whose diameter reflects the degree of defocus, or, in the general case of an astigmatic eye, as a blur ellipse whose long axis is parallel to the meridian of greatest dioptric error. If the eye is focused in front of the camera, a larger extent of blur is produced when the camera is defocused beyond the subject than if it is equivalently defocused in front, so comparison of two images with these camera settings allows the sign as well as the magnitude of the error to be determined. In practice a third image is required, since the calibration depends on the pupil size, which is determined from a picture focused on the subject in which the pupils appear as brightly illuminated discs. The method measures within ± 0.5 D over a range of ± 5 D, and has been calibrated against retinoscopic measurements (Atkinson *et al.*, 1984).

7.5.4.3 Eccentric photorefraction

Variants of this method have been called 'photographic static skiascopy' (Kaakinen, 1981), 'photoretinoscopy' (Howland, 1985; Shaeffel *et al.*, 1987), 'paraxial photorefraction' (Abramov *et al.*, 1990), and 'knife-edge photorefraction'. The camera is always in focus on the subject's eyes. The source lies to one side of the camera aperture. If the defocus of the eye is small, little or none of the light cone returning to the source will enter the camera; if it is large, most of the camera aperture will be illuminated. At the image plane, the returning light appears as a crescent within the pupil. The width of the crescent varies with the amount of defocus, and the sign of the defocus (myopic or hyperopic relative to the camera) determines on which side of the pupil the crescent appears.

Table 7.2 compares some of the features of the isotropic and eccentric methods. It should be noted, that while the optical theory is quite well worked out (Howland *et al.*, 1983; Bobier and Braddick, 1985; Howland, 1985; Wesemann *et al.*, 1991), empirical results follow the theory qualitatively rather than quantitatively; discrepancies may arise from aberrations, the nature of the fundal reflection, and the failure to detect the faintest regions of the returned image. Numerical results, therefore, should only be taken from an instrument which has been empirically calibrated from eyes having known refractive errors.

Table 7.2

Isotropic photo/video-refraction	Eccentric photo/video-refraction
Shows axis of astigmatism directly	Measures only the meridian along which the source is displaced from the lens, so multiple pictures with different source positions are necessary to assess astigmatic eyes
Sign of error derived from comparison of two images with different camera focus (and third image required for pupil size)	Sign of error derived from crescent position in a single image
Small 'dead zone' in which small errors cannot be estimated (typically less than ± 1 dioptre, and sign can be derived within this range)	'Dead zone' where no crescent appears depends on source eccentricity, but hard to reduce below ± 2 dioptres—no information on sign in this zone
Errors above ± 5 dioptres give diffuse images which can be recognized but not accurately quantified except to say they are ± 5D	For large errors crescent size varies little; the upper limit of measurement increases with source eccentricity (but so does the lower limit)

Capturing the image digitally has several advantages over film: the image is immediately available so that measurements which are uncertain or failed because of blinks, fixation changes, etc. can be repeated; in isotropic videorefraction this can be useful when a subject without cycloplegia is suspected to have changed accommodation in the interval (typically 5 s) between images. Digital capture also means that the image can be processed by computer. Commercial computer-based instruments have been developed: the isotropic Clement Clarke VPR-1 and the eccentric Tomey ViVA. The isotropic method provides a more satisfactory measurement of astigmatic errors, but the image structure is less amenable to automatic measurements. Currently, automatic measurement of eccentric videorefractive images is only reliable over a medium range of errors, and only for the meridian(s) on which the flash source is mounted. However, these limitations are likely to be overcome in future instruments.

7.5.5 'Dynamic retinoscopy' and 'near retinoscopy'

These are variants of the standard retinoscopic method which have been used in investigations of infants.

In *dynamic retinoscopy*, the source is made conjugate with the subject's fundus, not by inserting lenses in the beam, but by the retinoscopist moving forwards or backwards to locate the nulling point in space. Obviously this is only feasible for eyes which are focused in a near range of distances, and it cannot measure hyperopic refractions. However, this near range is that within which infants most often actively accommodate, so the method has been able to provide information on the accuracy of infants' accommodation to targets at different distances (Banks, 1980; Brookman, 1983).

Near retinoscopy (Mohindra, 1975; Thorn, 1996) is a variant of conventional retinoscopy using lenses, which aims to assess the structural refraction of the infant's eyes without the need for cycloplegic eye drops. Under conditions of darkness, with no target except the retinoscope beam itself, subjects take up a 'dark focus'. The finding that this refractive state averages 0.75 D of accommodation (Mohindra, 1977) means that an estimate of cycloplegic refraction can be obtained by correcting the measurement by this amount. It is likely, however, that the degree to which accommodation relaxes depends strongly on the exact procedure (e.g. how long the infants spend in the dark) and on the infant's resulting state of arousal (Thorn *et al.*, 1996). Anyone attempting to use this method would be recommended to calibrate their own procedure against cycloplegic measurements with the age group concerned, and to check its reliability in their hands.

7.6 Visual evoked potentials (VEPs)

7.6.1 Introduction

Evoked potentials, recorded from the surface of the scalp and synchronized with visual stimulus transitions, indicate a neural response to a stimulus without a verbal or behavioural indicator of detection. They have therefore been attractive as a method of research on infant vision. However, anyone contemplating this method should be aware: (a) that the 'objectivity' of VEP recording is just as dependent as behavioural methods on the child's fixation behaviour (see Section 7.6.6 below); and (b) that an active child of 6 months or more can rapidly remove scalp leads.

The small amplitude and variability of the potentials recorded at the scalp mean that the VEP signal can only be extracted by digital signal averaging. Chapter 8 provides a general account of VEP and related techniques. The theory, techniques, and results of VEP measurements in adults and infants have also been comprehensively reviewed by Regan (1989).

The procedures of electrophysiological recording appear, to the infant's family, more unfamiliar and invasive than behavioural testing. A careful and sympathetic explanation of the procedure is especially important, especially to clarify the idea that electrical energy is being detected rather than applied to the child's head!

7.6.2 Recording procedures

Small gold or Ag/AgCl cup electrodes, which can be filled with electrode jelly, are commercially available. Many clinical practitioners attach them with collodion glue. However, this takes a long time to attach and remove, and in infant recording, time is the most precious commodity. Most infants have only thin hair, and we find that two strips of Micropore surgical tape forming a cross over the electrode are a perfectly satisfactory way of securing an electrode for the usual duration of an infant recording. A headband made out of elasticated bandage, or adjustable by a Velcro fastening, can help to hold one or more electrodes in place. The leads from the electrodes, with the first connector, can conveniently be bunched together and taped to clothing on the child's back, leaving enough slack for head movement. It is often also convenient to carry the leads from this first connector to the preamplifier over the holder's shoulder, where they can again be lightly secured with adhesive tape.

Low and balanced electrode impedances are important in securing good quality recordings. The impedance should be checked with an applied voltage at a frequency of, say, 1000 Hz rather than DC to avoid polarization of the electrodes; commercially available amplifiers usually provide this facility. The goal should be impedance certainly below 10 kΩ, and preferably 5 kΩ or less, between each pair of electrodes. To achieve this the skin at the intended electrode site needs to be cleaned with cotton wool soaked in surgical spirit; at the same time a little electrode gel can be rubbed into the skin with a fingertip. Infants in the first week of life have high skin resistance, and it may be necessary to gently abrade the skin to reduce this.

EEG and VEP workers have developed systems, related to head landmarks, for specifying electrode locations, and with adult subjects they frequently use montages of 16 electrodes or more to map responses across the scalp. Very little is known about how developmental changes in brain topography (or in the electrical properties of intracranial tissues) may affect these patterns, and the smaller size of the infant's head makes it harder to locate electrodes precisely. Setting up large montages is time-consuming, and most of the useful results in infant VEP studies have been obtained with very few electrodes. The usual minimum is three: bipolar recording between an electrode over the occipital area (Oz on the 10–20 notation, about 1 cm above the infant's inion) and one on the forehead, with a third, reference or 'ground' electrode, either on the vertex or attached to the earlobe. For some purposes (e.g. Apkarian and Tijssen's (1992) studies of albinism and related pathologies) the lateralization of the signal is important and requires further recording sites.

7.6.3 Equipment

Before attaching infants, or any other human subjects, to electrical recording equipment it is necessary to be sure that local legal requirements concerning health and safety are satisfied, and that research procedures have the approval of the relevant ethical committee. In general, clinical physiological amplifiers will have to comply with established safety standards. These are likely to include a requirement that the subject is in electrical contact only with a low-voltage preamplifier, which is electrically isolated from the mains-powered amplifier.

Clinical installations will usually have an integrated suite of amplifiers and a special-purpose computer with software for VEP recording, stimulus control, signal averaging, and analysis. Unfortunately, as well as being expensive, such systems often lack minor but critical features for infant vision research. Examples are: (a) the facility to interrupt recording if the child is fretful or inattentive, and to discard the signal collected on the current sweep (most commercial systems simply allow the operator to prevent triggering on the next sweep); (b) the ability to accumulate separately averaged signals from interleaved presentations of different stimuli.

'Artefact rejection' is a common and useful feature, which excludes sweeps containing any voltage excursion beyond some specified value. Some care is needed in setting the cut-off so that it reliably rejects large voltages resulting from head movements, etc., but does not reject the signals of interest.

7.6.4 Transient and steady-state VEPs

If stimulus events occur at a rate of 1 per second or lower, the measurable electrical response to one event is complete before the next event occurs. This 'transient' evoked potential has a complex waveform which changes markedly with age. This waveform must be a sum of components arising from different stages of visual processing, and identifications of these stages have been proposed (e.g. by Maier *et al.*, 1987). Potentially the transient waveform should help in understanding the development of the different stages. However, the changes in the waveform, which include changes in the latency of particular features, can make it difficult to identify a particular component of the infant VEP across age.

If the stimulus events are repeated at 2 per second or faster, the responses to successive events overlap, producing a periodic waveform at the frequency of the stimulus. This waveform is known as the steady-state VEP. The higher the repetition rate, or the younger the infant, the more of the signal power is in the fundamental and low harmonic frequencies; in some cases the waveform approximates a pure sine wave quite closely.

Much research using VEPs with infants is concerned simply with the presence or absence of a VEP response to a particular stimulus, or with the relative amplitudes produced by different stimuli. For these purposes, the steady-state VEP has advantages over the transient VEP. First, steady-state recording is usually more sensitive in practice. The power to detect weak signals depends on the number of repeated presentations ('sweeps') over which the signal is averaged. Since each sweep is shorter at the high repetition rates used in steady-state recording, more averaging is possible within the limited time for which an infant remains in a good state. The value of this will depend, of course, on the frequency response of the response in question. While infants' high-frequency responses are weak, they rarely get much larger for frequencies below 2 Hz, so the useful frequencies are mostly in the steady-state range. Further, the noise power generally falls off with frequency so the optimum is not necessarily where the signal is largest. For many purposes, stimulus repetition rates between 2 and 12 Hz are convenient. Typically, signal averaging is performed on between 25 and 300 sweeps; the square root improvement in signal/noise with the number of sweeps, combined with the limited tolerance of infants for extended testing (Section 7.2.4), brings diminishing returns for longer measurement periods. Second, the quantitative analysis of steady-state signals is more straightforward (Section 7.6.7).

7.6.5 The stimulus event

The VEP response can only be identified in relation to a discrete, periodic stimulus event. Different types of event are needed to investigate different levels of the visual process. Commonly used examples are: an increase and/or decrease of field luminance ('flash VEP'); transitions between a pattern (e.g. grating or checkerboard) and a uniform field of equal space-average luminance ('pattern onset/offset'); and reversal of the contrast of a grating or checkerboard, i.e. substituting black for white and vice versa ('phase- or pattern-reversal').

Researchers studying infant vision would often like to examine VEPs elicited by changes in higher order properties, such as stereo disparity, orientation, or direction of movement. However, it is usually impossible to achieve such changes without local alterations in luminance and contrast, and so the VEP will include the response to these lower order transitions. A good way to isolate the effects of the higher order property is to make steady-state recordings in which the transition is embedded in a series of events that share its lower order effects. For instance, changing the binocular correlation of a random-dot pattern necessarily requires changes to some or all of the dots, an event which would elicit a VEP even monocularly. The specifically binocular component can be isolated by changing the correlation 4 times per second (say) and by replacing the whole random pattern at a multiple of this frequency (say 24 per second). The VEP component at 24 Hz must be assumed to arise from the dot changes, but components at 4 Hz and at the low harmonics such as 8, 12, and 16 Hz must be specific to the change in binocular relationships and so arise from neural processes combining the two eyes' signals (Braddick *et al.*, 1980; Julesz *et al.*, 1980). Analogous principles have been used to design stimulus sequences that isolate responses from orientation-specific (Braddick *et al.*, 1986*b*; Braddick, 1993) and direction-specific (Wattam-Bell, 1991) mechanisms.

Many stimulus sequences involve two opposite and intrinsically symmetrical events. For example, one complete cycle of a checkerboard-reversal stimulus includes two reversals, of which the second returns to the original pattern. On each of these reversals, 50% of the field switches from white to black with the opposite change for the other 50%. The VEP responses to each reversal should be equivalent, and so the response should appear at $F2$, twice the fundamental frequency of the stimulus cycle ($F1$), and perhaps at other even-numbered harmonics. (Care must be taken to distinguish this fundamental, expressed in cycles s^{-1}, with the rate of transitions expressed in reversals s^{-1} which is twice as high.) Responses at $F1$ (and other odd-numbered harmonics) imply a lack of symmetry in the response generation which may reveal information about the underlying process. For example, if a grating is oscillated with displacements equal to one-quarter of its spatial cycle, the VEP at $F2$ (the reversal rate) may include contributions both from motion-sensitive mechanisms, and from the appearance and disappearance of spatial contrast at particular locations. However, any $F1$ component implies a different response to leftward and rightward displacements. Such a response, which is found with monocular stimulation of young infants and in congenital esotropia (Norcia *et al.*, 1991), (a) is more certain than $F2$ to be motion-dependent, and (b) implies that the motion-sensitive mechanism has a directional asymmetry in development.

7.6.6 Stimulus control

A normal adult subject for VEP recording can be instructed to fixate a specified point on the screen. To get infant results of any value, we need to approximate this condition as far as possible. There are various techniques that help here:

1. Monitor the infant's fixation continuously, and interrupt recording when the infant looks away from the screen or has closed eyes. This is essential.
2. Attract the infant's attention with a visual stimulus superimposed on the display. We have used a large half-reflecting mirror in which an experimenter's face was visible to the infant (Harris *et al.*, 1976). A simpler method is to jiggle a small rattle or attractive toy against the screen. Provided that the attention-holding stimulus bears no temporal relationship to the test stimulus events, it should not contribute any systematic signal to the averaged VEP.
3. Use a homogeneous display that subtends as large a visual angle as possible, so that the pattern of stimulation is largely independent of the infant's fixation. Large screen (e.g. projection video) displays and/or close fixation distances (e.g. 40 cm) can achieve this; but inevitably limit the high end of the spatial frequency range.

7.6.7 Quantitative analysis

There are two basic measures from the VEP signal that may be of interest: amplitude and latency.

Amplitude measurements in transient VEPs depend on reliably identifying specified peaks and troughs in the waveform, and are very susceptible to noise at these extrema. Steady-state recording allows measurements that use more of the information in the signal. Specifically, the amplitude and phase of the component at the stimulus frequency ($F2$), and its harmonics, can be extracted. For a small number of frequencies this can be done by simply multiplying the averaged waveform with sine and cosine terms; alternatively a Fast Fourier Transform can be performed.

With increasing age, the higher harmonics of the VEP response generally become more prominent. It must be recognized, therefore, that a measure based solely on the fundamental frequency does not necessarily reflect the strength of the signal, either in describing developmental change or in assembling data from different individuals of the same age who may show differences in balance between the harmonics.

The amplitude measured at any frequency does, of course, include an unknown noise contribution. It is always desirable to have an indicator of statistical confidence to show that a stimulus-related signal is in fact present, and in some infant experiments the primary form of the conclusion may be that 'a response to stimulus X can be reliably detected at age A'. Such a conclusion must be based on the consistency of the signal across the samples which have been averaged together. A stimulus-related signal will have a constant phase, while the noise component at the same frequency will have a phase that varies randomly across 360°. (Another way of expressing this is that the phase-amplitude vector of the noise should average to zero.) The circular variance test (Moore, 1980; Wattam-Bell, 1985) provides a significance measure on the departure from random phase, weighted according to the amplitude of the signal in each sample. Alternatively, Hotelling's T^2 (Winer, 1971) provides a multivariate test of whether a vector is significantly different from zero, and so may be used taking the sine and cosine components at the signal frequency of each sample.

Latency is less affected than amplitude by recording conditions, which makes it attractive, e.g. for comparing responses of the same individual to different wavelengths (Dobson, 1976). However, latency measurements in transient recording, like amplitude, depend on identifying a particular feature of the waveform, such as the first negative peak. Variations with age and across individuals may complicate this. Latency variations are also reflected, arguably more reliably, in the phase of a steady-state signal (Porciatti, 1984). However, the 360° cyclic nature of phase measurements can create serious ambiguities of interpretation. It is not usually possible to derive an absolute value of latency from phase measurements; if the aim is to examine changes in latency within an individual as a function of a variable such as stimulus contrast, the continuity of the function may make it clear which of two phase values, 360° apart, is appropriate for comparison.

7.6.8 Comparing responses across stimuli

Many VEP investigations require the signals evoked by two or more stimuli to be compared. For instance, to show that the response to a complex stimulus is absent at an early stage of development (e.g. Braddick *et al.*, 1983; Braddick *et al.*, 1986*b*), it is important to include a control condition showing that this is not simply a limitation of recording technique or of overall temporal response, and that a VEP at the same frequency *can* be elicited by a simpler stimulus. A more specific requirement is the use of VEPs to estimate a threshold. For this purpose, signals must be recorded with a set of stimuli graded, e.g. in contrast or spatial frequency (Harris *et al.*, 1976; Sokol, 1978). A threshold can then be inferred from plotting VEP amplitude against the stimulus variable, and taking the intercept (Campbell and Maffei, 1970) (although there may be problems in selecting the range of this function to be fitted, given that the signal may saturate at high values).

Given the variability of responses due to infant state and electrode condition, the signals to be compared should be recorded as close together in time as possible. This can be achieved by interleaving blocks of, for example, 5- to 10-s recording with the different stimuli, and cumulating the responses from each as a separately averaged signal in the computer. The transition between stimuli may produce its own specific response, and it is good practice to prevent this contributing to the averaged steady-state signal by excluding the first 0.5 s of each block from the averages.

The 'sweep' method (Norcia and Tyler, 1985; Regan, 1989) can be regarded as taking the interleaving approach to its limit. A stimulus variable such as spatial frequency is varied either continuously or in very small steps over a period of 10–20 s, and the signal from repetitions of this whole sweep is averaged. The amplitude and phase of the averaged signal at the stimulus repetition frequency is analysed as a function of time in the sweep (and hence spatial frequency). A threshold value (acuity in the case of a spatial frequency sweep) is derived from the intercept where a linear fit to this plot reaches zero amplitude. This linear fit will be over some region of the amplitude vs. sweep–time function, and the criteria for selecting this fitted region are critical in achieving meaningful results. Criteria that have been used include: signal:noise ratio above a threshold value (where noise is estimated from a nearby frequency unrelated to the stimulus); and signal phase that remains consistent over the fitted region (Norcia and Tyler, 1985).

7.7 Infant eye movements

7.7.1 Development of oculomotor control

It will be apparent from earlier sections that much of our knowledge about the development of human vision rests on evidence from eye movements. Findings and methods in infant eye movement research have been reviewed by Hainline (1993) and by Shupert and Fuchs (1988).

The preferential-looking and habituation-recovery methods depend on observing saccadic shifts of fixation. Saccadic dynamics seem to be essentially adult-like at an early age (Hainline *et al.*, 1984; Hainline, 1993), although at 1 month, saccades frequently fall short of their targets (Aslin and Salapatek, 1975). However, the ease with which competing stimuli can elicit a saccade, and the latency with which they do so, change markedly with age and are revealing about the development of systems for the control of visual attention (Atkinson *et al.*, 1992; Hood, 1994)

Smooth pursuit elicited by a discrete moving target is almost absent before about 2 months of age (Aslin, 1981; Hainline, 1985, 1993). This is probably related to the immaturity of mechanisms which also subserve directional discrimination (Braddick, 1993; Wattam-Bell, 1996). In contrast, optokinetic nystagmus elicited by large-field motion is present from birth; the relationship to the later onset of pursuit, and the initial OKN asymmetry to monocular stimulation, may be interpreted in terms of the relationship between subcortical and cortical directional systems (Atkinson and Braddick, 1981; Braddick *et al.*, 1996).

Finally, the development of vergence movements of the eyes is important for understanding the development of binocularity. A substantial degree of vergence control appears to be present before the onset of sensitivity to binocular correlation and disparity around 10–16 weeks of age (Aslin 1993; Hainline and Riddell, 1996), raising the unanswered question: 'What drives infant vergence, if not disparity?'

Given the importance of inferences from infants' eye movements, it is unfortunate that current methods of recording these, summarized below, are so difficult or limited.

7.7.2 Direct observation

Almost all preferential-looking and habituation-recovery experiments depend on an observer judging the infant's direction of gaze, either directly or by video observation. Fortunately, a human observer is highly sensitive to another's gaze direction, especially with respect to a straight-ahead position (so it is important that the observation position should be centred with respect to the display). The minimum size of eye and head movement to be reliably judged is around 5° visual angle. The judgement is greatly aided by observing the reflection of the stimulus in the cornea. For video observation, this can be enhanced by positioning infrared light-emitting diodes symmetrically with respect to the display.

Infants have a larger angle kappa than adults, so the corneal reflection of a fixated target in a single eye appears off-centre (Slater and Findlay, 1975; Riddell, Hainline and Abramov, 1992). This is not usually misleading if both eyes are viewed together, since the corneal reflexes appear symmetrically placed in the two eyes in the case of central fixation. However, accurate observation is more difficult for infants viewing monocularly, or for strabismic subjects.

Direct observation yields information on fixations that is quite adequate for preference and habituation studies, but it can tell us little about the quantitative precision, timing, or kinematics of eye movements. Examination of a videotaped record can give information about the onset time and duration of movements (if the 50/60-Hz time resolution of a standard video is adequate), although the analysis is tedious, especially if the movements of interest have to be found by searching the taped records.

7.7.3 Automatic optical tracking

Methods of measuring eye movements, appropriate for adult subjects, are discussed in detail in Chapter 9. There are several alternative optoelectronic systems:

(1) the simplest, but with the lowest precision, tracking the position of the limbus;

(2) tracking the corneal reflection of an infrared source relative to the illuminated pupil;

(3) the most accurate and complex, tracking the corneal reflection (1st Purkinje image) relative to the 4th Purkinje image.

Instruments using method 3 require locating the head with a precision that is impractical with infants. Method 1 is seriously contaminated by head translations unless either the head is similarly restrained, or the LED sources and sensors are mounted on spectacle frames and so move with the head. Suitably lightweight mountings might be feasible although difficult to use with young infants; however, this does not seem to have been attempted.

Several laboratories have used instruments of type 2 (e.g. Hainline, 1981). The relative nature of the measurement means that it is not distorted by head translations. However, the systems require an enlarged video image of the eye, which typically requires the eye to remain within 1 cm of a fixed location so that it does not go out of the video frame or out of focus. This has proved possible to attain, but becomes increasingly hard with infants over 3 months of age, and even in the younger age range a large proportion of recordings are unsuccessful.

Such systems need to be calibrated with the individual infant in the test situation, by attracting the child's attention to known locations. The uncertainty about fixation means that measurements of several fixations at each location need to be averaged.

For some purposes, measurements of corneal reflex position within the pupil in still photographs may provide useful information (e.g. on the vergence angles that infants take up with targets at different distances (Hainline and Riddell, 1996)).

7.7.4 Electro-oculography (EOG)

Electrodes placed on the skin beyond the outer canthus of each eye pick up a voltage which varies with eye position, due to the standing corneoretinal potential across the eye. This method is possible with young infants, although their tolerance of attachments close to the face is typically less than for the scalp electrode of VEP recording. It measures the position of the eye in the orbit, so without some separate means of measuring head position (Regal *et al.*, 1983) it does not yield gaze direction. It is therefore more useful for studying infant oculomotor dynamics than for measuring the fixation position. For this purpose, it has the advantage over video-based methods that its time resolution is not limited. Even for studying dynamics, if a quantity such as pursuit gain is to be assessed, the signals need to be individually calibrated in the way mentioned in Section 7.7.3 (Finocchio *et al.*, 1990).

References

Abramov, I., Hainline, L., and Duckman, R. H. (1990). Screening infant vision with paraxial photorefraction. *Optometry and Vision Science*, **67**, 538–45.

Anker, S. E., Atkinson, J., Braddick, O. J., Ehrlich, D. L., Weeks F., and Wade, J. (1995). Accommodative measures of ametropia from video-refractive screening of a total infant population. *Investigative Ophthalmology and Visual Science* (Suppl.), **36**, S48 .

Apkarian, P. and Tijssen, J. (1992). Detection and maturation of VEP albino asymmetry: an overview and a longitudinal study from birth to 54 weeks. *Behavioural Brain Research*, **49**, 57–67.

Aslin, R. N. (1981). Development of smooth pursuit in human infants. In *Eye movements: cognition and visual perception* (ed. D. F. Fisher, R. A. Monty, and J. W. Senders), pp. 31–51. Lawrence Erlbaum Associates, Hillsdale, N J.

Aslin, R. N. (1993). Infant accommodation and convergence. In *Early visual development: normal and abnormal* (ed. K Simons), pp. 30–8. Oxford University Press, New York.

Aslin, R. N. and Salapatek, P. (1975). Saccadic localization of targets by the very young human infant. *Perception and Psychophysics*, **17**, 293–302.

Atkinson, J. and Braddick, O. J. (1981). Development of optokinetic nystagmus in infants: an indicator of cortical binocularity? In *Eye movements: cognition and visual perception* (ed. D. F. Fisher, R. A. Monty, and J. W. Senders), pp. 53–64. Lawrence Erlbaum Associates, Hillsdale, N J.

Atkinson, J. and Braddick, O. J. (1992). Visual segmentation of oriented textures by infants. *Behavioural Brain Research*, **49**, 123–31.

Atkinson, J., Braddick, O. J., and Moar, K. (1977). Development of contrast sensitivity over the first three months of life in the human infant. *Vision Research*, **17**, 1037–44.

Atkinson, J., Braddick, O. J., and Pimm-Smith, E. (1982). 'Preferential looking' for monocular and binocular acuity testing of infants. *British Journal of Ophthalmology*, **66**, 264–8.

Atkinson, J., Pimm-Smith, E., Evans, C., and Braddick, O. J. (1983). The effects of screen size and eccentricity on acuity estimates in infants using preferential looking. *Vision Research*, **23**, 1479–83.

Atkinson, J., Braddick, O. J., Durden, K., Watson, P. G., and Atkinson, S. (1984). Screening for refractive errors in 6–9 month old infants by photorefraction. *British Journal of Ophthalmology*, **68**, 105–12.

Atkinson, J., Pimm-Smith, E., Evans, C., Harding, G., and Braddick, O. J. (1986*a*). Visual crowding in young children. *Documenta Ophthalmologica Proceedings Series*, **45**, 201–13.

Atkinson, J., Wattam-Bell, J., Pimm-Smith, E., Evans, C., and Braddick, O. J. (1986*b*). Comparison of rapid procedures in forced choice preferential looking for estimating acuity in infants and young children. *Documenta Ophthalmologica Proceedings Series*, **45**, 192–200.

Atkinson, J., Anker, S., Evans, C., Hall, R., and Pimm-Smith, E. (1988*a*). Visual acuity testing of young children with the Cambridge Crowding Cards at 3 and 6 metres. *Acta Ophthalmologica*, **66**, 505–8.

Atkinson, J., Hood, B., Wattam-Bell, J., Anker, S., and Tricklebank, J. (1988*b*). Development of orientation discrimination in infancy. *Perception*, **17**, 587–95.

Atkinson, J., Hood, B. M., Wattam-Bell, J., and Braddick, O. J. (1992). Changes in infants' ability to switch visual attention in the first three months of life. *Perception*, **21**, 643–53.

Banks, M. S. (1980). The development of visual accommodation during early infancy. *Child Development*, **51**, 646–66.

Bobier, W. R. and Braddick, O. J. (1985). Eccentric photorefraction: optical analysis and empirical measures. *American Journal of Optometry and Physiological Optics*, **62**, 614–20.

Bornstein, M. H. (1985). Habituation of attention as a measure of visual information processing in human infants: Summary, systematization, and synthesis. In *Measurement of audition and vision in the first year of life* (ed. G. Gottlieb and N. Krasnegor), pp. 253–300. Ablex, Norwood, NJ.

Braddick, O. J. (1993). Orientation- and motion-selective mechanisms in infants. In *Early visual development: normal and abnormal* (ed. K. Simons), pp. 163–77. Oxford University Press, New York.

Braddick, O. J., Atkinson, J., French, J., and Howland, H. C. (1979). A photo-refractive study of infant accommodation. *Vision Research*, **19**, 319–30.

Braddick, O. J., Atkinson J., Julesz B., Kropfl W., Bodis-Wollner I., and Raab, E. (1980). Cortical binocularity in infants. *Nature*, **288**, 363–5.

Braddick, O. J., Wattam-Bell, J., Day, J., and Atkinson, J. (1983). The onset of binocular function in human infants. *Human Neurobiology*, **2**, 65–9.

Braddick, O. J., Atkinson, J., and Wattam-Bell, J. (1986*a*). Development of the discrimination of spatial phase in infancy. *Vision Research*, **26**, 1223–39.

Braddick, O. J., Wattam-Bell, J., and Atkinson, J. (1986*b*). Orientation-specific cortical responses develop in early infancy. *Nature*, **320**, 617–19.

Braddick, O. J., Atkinson, J., Wattam-Bell, J., Anker, S., and Norris, V. (1988). Videorefractive screening of accommodative performance in infants. *Investigative Ophthalmology and Visual Science* (*Suppl.*), **29**, 60.

Braddick, O. J., Atkinson, J., and Hood, B. M. (1996). Striate cortex, extrastriate cortex, and colliculus: some new approaches. In *Infant vision* (ed. F. Vital-Durand, O. Braddick, and J. Atkinson), pp. 203–20. Oxford University Press.

Brookman, K. (1983). Ocular accommodation in human infants. *American Journal of Optometry and Physiological Optics*, **60**, 91–9.

Campbell, F. W. and Maffei, L. (1970). Electrophysiological evidence for the existence of orientation and size detectors in the human visual system. *Journal of Physiology*, **207**, 635–52.

Caron, A. J., Caron, R. F., and Carlson, V. R. (1978). Do infants see objects or retinal images? Shape constancy revisited. *Infant Behaviour and Development*, **1**, 229–43.

Dobson, V. (1976). Spectral sensitivity of the 2-month infant as measured by the visual evoked potential. *Vision Research*, **16**, 367–74.

Fantz , R. L., Ordy, J. M., and Udelf, M. S. (1962). Maturation of pattern vision in infants during the first six months. *Journal of Comparative and Physiological Psychology*, **55**, 907–17.

Finocchio, D. V., Preston, K. L., and Fuchs, A. F. (1990). Obtaining a quantitative measure of eye movements in human infants: a method of calibrating the electroretinogram. *Vision Research*, **30**, 1119–28.

Fox, R., Aslin, R. N., Shea, S. L., and Dumais, S. T. (1980). Stereopsis in human infants. *Science*, **207**, 323–4.

Gwiazda, J., Brill, S., Mohindra, I., and Held, R. (1978). Infant visual acuity and its meridional variation. *Vision Research*, **18**, 1557–64.

Hainline, L. (1981). An automated eye movement recording system for use with human infants. *Behavioral Research Methods and Instrumentation*, **13**, 20–4.

Hainline, L. (1985). Oculomotor control in human infants. In *Eye movements and human information processing* (ed. R. Groner, G. W. McConkie, and C. Menz), pp. 71–84. Elsevier-North Holland, Amsterdam.

Hainline, L. (1993). Conjugate eye movements of infants. In *Early visual development: normal and abnormal* (ed. K Simons), pp. 47–79. Oxford University Press, New York.

Hainline, L. and Riddell, P. (1996). Eye alignment and convergence in young infants. In *Infant vision* (ed. F. Vital-Durand, O. Braddick, and J. Atkinson), pp. 221–47. Oxford University Press.

Hainline, L., Turkel, J., Abramov, I., Lemerise, E., and Harris, C. (1984). Characteristics of saccades in human infants. *Vision Research*, **24**, 1771–80.

Hainline, L., De Bie, J., Abramov, I., and Camenzuli, C. (1986). Eye movement voting: a new technique for deriving spatial contrast sensitivity. *Clinical Vision Sciences*, **2**, 33–44.

Harris, L., Atkinson, J., and Braddick, O. J. (1976). Visual contrast sensitivity of a 6-month-old infant measured by the evoked potential. *Nature*, **264**, 570–1.

Held, R., Birch, E. E., and Gwiazda J. (1980). Stereoacuity of human infants. *Proceedings of the National Academy of Sciences USA*, **77**, 5572–4.

Hood, B. M. (1994). Visual selective attention in infants: a neuroscientific approach. In *Advances in infancy research*, Vol. 9 (ed. L Lipsitt and C. Rovee-Collier), pp. 163–216. Ablex, New Jersey.

Hood, B. M., Murray, L., King, F., Hooper, R. Atkinson, J., and Braddick, O. J. (1996). Habituation changes in early infancy: longitudinal measures from birth to six months. *Journal of Reproductive and Infant Psychology* **14**, 177–85.

Horowitz, F. D., Paden, L., Bhana, K., and Self, P. (1972). An infant-controlled procedure for studying infant visual fixations. *Developmental Psychology*, **7**, 90.

Howland, H. C. (1985). Optics of photoretinoscopy: results from ray tracing. *American Journal of Optometry and Physiological Optics*, **62**, 621–5.

Howland, H. C. and Howland, B. (1974). Photorefraction, a technique for the study of refractive state at a distance. *Journal of the Optical Society of America*, **64**, 240–9.

Howland, H. C., Atkinson, J., Braddick, O. J., and French, J. (1978). Infant astigmatism measured by photorefraction. *Science*, **202**, 331–3.

Howland, H. C., Braddick, O. J., Atkinson, J., and Howland, B. (1983). Optics of photorefraction: orthogonal and isotropic methods. *Journal of the Optical Society of America*, **73**, 1701–8.

Julesz, B., Kropfl, W., and Petrig, B. (1980). Large evoked potentials of dynamic random-dot correlograms and stereograms permit quick determination of stereopsis. *Proceedings of the National Academy of Sciences USA*, **77**, 2348–51.

Kaakinen, K. (1981). Simultaneous two-flash static photoskiascopy. *Acta Ophthalmologica*, **59**, 378–86.

McDonald, M. A., Sebris, S. L., Mohn, G., Teller, D. Y., and Dobson, V. (1985). The acuity card procedure: a rapid test of infant acuity. *Investigative Ophthalmology and Visual Science*, **26**, 1158–62.

McKee, S. P., Klein, S. A., and Teller, D. Y. (1985). Statistical properties of forced-choice psychometric functions: implications of probit analysis. *Perception and Psychophysics*, **37**, 286–98.

Maier, J., Dagnelie, G., Spekreijse, H., and van Dijk, B. W. (1987). Principal components analysis for source localization of VEPs in man. *Vision Research*, **27**, 165–77.

Mash, C., Dobson, V., and Carpenter, N. (1995). Interobserver agreement for measurement of grating acuity and interocular acuity differences with the Teller Acuity Card procedure. *Vision Research*, **35**, 303–12.

Mayer, D. L. and Dobson, V. (1980). Assessment of vision in young children: a new operant approach yields estimates of acuity. *Investigative Ophthalmology and Visual Science*, **19**, 566–70.

Mohindra, I. (1975). A technique for infant examination. *American Journal of Optometry and Physiological Optics*, **52**, 867–70.

Mohindra, I. (1977). Comparison of 'near retinoscopy' and subjective refraction in adults. *American Journal of Optometry and Physiological Optics*, **54**, 319–22.

Moore, B. R. (1980). A modification of the Rayleigh test for vector data. *Biometrika*, **67**, 175–80.

Norcia, A. M. and Tyler, C. W. (1985). Spatial frequency sweep VEP: visual acuity during the first year of life. *Vision Research*, **25**, 1399–405.

Norcia, A. M., Garcia, H., Humphry, R., Holmes, A., and Orel-Bixler, D. (1991). Anomalous motion VEPs in infants and in infantile esotropia. *Investigative Ophthalmology and Visual Science*, **32**, 436–9.

Porciatti, V. (1984). Temporal and spatial properties of the pattern-reversal VEPs in infants below 2 months of age. *Human Neurobiolology*, **3**, 97–102.

Prechtl, H. F. R. (1974). The behavioural states of the newborn infant (a review). *Brain Research*, **76**, 185–212.

Regal, D. (1981). Development of critical flicker frequency in human infants. *Vision Research*, **21**, 549–55.

Regal, D. M., Ashmead, D. H., Salapatek, P. (1983). The coordination of eye and head movements during early infancy: a selective review. *Behavioural Brain Research*, **10**, 125–32.

Regan, D. (1989). *Human brain electrophysiology: evoked potentials and evoked magnetic fields in science and medicine*. Elsevier, New York.

Riddell, P. M., Hainline L., and Abramov, I. (1992). Measurement of the Hirschberg test in human infants. *Investigative Ophthalmology and Visual Science*, **35**, 538–43.

Rieth, C. and Sireteanu, R. (1994). Texture segmentation and 'pop-out' in infants and children: a study with the forced-choice preferential looking method. *Spatial Vision*, **8**, 173–91.

Schupert, C. and Fuchs, A. F. (1988). Development of conjugate human eye movements. *Vision Research*, **28**, 585–96.

Shaeffel, F., Farkas, L., and Howland, H. C. (1987). Infrared photoretinoscope. *Applied Optics*, **26**, 1505–9.

Sheridan, M. D. (1976). *Manual for the STYCAR vision tests*. NFER, Slough.

Shimojo, S., Birch, E. E., Gwiazda, J., and Held, R. (1984). Development of vernier acuity in infants. *Vision Research*, **24**, 721–8.

Sireteanu, R., Fronius, M., and Constantinescu, D. H. (1994). The development of visual acuity in the peripheral visual field of human infants: binocular and monocular measurements. *Vision Research*, **34**, 1659–71.

Slater, A. M. and Findlay, J. (1975). The corneal reflection technique and the visual preference method: sources of error. *Journal of Experimental Child Psychology*, **20**, 248–73.

Slater, A. M., Morison, V., and Somers, M. (1988). Orientation discrimination and cortical function in the human newborn. *Perception*, **17**, 597–602.

Smith, J. C., Atkinson, J., Anker, S., and Moore, A. T. (1991). A prospective study of binocularity and amblyopia in strabismic infants before and after corrective surgery: implications for the human critical period. *Clinical Vision Sciences*, **6**, 335–53.

Sokol, S. (1978). Measurement of infant visual acuity from pattern-reversal evoked potentials. *Vision Research*, **18**, 33–40.

Swanson, W. H. and Birch, E. E. (1992). Extracting thresholds from noisy threshold data. *Perception and Psychophysics*, **51**, 409–22.

Teller, D. Y. (1979). The forced-choice preferential looking procedure: a psychophysical technique for use with human infants. *Infant Behavior and Development*, **2**, 135–53.

Teller, D. Y. (1983). Measurement of visual acuity in human and monkey infants: the interface between laboratory and clinic. *Behavioural Brain Research*, **10**, 15–23.

Teller, D. Y. and Palmer, J. (1996). Infant color vision: motion nulls for red/green and luminance modulated stimuli in infants and adults. *Vision Research*, **36**, 955–74.

Teller, D. Y., Morse, R., Borton, R., and Regal, D. (1974). Visual acuity for vertical and diagonal gratings in human infants. *Vision Research*, **14**, 1433–9.

Thorn, F. (1996). Basic considerations when refracting infants. In *Infant vision* (ed. F. Vital-Durand, O. Braddick, and J. Atkinson), pp. 97–112. Oxford University Press.

Thorn, F., Gwiazda, J., and Held, R. (1996). Using near retinoscopy to refract infants. In *Infant vision* (ed. F. Vital-Durand, O. Braddick, and J. Atkinson), pp. 113–24. Oxford University Press.

Wattam-Bell, J. (1985). Analysis of infant visual evoked potentials (VEPs) by a phase-sensitive statistic. *Perception*, **14**, A33.

Wattam-Bell, J. (1991). The development of motion-specific cortical responses in infants. *Vision Research*, **31**, 287–97.

Wattam-Bell, J. (1992). The development of maximum displacement limits for discrimination of motion direction in infancy. *Vision Research*, **32**, 621–30.

Wattam-Bell, J. (1996). The development of visual motion processing. In *Infant vision* (ed. F. Vital-Durand, O. Braddick, and J. Atkinson), pp. 79–94. Oxford University Press.

Weatherill, G. B. and Levitt, H. (1965). Sequential estimation of points on a psychometric function. *British Journal of Mathematical and Statistical Psychology*, **18**, 1–10.

Wesemann, W., Norcia, A. M., and Allen, D. (1991). Theory of eccentric photorefraction (photoretinoscopy): astigmatic eyes. *Journal of the Optical Society of America*, **A8**, 2038–47.

Winer, B. J. (1971). *Statistical principles in experimental design* (2nd edn). McGraw-Hill, New York.

Yonas, A. and Granrud, C. (1985). Reaching as a measure of infants' spatial perception. In *Measurement of audition and vision in the first year of life* (ed. G. Gottlieb and N. Krasnegor), pp. 301–22. Ablex, Norwood, NJ.

8

Gross potential recording methods in ophthalmology

H. SPEKREIJSE and F. C. C. RIEMSLAG

8.1 Introduction

The recording of electro-oculograms (EOGs), electroretinograms (ERGs), and visual evoked potentials (VEPs) in a clinical setting can be performed using various protocols, adapted to the question at hand. In this chapter we will discuss schemes that are an extension of the ISCEV (International Society for Clinical Electrophysiology of Vision) standard recording routines which have become available recently (EOG: Marmor and Zrenner, 1993; ERG: Marmor *et al.*, 1989, 1994; VEP: Celesia, *et al.*, 1993; Harding, 1995; visual electrodiagnostics: Galloway *et al.*, 1995). While the standard routines guarantee comparability of recordings between different laboratories, they do not optimally probe clinical or research questions. For EOGs, the standardized method, proposed by the ISCEV committee (Marmor and Zrenner, 1993), will be discussed; in addition, a method will be presented that can improve the accuracy of the measurement and which allows the EOG to be measured with arbitrary eye movements. In the case of the ERG we will discuss the standardized protocol (Marmor *et al.*, 1989, 1994), some applications, and an extension of the protocol with coloured light stimuli for distinguishing rod- and blue (S)-cone input. We will end with a method to be used in cases of severe absorption by media opacities, where a correction for absorption is necessary to judge retinal function. As for the VEP, we will discuss several methods addressing visual pathway conduction, misrouting, and objective visual acuity measurement, especially in malingering subjects. The stimuli in VEP studies are the flash, the pattern reversal, the pattern onset, and pattern motion. For the latter we will give an example to demonstrate the usefulness of combining brain-imaging techniques, such as magnetic resonance imaging (MRI) and single photon emission computed tomography (SPECT), with VEP brain maps for the interpretation of these recordings.

8.2 Instrumentation

8.2.1 Signal amplification

For recording gross potentials the signals have to be converted from analog signals into a time series of numbers in order to represent the signals in the computer (analog to digital, or A/D, conversion). Usually these A/D converters work with ranges in the order of volts. The signals of interest are in the range of microvolts

(μV) for the VEP and up to millivolts for the ERG and the EOG. Therefore the gross potentials have to be amplified with commercially available EEG amplifiers into the A/D conversion range. This means amplification with values of 10^5–10^6 for the VEP and 10^3–10^4 for the ERG and the EOG. Many EEG amplifiers are used directly connected to some plotting device, with amplification specified in terms of sensitivity: e.g. $50\,\mathrm{mV\,mm^{-1}}$. Usually, the manual gives the conversion by stating how many volts at the pen are needed to produce, for example, a 1-cm deflection. To obtain such levels of amplification without external interference these systems usually consist of separate preamplifiers that amplify the signal moderately and transport it through a long, shielded thick cable (or better: a glass optic fibre) to the main amplifier set, where the signals are filtered and amplified further. Modern practice dictates that preamplifiers should physically be kept as small as possible in order to reduce the capacitive coupling of the equipment (Fig. 8.1).

If these requirements are fulfilled it even becomes possible to record relatively artefact-free responses evoked by electrical pulse stimulation of the eye. Such recordings are useful for evaluating the integrity of the visual pathways behind a dense cataract. The most recent development in this context concerns A/D conversion immediately in the preamplifier system, and sending the multiplexed digital signals by optical fibre or wireless transmission up to the computer. One should be aware that these high-frequency signals may interfere with other types of equipment, especially

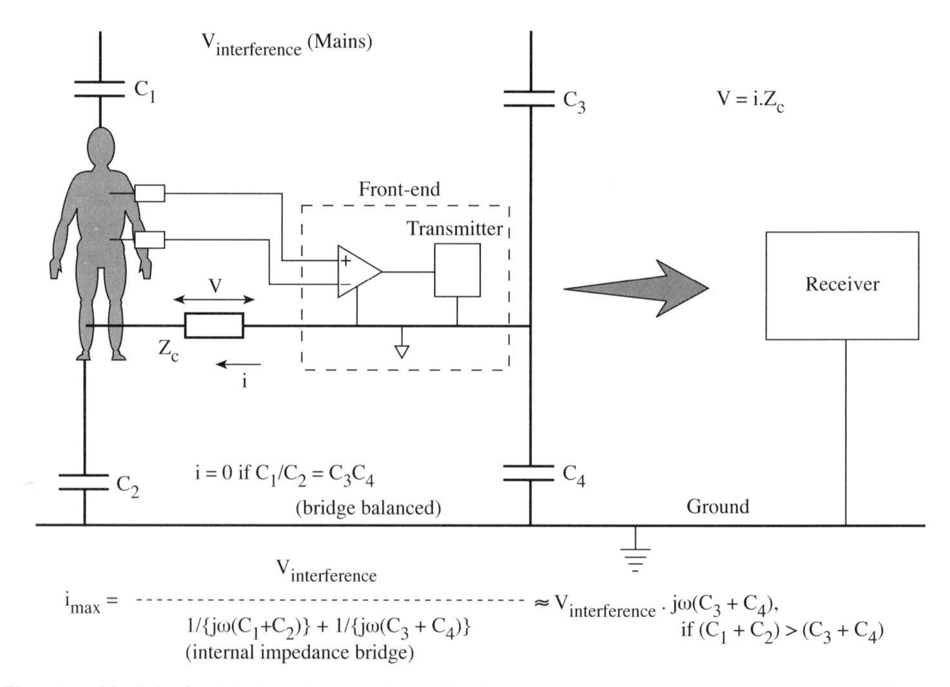

Fig. 8.1 Model of a bioelectric recording. C_1, C_2 capacitive coupling of the subject with interference source and ground; C_3, C_4 capacitive coupling of front-end with interference source and ground. If the front-end is kept physically small, then C_3 and C_4 remain small in comparison with C_1 and C_2.

brain scanners for functional MRI. So for simultaneously recording MRI images and VEPs there remains a preference for analog transmission and digitalization far away from the recording site.

While systems nowadays provide very stable amplification, regular calibration remains advisable for a standard signal (sine or square wave) with a frequency of 10 Hz and an amplitude of about 10 mV for the VEP, or about 100 mV for the EOG and the ERG. Since ERGs contain frequencies up to 125 Hz (oscillatory potentials, OPs) additional calibration with such a high-frequency signal is recommended. The analysis of VEPs in clinical practice is restricted at the high-frequency end to 70 Hz, although higher frequencies may be of interest in basic research. Therefore the bandwidth of the amplifiers should be adjustable. In the near future, general standard settings for amplifiers with 10-kHz digitalization in the 60-mV range and with a 16-bit resolution will provide band limitation *after* the signal has been digitized. At the high frequency end, the limit proposed for standard recording is 250 Hz (roll-off slope not exceeding 24 dB/octave). This is appropriate for ERG recordings, but may be disadvantageous for the signal-to-noise ratio (S/N ratio) in VEP recordings, and dictates higher rates of A/D conversion than really necessary; therefore in practice for VEPs we use 70 Hz as the high-frequency limit. For the EOG, one is not really interested in the actual eye movements made, which would require recording frequencies up to 250 Hz at least. More important for the EOG is the low-frequency cut-off, because this determines the recording of the potential difference during the fixation periods, which is the parameter of interest. In practice we use a bandwidth of 0.03 Hz (time constant of 5 s) to 70 Hz.

For VEP and ERG recordings the low-frequency (F) cut-off should be as low as 0.3 Hz with a roll-off slope not exceeding 12 dB/octave. Many commercial amplifiers specify the high-pass cut-off in terms of a time constant: 0.3 Hz corresponds to a time constant (τ) of 0.5 s ($\tau = 1/(2\pi F_{\text{cut-off}})$. For appropriate recording of the OPs it is necessary to raise this high-pass cut-off frequency to 100 Hz in order to eliminate the a- and b-wave components.

These low- and high-pass filter parameters always influence the parameters measured. For instance, peak latencies will increase when the low-pass cut-off frequency is lowered. If a low-pass filter is used with a 35-Hz cut-off point, and this filter is of the first order, an extra peak latency of about 7 ms is added to the physiological peak latency. Since the actual latency shift depends on the frequency content of the signal recorded, every laboratory should establish its own latency (and amplitude—for the ERG) standard, and published data should be accompanied by information about the filtering used.

8.2.2 Electrodes

For the derivation of signals from the skin, standard EEG cup electrodes (1-cm diameter) can be attached with electrode paste, after careful rubbing of the skin with alcohol to ensure that the resistance between the electrode and the skin is less than 5 kΩ. This, by the way, may be difficult in neonates and children, for example, and should be considered as an optimal value. However, clear responses can sometimes be recorded, even when the contact is not optimal, as in the example shown in

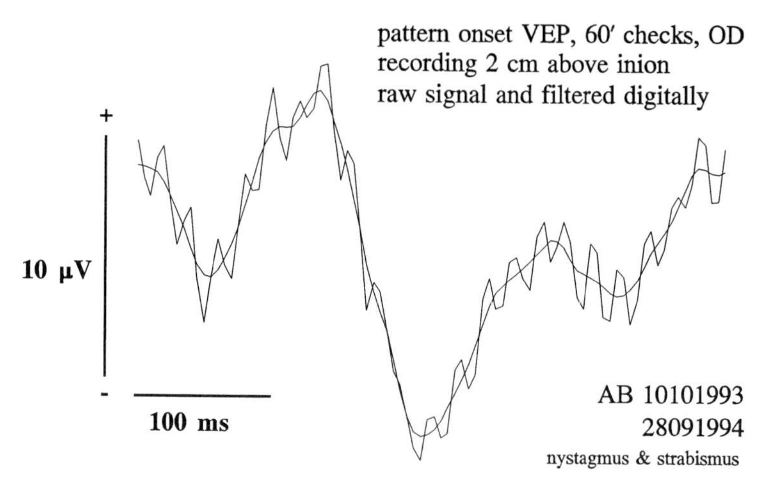

Fig. 8.2 Recording with non-optimal impedance in a 1-year-old child, heavily obstructing the attachment of the electrodes. A bad recording can be filtered, and sometimes must be accepted instead of upsetting the child and ending up with no response at all.

Fig. 8.2. When actively shielded electrodes are used, less stringent values of the impedance can be adopted. This, however, requires the use of special amplifier systems that are now becoming available commercially.

The contact of metal to electrode paste (an electrolyte) will always produce a small standing chemical potential. In principle, this DC potential varies in time, follows changes in temperature, etc., and thus adds noise to the signal. Furthermore, if this potential is substantial (e.g. 100 mV), because electrodes of different material (e.g. disposable ECG electrodes together with silver chloride EEG electrodes) are used, and if the signal is preamplified before filtering (e.g. 50 ×) then clipping of the amplifiers can easily occur. Therefore, care should be taken to reduce this chemical potential. For this reason the use of silver (gold is even better but more expensive) electrodes is common, and especially silver chloride (Ag/AgCl) electrodes, since the chemical potential is then even less. The modern amplifiers mentioned above have special features, such as low-voltage preamplification and guided-shield amplification, that take care of these problems. This means that different types of electrodes may be used simultaneously; however, this is still not advisable if the problem is avoidable. Many versions of contact lenses with silver electrodes are available for ERG, and these are being used routinely in the clinic. However, alternative electrodes that can be attached under the eyelid are becoming more and more popular, for instance gold foil and DTL electrodes. The latter consist of a thin nylon fibre coated in silver, which is attached with metallic glue to a copper wire connected to the preamplifier. We prefer the wire to be connected to two copper-wire terminals that are connected together to the final electrode lead attached to the amplifier (Fig. 8.3). Since the nylon fibre cannot be sterilized, each patient has to be tested with new electrodes.

Especially with gold foil and nylon fibre electrodes, the DC chemical potential artefacts may play an important role, although they are small when the recording is referenced to Ag/AgCl electrodes. Their main advantage is that they are much more

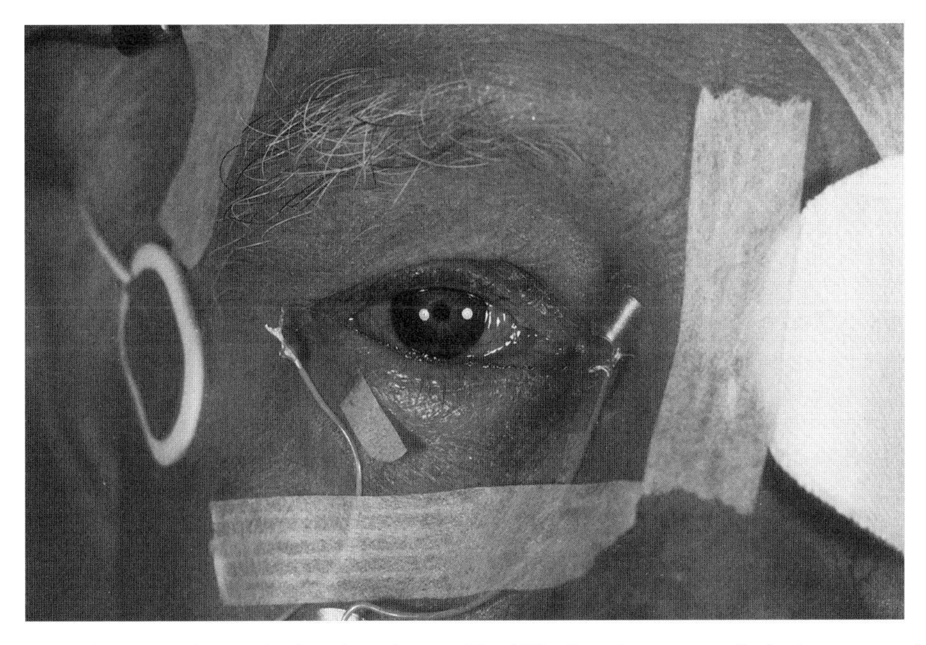

Fig. 8.3 DTL electrode attached against the eye. The DTL silver-impregnated wire is connected at both ends to flexible copper wires. These two wires are brought together to the amplifier. The, in principle, uncomfortable contact between skin and wire endings is prevented by the use of two-sided, EEG electrode stickers at both canthi of the eye. The electrode is very comfortable indeed: generally, anaesthesia of the cornea is not required.

comfortable than the standard contact lens and can be used easily in long recording sessions lasting several hours, nor do they require anaesthetic drops to render the cornea insensitive to foreign bodies. They have the further advantage of leaving the optics of the eye unaffected, and are therefore indicated for the recording of the ERG to pattern stimulation. A disadvantage is that they do not prevent the patient from blinking, which—although not troublesome in steady-state recordings, like pattern reversal—may cause strong artefacts with transient flash stimulation. In this case it is worthwhile skipping the first three to four sweeps, after which the subject becomes used to the flashes and can refrain from blinking and rotating the eyeball.

8.2.3 Electrode positions

For positioning the EEG electrodes at the scalp the 10–20 international system is usually employed. The position names of this system are often quoted on the input box of the EEG amplifiers. In the 10–20 system, distances are measured in percentages of the distance between nasion and inion in the sagittal direction. The first location above the inion is named Oz (occipital), and is located 10% above the inion. In steps of 20% there then follow, respectively, parietal (Pz), central (Cz), frontal (Fz), and a frontal pole (Fp). In the lateral direction the distances are referred to the distance between the preauricular depressions, with 10% above the left ear the temporal lead (T3). Then again in steps of 20% one finds central left (C3), central centre (Cz), and central right (C4), and finally temporal (T4). Analogously, further backwards, parietal (P) and occipital (O) positions are labelled; and forward, the frontal (F) positions. Reference to this scheme gives a practical way of localizing electrodes at the scalp, even when intermediate positions are being used. One should, however, not attribute too great a value to this system: distances between the landmarks of the skull tend to show substantial variability, and MRI has shown considerable asymmetries between the hemispheres and the way the brain is oriented within the skull. As for the visual evoked potential used standardly, the part of the skull in the area where the most important components arise can be best approximated by a pure sphere with its midplane through a point approximately 3 cm above the inion and a diameter equal to the distance between the two ears. This being the case, one might practically prefer to choose a positioning system that identifies positions in centimetres relative to the inion. This is especially advantageous when used in dipole source-localization procedures of the visual response, where one is only interested in the back part of the skull, with multiple electrodes just in that region. If one is interested in hemispheric contributions, a row of five electrodes is the minimum advisable number: one on the midline, 2 cm above the inion, and the other two pairs at each side (left and right) at 3-cm and 6-cm distances. For source localization one should record the actual electrode positions after montage, and these positions should be used for constructing the potential fields. For this position measurement we use the Isotrak III (Polhemus Inc.). The system consists of a stylus transmitter and a receiver, such that the position of the electrodes can be pinpointed by the stylus and its position measured accurately in space referenced to the position of the fixed receiver.

8.2.4 Derivations

The recording of gross potentials consists of recording potential *differences* at the skull. Most of these recordings are referred to a position at the skull where relatively low activity is expected (so-called inactive positions). This type of recording is called a monopolar derivation, since only one of the two electrodes is assumed to be active ('the hot electrode').

The inactive position for the ERG is considered to be a site on the skin near the lateral or medial canthus of the eye (inactive since most of the current involved is restricted to the eye and does not enter the orbit); the inactive location for the VEP is considered a central frontal position (Fz). The so-called 'connected ears' is often used as the reference for the VEP: two electrodes positioned at the lower lid of the ears (the earlobe clearly inactive) are connected together to the amplifier. If one is interested mainly in the temporal behaviour of the response (the waveform) this may be applicable, but the moment one is interested in the distribution of the response across the skull, the use of linked ears as a reference may result in uninterpretable data, since in this way the two ears are forced to the same potential.

So-called bipolar derivations, i.e. the recording of potential differences between two active positions, are used for recording the EOG, with electrodes placed at both canthi of the eyes. VEP and bipolar derivations, which gives a better S/N ratio and shows specific advantages for the interpretation of the data, will be discussed below.

Furthermore, a more specialized derivation (Laplacian), which makes use of the potentials at five different positions on the head, will also be discussed later.

8.2.5 Grounding

For every biological recording one needs to connect the body to the ground of the amplifier in order to reduce interference with the mains, which can add large 50- (60-) Hz signals ('hum') to the biological signals and will hamper the recording of the signals of interest. The position of the ground electrode is not critical, but ideally should be in the centre of all electrodes attached—this will reduce the total surface between the wires, and thus the flux produced by the field of the mains. Area reduction between the wires, by twining the electrode wires, also helps to reduce the interference. The use of actively driven shielded electrodes yields even better interference reduction.

8.2.6 Signal processing

After amplification the signal is converted into a digital signal, for which the sample frequency for the VEP should be at least 200 Hz. (To avoid aliasing the sample frequency should be at least twice the highest frequency of interest.) For the ERG a sample rate of 1000 Hz is advisable, since the interindividual variability of the peak latencies for the OPs is reported to be of the order of 2 ms. After A/D conversion the signal can be stored, averaged, or digitally filtered and processed in various ways depending on the question at hand. On-line display of the averaged signal will enhance the quality of the records for the ERG and is imperative for VEP recording. An interrupt mode for the averaging process, with the examiner continuously checking (by visual inspection) the attention of the subject, allows recordings to be made when the subject is actually looking. Automatic level-crossing artefact rejection improves the quality of the records greatly. Especially for the ERG, where artefacts are present many-fold, it is practical to have the opportunity to skip on-line sweeps that contain artefacts, e.g. due to blinking. Finally, on-line Fourier analysis of a signal is imperative for steady-state recordings.

A 2-channel amplifier system is sufficient for the EOG; but if a recording of the actual eye movements during the EOG recording (as we will discuss later in this chapter) is wanted, a 4-channel system is required. For the ERG two channels are usually sufficient, but a 4-channel system may provide simultaneous OP processing. For the VEP, at least a 5-channel system is required, but a 7-channel system is better, since it allows for the bipolar derivations to be displayed on-line. For on-line Laplacian recording six channels are needed. In practice, we often use 10 channels to provide both the row of the five derivations mentioned, two bipolar derivations, and the channels needed for the Laplacian derivation (see later in this chapter) over the inion. Finally, 32- or 64-channel systems are available for brain mapping. These systems are used for the localization of the sources of evoked activity in MRI scans of the subject.

8.2.7 Subject instruction

Subjects should be made comfortable, at ease, and adequately instructed as to what is expected to happen and what they should or should not do. Talking during the actual measurement, chewing gum, or making sudden movements, should especially be avoided. Attention should be paid to the fixation mark in the stimulus field, and instruction as to eye blinking can be helpful. Before we start recording a VEP we often ask the patient to chew quickly and to blink while they observe their own gross potentials displayed on a monitor. In this way they can see that chewing and blinking provoke signals moving all over the screen. These instructions serve two purposes: they provide the experimenter with a check as to whether the connections are patent, and explain to the subject why he should refrain from moving his jaws during the actual measurement. Infrared video monitoring of the patient can be helpful, especially when a patient with a dubious visual acuity loss is being tested. To attract the attention of children we interrupt stimulation (by pressing a button) and replace the stimulus with a cartoon movie when they are not looking. As soon as the child looks again, stimulation is restarted with the sound of the cartoon (preferably music) continued.

8.2.8 Stimulation equipment

The standard protocols recommend flash stimulation for both ERG and VEP. For the ERG a Ganzfeld dome is strongly recommended to illuminate the full retina as homogeneously as possible. This requirement is less pertinent for the flash VEP, which is dominated by macular input and not by contributions from the peripheral retina as in the ERG. Steady illuminated backgrounds can easily be produced in such a dome, in addition to well-controlled short-duration ($< 5\,\mu s$, usually of the order of μs) intense flashes by gas discharge tubes (colour temperature 7000 °K). According to the ISCEV standardization for the flash ERG, a standard flash should have a strength of 1.5–3.0 cd m^{-2} • s(luminance × time) as measured with a photometer at the position of the subject's eye. This calibration is not a straightforward task, since it needs a photometer that can integrate the strength of a number of flashes (e.g. 10). Nevertheless, flash strength should be checked regularly, since these flash sources tend to show extensive variability with time. The flash source chosen should be capable of producing intensities up to 1–2 log units above the level of the standard flash, to overcome absorption in patients with opacities of the media, and should be able to produce 30-Hz flicker at the standard intensity. These two requirements cannot be fulfilled by most commercially available flashers. Usually the flash intensity goes down with increasing frequency of stimulation. This decrease of intensity can go up to 50% and sometimes comes on top of instability of the flash frequency for the higher intensity conditions of the system.

The intensity and colour of the flash and background in the dome can be adjusted by means of optical filters—preferably in electronically driven systems, so that the subject can remain in the dark without the examiner having to enter the examination room in order to change filters.

When a Ganzfeld dome cannot be used because the ERG recording has to be done outside the electrophysiology unit (for example in the surgical theatre or a neonatal intensive care unit), one can approach homogeneously distributed luminance stimulation by covering the eye with part of a table tennis ball, by placing a sheet of drawing paper in front of the subject, or by using contact glasses with built-

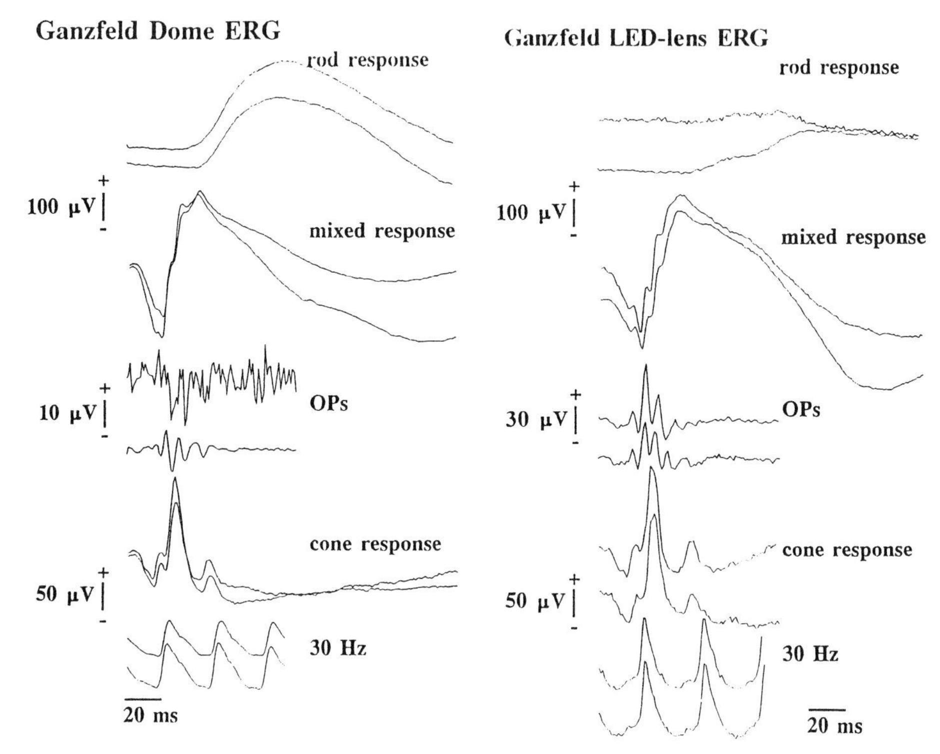

Fig. 8.4 A comparison of the responses obtained with a Ganzfeld dome and with a Ganzfeld LED-lens stimulator. In both, the standard recording conditions of the ISCEV standard were used and comparable responses obtained (Kooijman, personal communication).

in LED stimulators. The latter is especially suitable for use in the surgical theatre since the blackened contact glass prevents the strong theatre lights from interfering with the stimulation. The ISCEV committee leaves room for these kinds of solutions, but recommends comparison with the standard stimulation in a number of controls before application in the patient environment. Such a comparison can be found in Fig. 8.4.

Apart from flash stimulation, VEPs (and for certain applications also the ERG) require patterned stimulation. For the VEP, pattern stimulation has replaced, to a large extent, flash stimulation; but we still regard the flash as the prime stimulus for the ERG. The pattern stimulus consists of a regular pattern of black and white elements (checkerboards, bars, diamonds, etc.) presented in various ways: reversing, appearing–disappearing, moving, etc. Pattern stimulation should never be contaminated by changes in overall luminance: thus in the pattern onset–offset mode, when the pattern is appearing from a homogeneous grey field, it should be checked that there is no net luminance change. This is obviously obtained when the luminance of the grey is equal to the mean of the luminances of the white and black elements. It should be realized that an equal coverage of area by the two sets of elements is required to produce pure contrast stimulation. Furthermore, correction is needed for the non-linear intensity curve (z-axis) of the monitor used, though this does not play a role for pattern reversal. An easy check for luminance changes during pattern-onset stimulation can be obtained by blurring the image. This can be achieved by inserting frosted glass or a piece of white paper in front of the monitor, or by positioning the TV monitor such that its light is reflected on the wall of the room. When this is done in a dark room, at a stimulation rate around 10 Hz, very small luminance modulations (about 1%) can be perceived easily, since the flicker sensitivity of the human eye is the highest around this frequency.

Several ways of producing pattern stimuli have been proposed. The most feasible method consists of generating the pattern on a TV monitor under computer control, for which computer graphics are now widely available. Great care should be taken that triggering and luminance calibration are appropriately fulfilled. The triggering, that is the change from one image to another, should take place locked to the TV refresh rate, and especially only in the 'fly-back' of the monitor, such that the averaging process can take place coherently locked to the stimulus. If not, as is still the case in some commercially available monitors, trigger jitter will substantially change the waveform of the responses and hence the peak latencies that are frequently used as diagnostic criterion. The luminance calibration is another technical detail, since the z-axis (intensity) of most TV monitors is not a linear voltage-intensity transformer. Therefore one should adjust the voltages used within the linear range of the system, or make use of a look-up table to correct for the non-linearity. The latter is required when large contrast steps are employed, as is generally the case. High-contrast stimulation is also advisable to avoid latency increases due to reduced imaging quality by the eye media. Since the voltage-intensity characteristics change in time they should be checked regularly. Furthermore, they differ for fine and coarse patterns, so that the look-up tables should be adjusted to the element size within the stimulus. An intensity calibration procedure should be part of the equipment. In our lab-made system the calibration tests not only the full screen, but also the intensity when only one half of the TV screen has a high intensity and the other half is dark, since in the presence of such large contrast steps the voltage intensity curve may change substantially.

8.3 Electro-oculogram

8.3.1 Standard

The electro-oculogram (EOG) is a registration of the increase in the corneoretinal potential caused by an increase of the illumination of the retina. This corneoretinal potential (i.e. the potential difference between the front and the back of the eye, front positive: Fig. 8.5(I)) originates in the retinal pigment epithelium (RPE).

The increase of the potential due to a sudden increase in global illuminance of the retina occurs only slowly: the potential increases to a maximum in about 10 minutes after onset of the illumination. Such slowly changing potentials cannot be registered easily by biological amplifier/electrode systems, and even so the registration would be considerably disturbed by ongoing movements of the eye. Therefore the recording is performed by converting the signal into an AC signal: the subject is instructed to

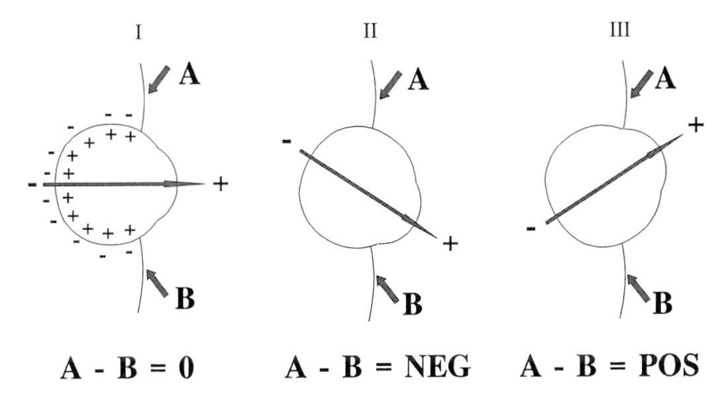

Fig. 8.5 Schematic drawing of the principle of recording an electro-oculogram. Across the retina, and more specifically at the retinal pigment epithelium layer, an only slowly changing potential step is present. The potential field produced by this layer dipole can be represented as a single dipole pointing outwards from the eye. Two electrodes attached at both sides of the eye will record a zero (I), negative (II), and positive (III) potential difference, respectively, when the eye is looking forward, right, and left.

perform fast eye movements between two fixation marks. When electrodes are attached at both sides of the eye (Figs 8.5(II) and (III)), these eye movements will result in abrupt changes in the potential difference between the two electrodes (Fig. 8.6A).

The size of this potential change is obviously dependent on many parameters, such as the distance between the electrodes, the size of the eye movement, the intensity

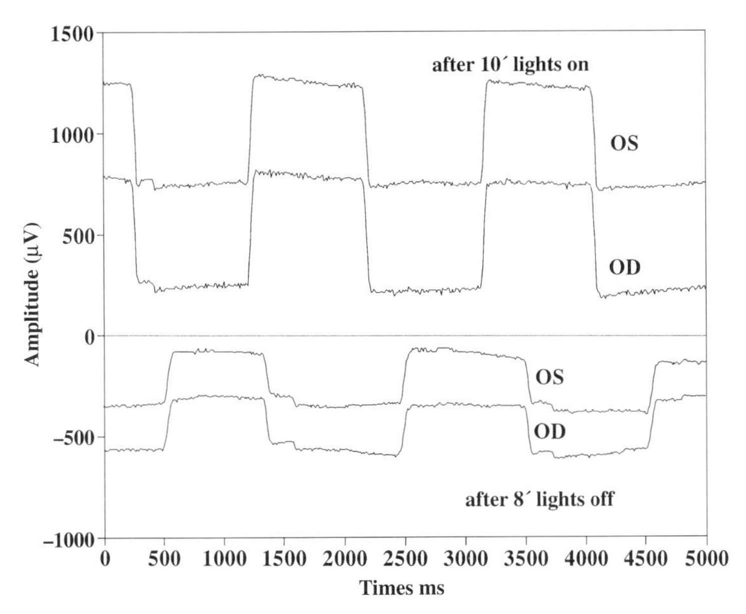

Fig. 8.6(A) Raw EOG recordings from electrodes attached at both sides of the eyes. Saccadic eye movements were generated upon alternating fixation LEDs after 8′ in the dark (lower panel) and after 10′ with background light on (upper panel). All traces are shifted for clarity.

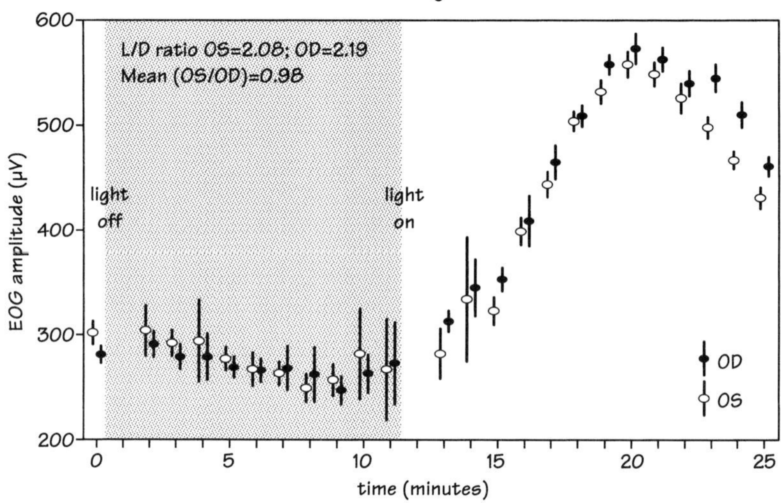

Fig. 8.6(B) Measured mean and SD of the amplitudes of the saccadic eye movements within every sweep for the left eye (open circles) and the right eye (closed circles), plotted as a function of time during the EOG procedure. Lights were switched off after the first recording, and switched on after 11 minutes. The L/D (light/dark) ratio is estimated by the mean of the top and its neighbouring two points divided by the mean of the lowest and its neighbours. The mean OS/OD (left/right eye) ratio is determined by the ratios of the mean at all points.

of the illumination, and, most important for us, the strength of the potential across the RPE. It is the behaviour of this potential that one wants to measure by the EOG procedure. Inasmuch as the potential step depends on so many parameters, the absolute value of this step is considered to be of little value. Therefore the procedure proposed by Arden and standardized by the ISCEV committee is to record these steps every minute for fixed eye movements, first during 12 minutes in the dark, and next during at least 12 minutes in bright light (Fig. 8.6(B)). By calculating the ratio between the peak value after about 8–10 minutes in the light and the value in the dark (about 8 minutes after switching off the light) the influence of parameters such as tissue conductivity, size of the eye movements, etc., which are supposed to be equal in the dark and in the light, is eliminated. In controls, a fairly constant value of 2 is obtained; when this value becomes less than 1.85 in patients it is considered pathological. The recording can be performed in the Ganzfeld dome with LED fixation marks at positions 15° left and right of the centre to guide the eye movements. The illumination should be at least 400–600 cd m^{-2} (undilated pupil).

The interpretation of the EOG is not as straightforward as often stated in literature: 'the Arden ratio is affected to the extent that retinal area is involved in the pathological process'. This is the reason why in vitelliform macula degeneration, where the EOG is flat, it is assumed that the retina as a whole is affected by the disease, although this is only visible in the macula. This may also be the case in panuveitis (Fig. 8.7).

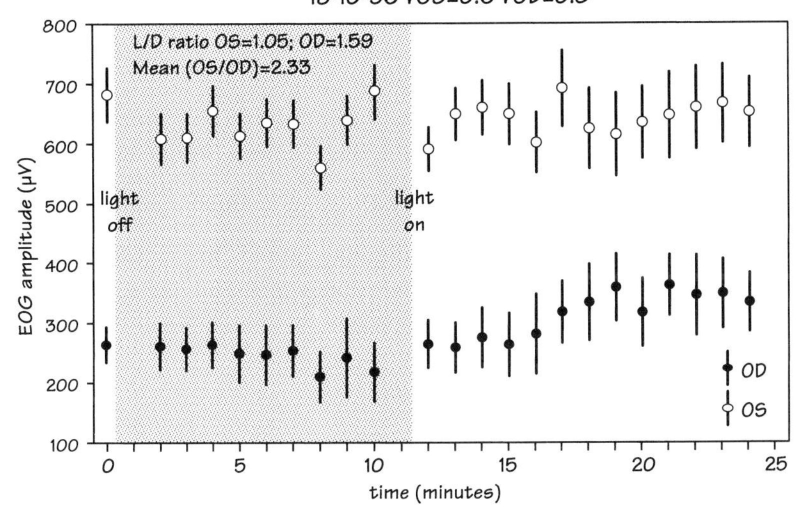

Fig. 8.7 EOG of a patient with panuveitis. There is almost no light rise for the left eye (open circles), and a too low rise for the right eye (closed circles). The mean OS/OD ratio is far above one and this value cannot be accounted for by differences between orbit tissue, etc.

However, it seems that the right eye of this patient is more affected than the left, since the signals of the two eyes differ in absolute amplitude by more than a factor of two, which intra-individually cannot be accounted for by geometrical differences between the orbits. Another example of interocular EOG differences is given in Fig. 8.8. This figure shows a patient with a relatively limited section of non-functioning retina (caused by a local toxoplasmosis chorioretinitis lesion) and with a considerable difference between the potential step amplitudes of the two eyes. Yet they both rise to a maximum at 10 minutes, and the ratio in both eyes remained greater than 2. It is as if part of the voltage is lost somewhere, like by a leaking current through the retinal pigment epithelium: the shape of the curve remains normal, but less of the voltage is tapped.

An improvement

Finally, in children or older people recording the EOG may not be easy since the subjects have to carry out constant eye movements during the whole procedure. To overcome this problem, one could record the actual eye movements using an infrared device that makes use of the infrared light reflected form the sclera (Fig. 8.9), and correct the measured steps for the variability of the eye movements. The variability of the ratio is reduced substantially in this way, and it allows even for recording the EOG with arbitrary eye movements (also, for example, nystagmus), using, for instance, a computer tennis game (Fig. 8.10, subject is 8 years old), or an optokinetic nystagmus stimulus like a rotating drum with high-contrast bars.

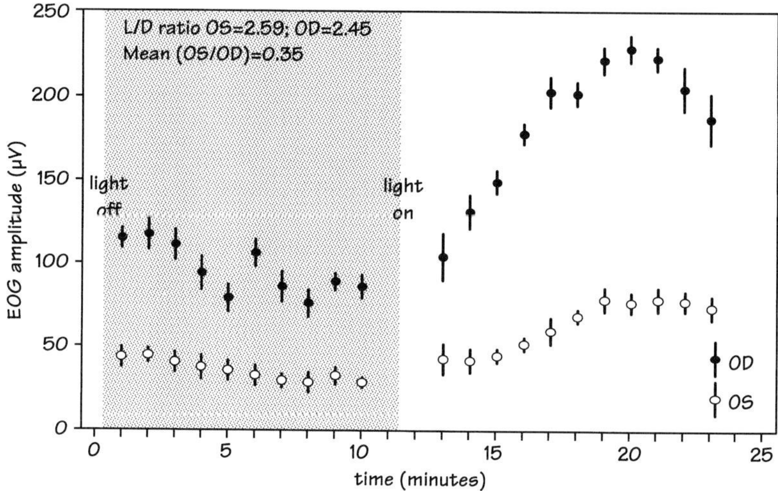

Fig. 8.8 EOG of a patient with a limited toxoplasmotic chorioretinitis lesion in the right eye. Here the light rise is seen to be present in both eyes with about normal L/D ratios. However, the mean OS/OD ratio is about one-third of normal.

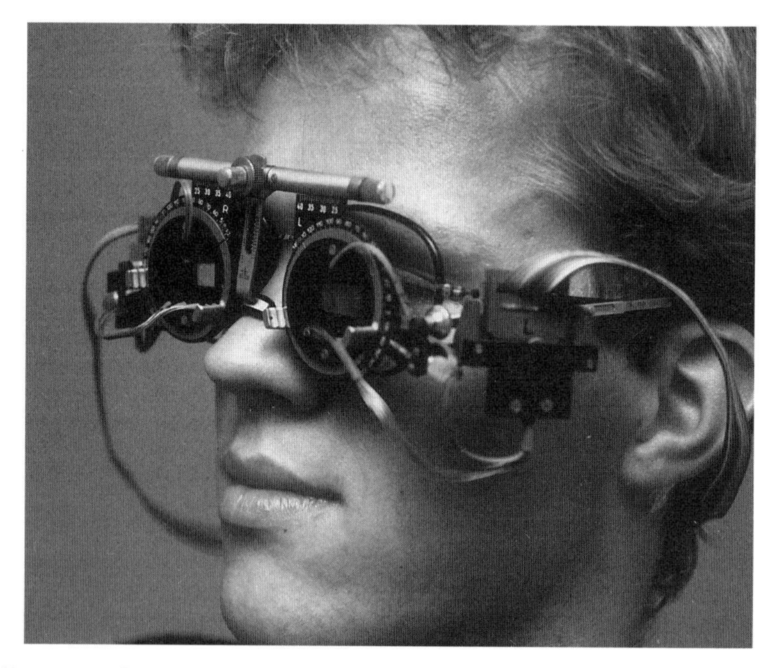

Fig. 8.9 Prototype of the IRIS eye-movement recording system. Infrared light-emitting diodes (LEDs) and light-dependent resistances (LDRs) are mounted in a trial frame. The reflected infrared light from the eye is measured and monitors the eye position (Reulen *et al.* 1988).

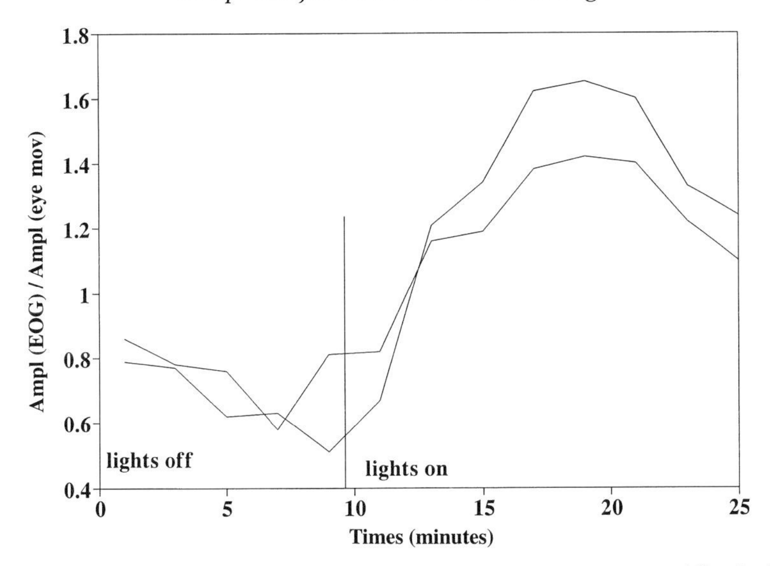

Fig. 8.10 EOG measured with the subject making arbitrary eye movements, while playing a computer tennis game. The voltage per degree of eye movement was calculated from a simultaneously registered, independent eye position signal obtained with an infrared reflection technique (IRIS). In this way one can correct for the actual eye movements made, instead of assuming that they are constant.

The RPE is only indirectly involved in visual processing: it is the outer layer of the retina, separating the receptors from the choroidal network, and regulates the metabolic exchange between the blood in the choroid and the receptors. The EOG probes the integrity of the RPE, and seems to require contact with an intact receptor cell layer. Although there is no consensus about the function of this potential, there are indications that this steady potential might play a role in guiding enzymes and oxygen, needed for the regeneration of rhodopsin, from the RPE towards the receptor outer segments. So if the contact between the two layers is disturbed, which expresses itself in a lowered EOG, then one also expects a prolonged dark adaptation and prolongation of the regeneration of the b-wave, as has been reported in Best's disease.

8.4 Electroretinogram

8.4.1 Standard

The shape of the flash response measured from the front of the eye depends strongly on the strength of the flash, and the background on which the flash is delivered. If a series of intensities is delivered to the dark-adapted eye (20 minutes in complete darkness) a series of responses, as plotted in Fig. 8.11, is observed. With a flash of 3.5 neutral-density log units (ND) below the standard flash intensity (not shown), only little changes of the potential can be seen. When the stimulus is increased by a factor of 10 (1 ND) a broad positive wave (called b-wave) is found, with an onset latency of about 60 ms and an implicit time of the positive peak of around 120 ms. Increasing the intensity further in about 0.35-ND steps gives the following results: the amplitudes increase, and the onset- and peak latencies decrease, until at about – 1.1 ND a negative deflection precedes the positive wave: the a-wave. A further increase reveals an even earlier a-wave. At that intensity a second b-wave can also be seen, on the trailing flank of the b-wave. The components thus found, i.e. at the intensity of the standard flash (I = 0 ND) are termed respectively a1, a2 and b1 and b2, and are generally being interpreted as contributions to the ERG mediated through the cones and the rods, respectively. At higher intensities (+ 0.35 ND and + 0.5 ND, respectively), two more responses, with deeper a-waves and a seemingly lowered b-wave, are given in the figure. Only the a-waves of these responses are

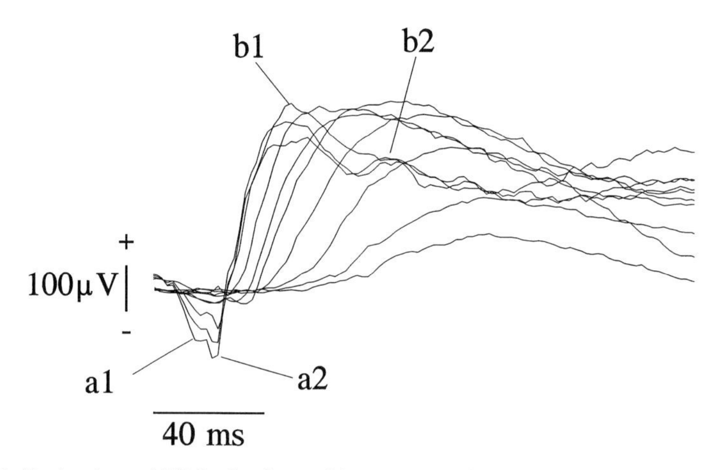

Fig. 8.11(A) Dark-adapted ERGs for intensities varying from – 2.5 ND in steps of 0.35 ND up to + 0.5 ND (relative to standard ISCEV). The responses change from purely rod-mediated (broad positive 'b'-wave for the lower intensities) to a mixed response in which two a-waves and two b-waves ((a_2, b_2) rod- and (a_1, b_1) cone-mediated) can be discriminated.

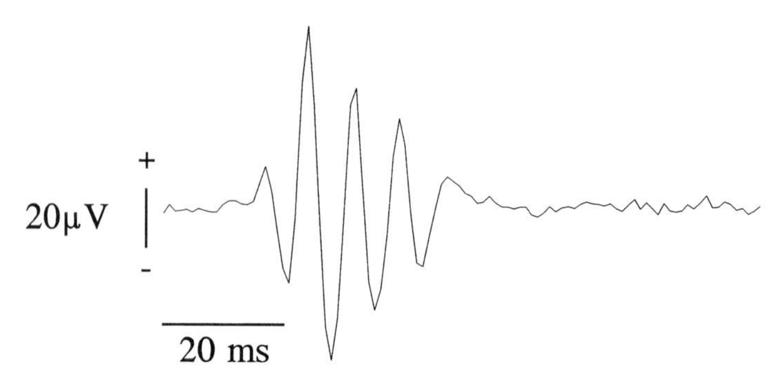

Fig. 8.11(B) Oscillatory potentials. Standard intensity (as the mixed response condition), filtered signal between 100 and 200 Hz.

directly produced by the cones and rods, whereas the b-waves are thought to have their origin in the Müller cells. They are supposed to be due to a radial current flow along the cell bodies that is generated by a depolarization of the Müller cells and mediated by potassium ions. So the b-wave is supposed to be an epiphenomenon of visual processing, and reflects only indirectly the neural activity. That is, by the way, the reason why the visual cortex can be active before the b-wave has developed. One can expect mainly to record activity from the receptors and the Müller cells with an electrode at the front of the eye, since these cells are positioned well aligned perpendicular to the retina. The other retinal cells, like ganglion cells, and even more so the horizontal cells, are oriented within the plane of the retina, and their responses can therefore not express themselves strongly in the algebraic sum made by the electrode at the front of the eye. It may be argued that this is not the case for the ERG to pattern reversal, in which the a- and b-waves are cancelled to a large extent by the nature of the stimulus, so that the weak (distortion) components get a chance to dominate the response. This is the so-called pattern ERG.

In the above account we have qualitatively described the behaviour of the ERG in dark-adapted (scotopic) conditions. For lower intensities the rods contribute solely, since the cone system is about three log units less sensitive than the rod system. But for intensities near the standard flash both systems contribute to the response. At these higher intensities another component in the response also becomes visible, as shown in Fig. 8.11B: the so-called oscillatory potentials (OPs). In the ERG to the standard flash, these OPs can be seen as wavelets on the rising slope of the b-wave. They can be made visible more clearly with appropriate filtering (high-pass frequency at 100 Hz, 2nd order; low-pass 250 Hz). The recording under standard conditions should contain five positive peaks at 8-ms intervals, with the first peak at 18 ms after flash onset. Note that this value for the peak latency is severely influenced by the specific filtering technique used. Several alternatives for measuring the amplitudes of these wavelets have been proposed in the literature. In our lab. only the absence of these OPs is considered as a pathological sign. These OPs have been described as originating in the 'deeper' (deeper in terms of signal transport) layers of the retina, that is the vitreous side of the retina and specifically the amacrine cell layer. Since this layer depends on the retinal circulation, OPs are affected when this circulation is hampered. The other components of the ERG, originating from receptors and Müller cells, obviously depend more on the other circulation of the eye: the choroidal blood supply. This blood supply is so superfluous that partial obstruction of it rarely affects the ERG.

In the light-adapted state, when the Ganzfeld dome is illuminated with a steady white intensity of about 15–30 cd m^{-2}, the responses obtained consist of (Fig. 8.12) a short latency a- (16 ms) and b-wave (35 ms). This brisk response is considered to be generated mainly through the cones, though rods may also contribute substantially (as illustrated by the light-adapted ERG of an achromat, see Figs 8.15 and 8.16). One should always be cautious about rod contamination of the ERG, since the rods so overwhelmingly outnumber the cones (100×10^6 versus 5×10^6). However, one way of eliminating this contamination is to use a high flash rate of, for example, 30 flashes per second (Fig. 8.13). The rods are unable to follow such a flash rate, as is also shown in the achromat response set of Fig. 8.15 (see below).

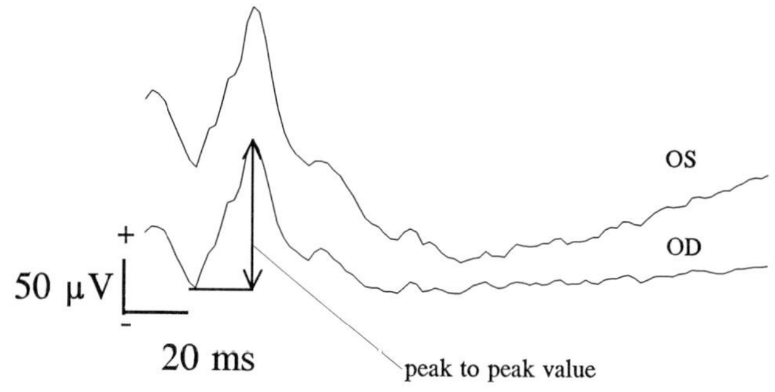

Fig. 8.12 Light-adapted ERGs at standard flash intensity upon a continuously illuminated white background.

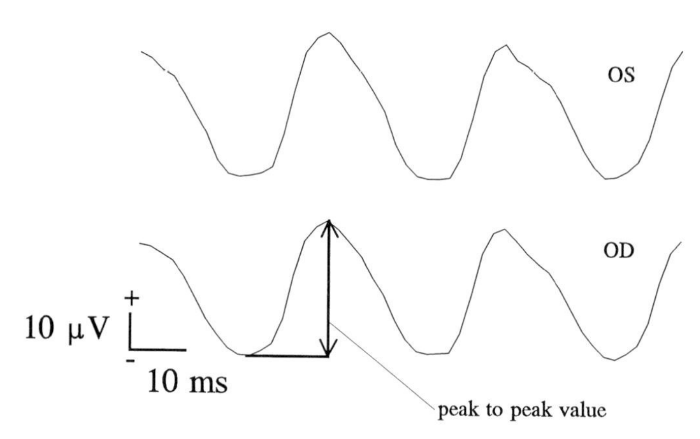

Fig. 8.13 30-Hz cone responses. A series of three consecutive responses is shown by averaging a 100-ms epoch.

For clinical use, a standard test has been prescribed by the ISCEV standardization committee that makes a selection of these responses. This choice is directed to optimally separate the activity developed through the two receptor systems and consists of:

(1) a cone ERG, with a standard flash upon a bright background;
(2) a 30-Hz flicker ERG with standard flashes upon a bright background;
(3) a rod ERG, after 20 minutes dark adaptation, in complete dark with a – 2.5-ND log flash;
(4) mixed ERG, complete dark, standard flash.;
(5) oscillatory potentials, complete dark, standard flash.

An example of a complete set of responses is given in Fig. 8.14.

The use of this standard set of stimuli has become widely accepted and is applied routinely. Space does not permit an extensive overview of the changes caused by different types of diseases, but some examples will be presented that serve to illustrate the interpretation of the responses in terms of their origin.

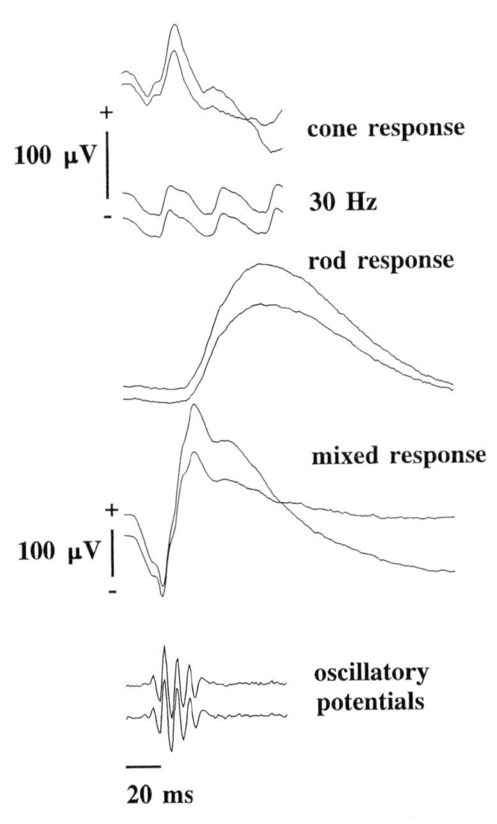

Fig. 8.14 Responses obtained in a standard ISCEV procedure, containing light-adapted, transient (1 flash per second) and steady-state (30 flashes per second), cone-mediated responses, rod-mediated, and mixed responses in the dark as well as oscillatory potentials (also in the dark). Responses from both eyes of the same subject are given.

8.4.2 Origin of the components of the ERG

Figure 8.15 gives the response of a complete *achromatopsia* patient. These patients lack cone activity completely, and only slow a- and b-waves are found. The responses under photopic conditions also show activity, of which the latency, however, is longer than for the usual cone-activity, but much shorter than that for rod activity measured in the dark-adapted state. The ultimate check for whether this activity is indeed mediated through the rods is made by comparing the responses to blue (Wratten no. 47B) and green (Wratten no. 58) flashes in the light-adapted condition. It can be concluded from the spectral sensitivity curves of the rod and the cone systems, that responses mediated by the rod system should be equal for these stimuli, whereas the cone system should produce quite different responses.

Figure 8.16 does indeed show that the blue and the green flashes yield quite different responses in the normal subject, but similar responses in amplitude and latency in the achromatic subject. The photopic response in Fig. 8.15 implies that in the light-adapted state, with the standard background illumination, a rod retina can produce

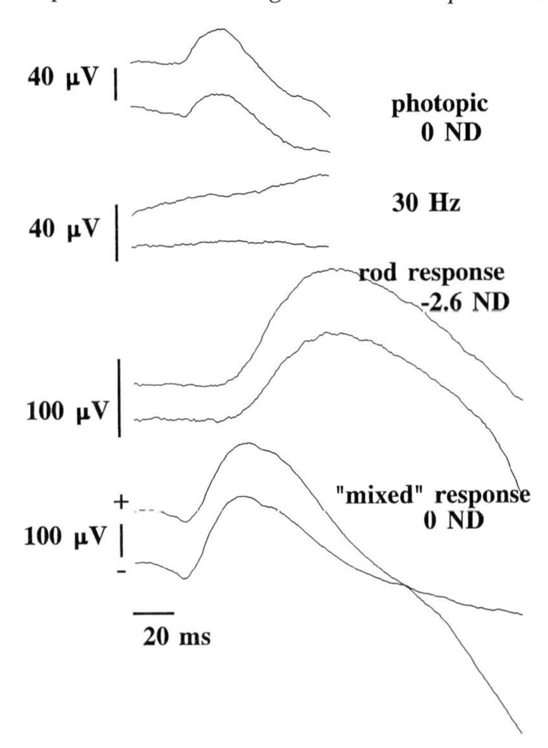

Fig. 8.15 ERG of a rod-achromat. The mixed response, measured at standard intensity in the dark, only contains the later a- and b- wave, the 30-Hz response is absent and the photopic response, measured at standard intensity with a light background, contains a slow wave, later than the normally obtained cone-mediated response but faster than the dark-adapted rod response. Note that although the background is intended to silence the rods, there still is a response.

activity of up to around 50 mV, which is quite unexpected since the background is meant to suppress the rod-mediated activity completely. It may be concluded that there is active suppression of rod activity by the cones, which is lacking in the achromat, or alternatively that the photopic normal ERG should be interpreted as a combination of cone- and rod-mediated activity. This seems, however, to be unlikely since the blue-flash ERG for the normal subject lacks the broad scotopic b-wave of the rod-achromat. Here the 30-Hz flicker response included in the standard becomes important: rods are known to follow activity up to about 10 Hz only. The responses to 30-Hz flicker are therefore completely determined by the cones. The isolation of cone and rod activity might be even better when different coloured stimuli and backgrounds are used. Many authors use, for example, a dim blue flash, sometimes on a red background (for which the rods are insensitive), to isolate the rods. Analogously, a red stimulus flash may be used on a blue background to isolate cone activity. However, it is rather difficult to obtain a sufficiently high-intensity blue background to suppress the rods completely (the blue background is usually obtained by a colour filter, which is nothing more then removing a considerable part of the intensity from the white background).

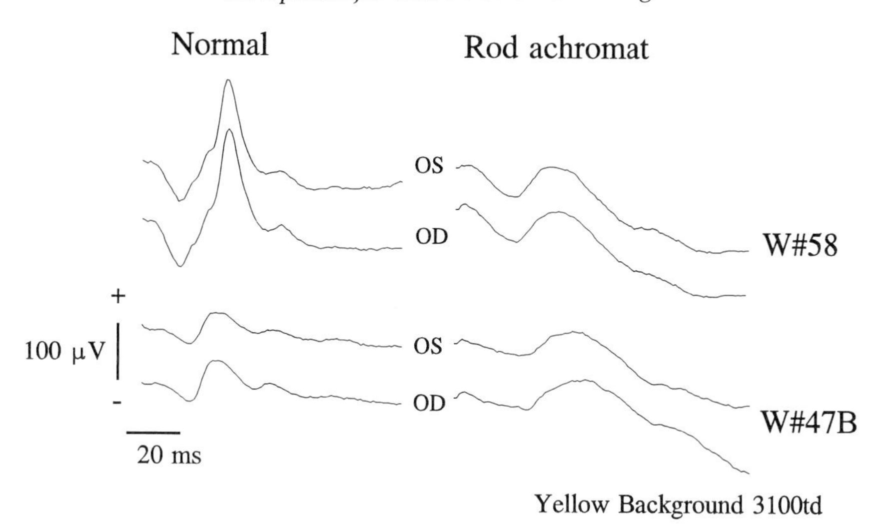

Fig. 8.16 To test whether the photopic responses obtained in the rod-achromat really should be ascribed to the rods, flash responses to blue (Wratten no. 47B) and green (Wratten no. 58) flashes (matched for the rods) have been measured in photopic condition. As can be seen from the normal responses shown at the left, these stimuli are quite different for the normal cone system. However, in the achromat, the obtained responses in this condition are fairly equal, compatible with the rod origin of these responses. It can be concluded that also with fairly high backgrounds (like the specified background intensity of ISCEV) rod contributions to the response still cannot be excluded, or alternatively, an active cone–rod interaction mechanism suppresses the rod activity in normals whereas this mechanism is absent in rod achromats. A yellow background was chosen to distinguish the rod contribution from possible S-cone contribution (see Fig. 8.22)

The behaviour of the responses in achromatopsia is quite different from that in *retinitis pigmentosa* (RP), of the rod (cone) type (Fig. 8.17). Here the responses are severely reduced (to some 25–50% of the normal mean) with, however, a preserved shape of the response, which is caused by the fact that this disease affects the very first elements in the retina: the receptors. If the amplitude reduction is less severe for the scotopic than for the photopic stimulation one speaks of RP of the cone (rod) type.

The responses in *congenital stationary night blindness* (CSNB) (Fig. 8.18) show a relatively preserved a-wave in the mixed response, measured in the dark-adapted state, with a strongly reduced b-wave. This disorder is generally considered as affecting not primarily the receptor layer itself but subsequent stages of retinal processing, since the receptor component (a-wave) is preserved in the response, whereas the Müller-cell component is reduced considerably. This type of ERG pattern is often termed a 'negative ERG', for which the ratio of the b-wave amplitude to the a-wave amplitude (b/a ratio) provides a sensitive index to quantify the effect. These examples show that on the basis of the ERG wave shape, conclusions can be reached about where and how strongly a disease has affected the retina.

This is also the case for the ERG of an *obligate carrier of CSNB* (Fig. 8.19). Here all the responses look normal except for the oscillatory potentials. This also illustrates that the OPs are poorly correlated with visual function since carriers show no visual complaints. With the OPs originating 'higher up' in the retina, they can be

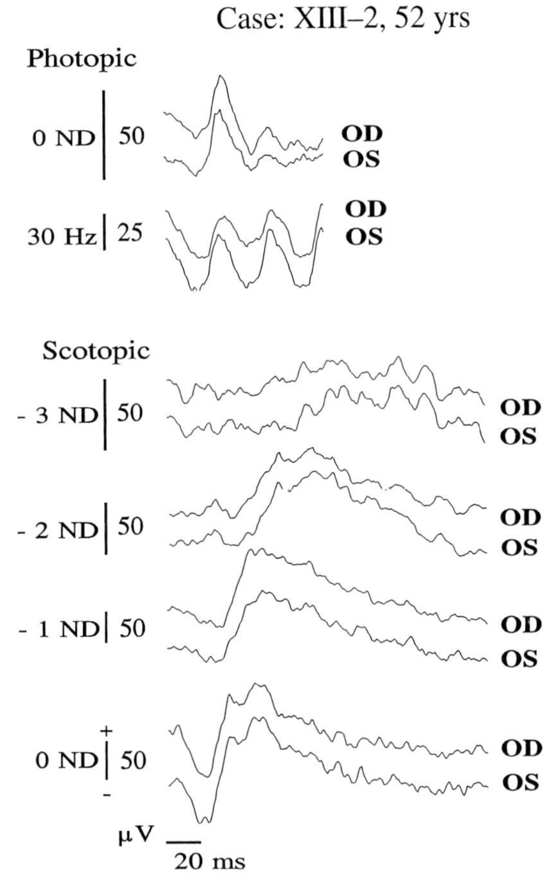

Fig. 8.17 ERG responses of a patient with retinitis pigmentosa of the rod (cone) type. Here the pathology is influencing the retina at the receptors itself, i.e. at the beginning of the process. Therefore the responses preserve their shape, but are reduced in amplitude as a whole. Also plotted are responses obtained at intensities in between the intensities of the 'rod' and 'mixed response of the standard.

used for detecting diseases that affect the retina from the corpus vitreous side, e.g. central retinal vein occlusion, diabetic retinopathy, glaucoma, etc. However, a standard analysis of the OPs is not available yet, and, since the high-pass filtering can be done in so many different ways, there is an imperative need for laboratories to establish their own standards for the OPs.

Local current source density analysis has shown that oscillatory potentials originate in the inner plexiform layer, most likely from amacrine cells and inner plexiform cells. There seems to be evidence that the origin of the first two peaks in the OPs differs from that of the last two peaks, as shown for example by neuropharmacological studies. So the first two peaks are depressed by γ-aminobutyric acid (GABA) and dopamine antagonists and the last two by strychnine and ethanol.

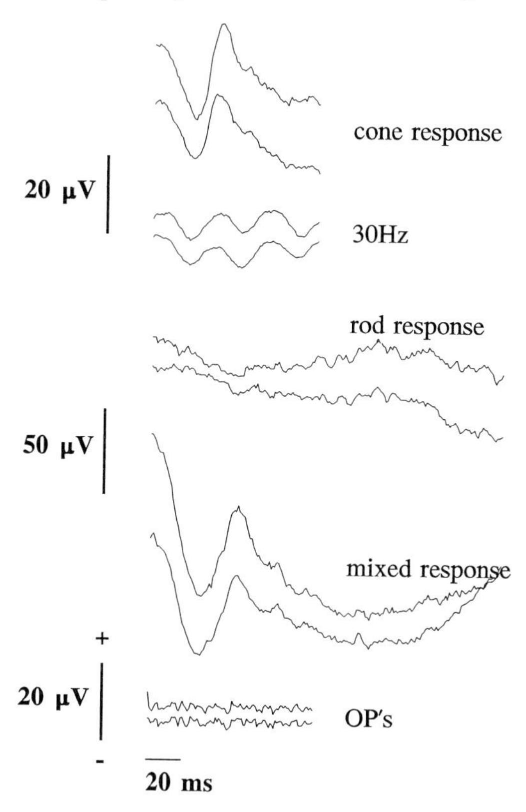

Fig. 8.18 ERG response of a patient with complete congenital stationary night blindness. The rod-mediated responses are absent most likely because of a defect in the connection between the rods and the bipolar cells. The mixed responses show a negative waveshape, containing a normal a-wave but lacking a b-wave. The cone responses in these subjects are only mildly affected.

8.4.3 Absorption within the eye

For the important category of patients with media opacities such as cataract, vitreous haemorrhage, etc., the interpretation of the ERG is hampered by the fact that part of the stimulus light is absorbed in the media, so that one cannot be sure how much of the flash reaches the retina. An example of overcoming this problem is given for a patient with vitreous clouding induced by the activity of toxoplasmotic chorioretinitis. Vitreous clouding prevents ophthalmoscopic evaluation of the retina. Yet the ophthalmological attendant needs information about how large the resulting lesion from this attack will be, and even more so, where it will be located. If in such a patient the amplitude of the ERG is severely reduced, it is difficult to decide whether this is caused by the absorption of the light in the media and/or by malfunctioning of part of the retina. In patients with old toxoplasmotic lesions, where the disease process is inactive, and the vitreous clear, we never observed an increase in ERG latency, but often found amplitude reductions. To interpret the ERG reductions we therefore adopted the strategy illustrated in Fig. 8.20. Note that in such a

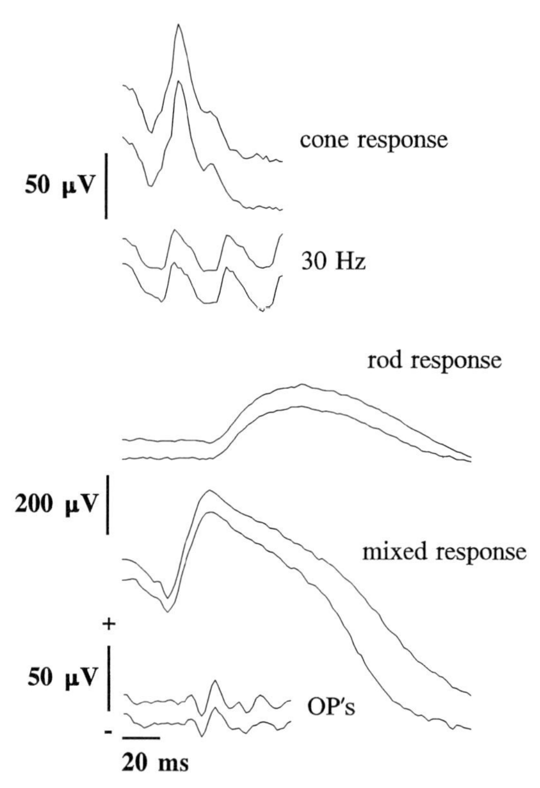

Fig. 8.19 ERG responses of an obligate carrier of congenital stationary night blindness. All responses are normal except for the oscillatory potentials.

condition the ERG standard does not provide sufficient information to be able to interpret the results.

Figure 8.20 shows a series of responses to different intensities of flashes in the dark-adapted state. It can be seen that the responses of the affected eye are systematically smaller, and prolonged in latency (compare with Fig. 8.11(A)). The latencies of the b-wave as a function of log intensity are given in Fig. 8.20(B) (lower panel), and it can be seen that the latencies of the affected eye, are systematically longer and follow a straight line over a considerable intensity range. If by linear regression these straight lines are replaced by the best-fitting regression lines, equal slopes are found for both eyes, but the line of the affected eye is shifted to the right along the intensity axis. From this shift, the absorption in the eye media can be estimated, and more importantly the amplitude can be corrected (Fig. 8.20(C) (upper panel)). Since after this correction, a considerable difference in ERG amplitudes between the two eyes remained, it had to be concluded that in this case considerable retinal damage had taken place. Furthermore, the ratio between the amplitude of the affected eye and that of the control eye was lowest for the lower intensities, where the response is determined by the rods. At the intensities where the cones begin to contribute to the response this ratio increases, as is further exemplified by the light-adapted responses

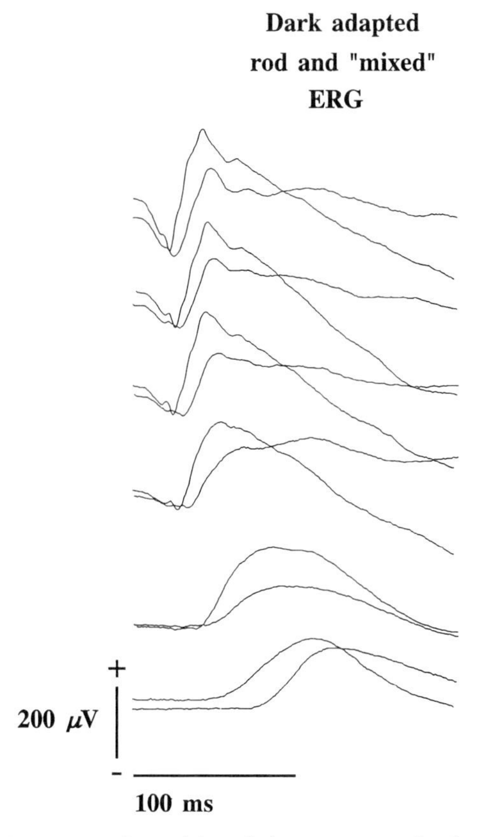

Fig. 8.20A Responses for various intensities of the two eyes of a toxoplasmosis patient with active chorioretinitis in the left eye. Responses shown are recorded with stimulus intensities – 3.25, – 2.25, – 1.25, – 1.0, – 0.5, and + 0.25 ND of the standard ISCEV intensity.

given in Fig. 8.20(D). Thus it could be concluded that the lesion affected mainly the rods, and probably would be restricted to the periphery. This was confirmed in this patient when the vitreous clouding had cleared away.

This case was chosen to illustrate a technique that can overcome the difficulties with the interpretation of the ERG when part of the stimulus light is absorbed. More recently, we have applied electrical stimulation to the eye, in order to probe the functioning of the optic nerve, thus bypassing absorption of the stimulus light by the media (Fig. 8.21). Obviously there are many diseases of the eye where absorption hampers vision considerably, and in those patients, where the ophthalmologist cannot observe the retina, information about retinal functioning is important. Note that in such cases the standard is not sufficient.

8.4.4 Separation of receptor systems

In disorders where only one of the *cone* populations is affected, tailored stimuli have to be applied in order to estimate the relative contributions of the different

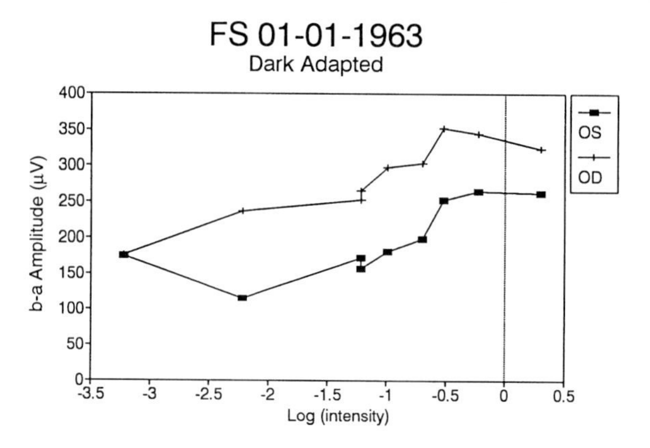

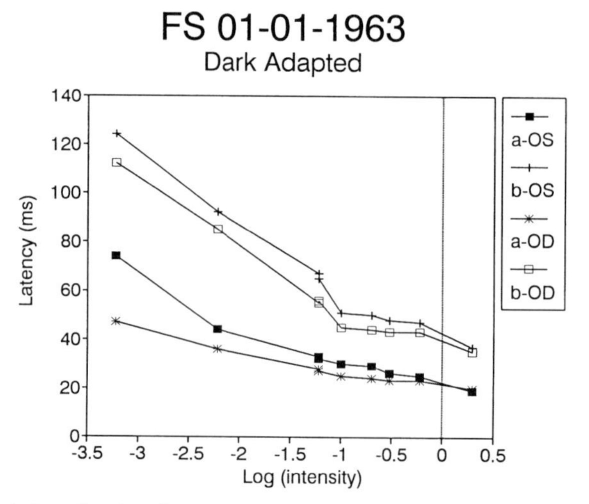

Fig. 8.20B (Upper) Amplitudes (b-wave to a-wave) of the responses of the left eye (black squares) and right eye (crosses) as a function of the (log) intensity of the stimulus flash. Note that there is a consistent difference in amplitude between the responses of the affected left eye, compared to those of the normal eye. (Lower) Latencies of the a-wave (OS, black squares; OD, asterisks) and the b-wave (OS, crosses; OD, open squares) of the ERG as a function of the log intensity of the stimulus flash. The latency of the b-wave depends strongly on the stimulus intensity for the four lowermost intensities. For higher intensities the latencies of the b-waves become more equal, probably by an intruding cone-mediated b-wave. The data for the left eye can be obtained by shifting those of the right eye 0.25 ND to higher intensities. This indicates that by absorption in the media, the actual intensity reaching the retina is reduced by about 0.25 ND.

cone systems to the ERG. Separation of the long wavelength (red, L cones) and medium wavelength (green, M cones) sensitive cones is very difficult with Ganzfeld flash stimulation, since their spectral sensitivities overlap considerably. For this a silent substitution technique seems indicated, in which two stimuli, that are of

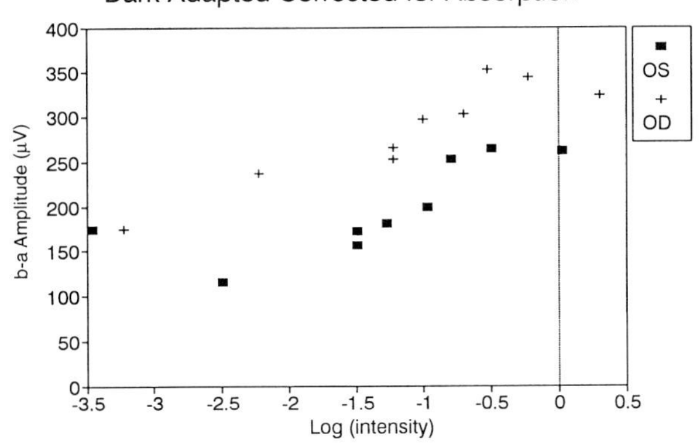

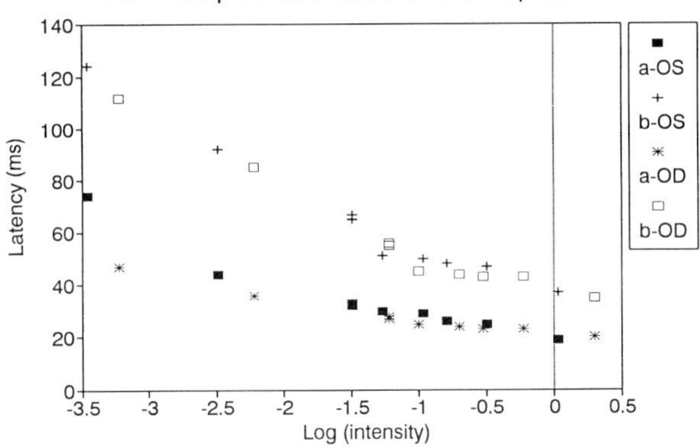

Fig. 8.20C Amplitude and latency data after a – 0.25-ND shift of the left eye data. The latency data can be seen to coincide fairly well. However, the amplitude data do not coincide, and especially so for the lower intensities, suggesting pathology for the rod system, i.e. in the peripheral field. Note that for this disease it is expected that part of the retina functions normally, and part is not functioning at all. If this would have been the case then the difference in amplitude should have grown with increasing intensity.

equal strength for one cone system and thus differ for the other, are exchanged continually (Estévez and Spekreijse, 1982). If the separation concerns the isolation of the short wavelength sensitive cones (blue, S cones), as in blue-cone monochromacy or the S-cone hypersensitivity syndrome, their contribution to the ERG may be revealed by very intense, blue-flash stimulation on a very strong white (or yellow) background. The response obtained is a small positive deflection riding on the trailing edge of the b-wave of the L and M system (Gouras, *et al.* 1989, Fig. 8.22).

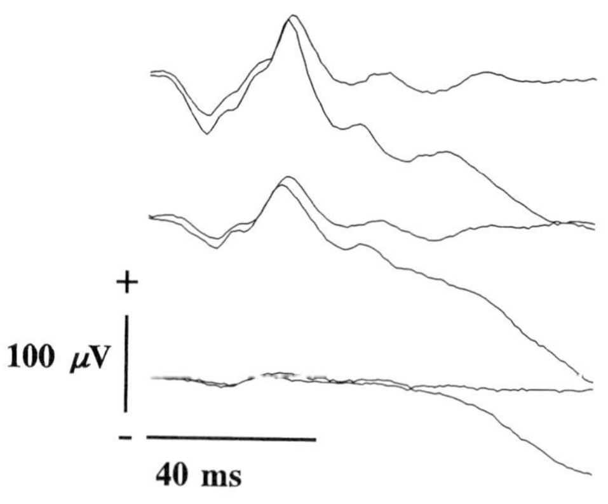

Fig. 8.20D Light-adapted cone-mediated responses, which can be seen to be fairly equal for the two eyes and especially so for the amplitudes of the responses (+ 0.5, 0, and – 0.5 ND).

The lower two traces of this figure show a method for isolating the S-cone mediated response from the responses mediated through the L and M cones. Since both the L system and M system seem to react with similarly shaped flash ERGs, one may

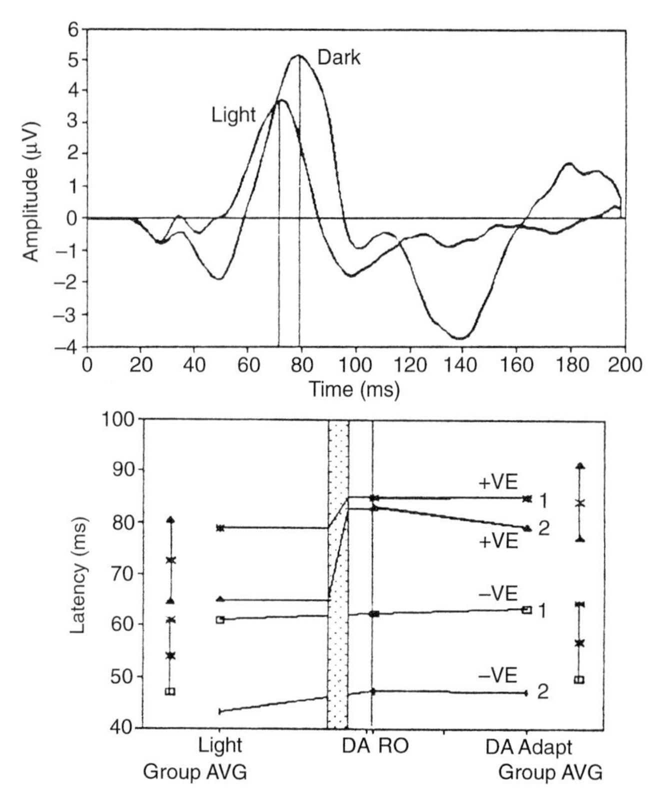

Fig. 8.21 For caption see p. 215.

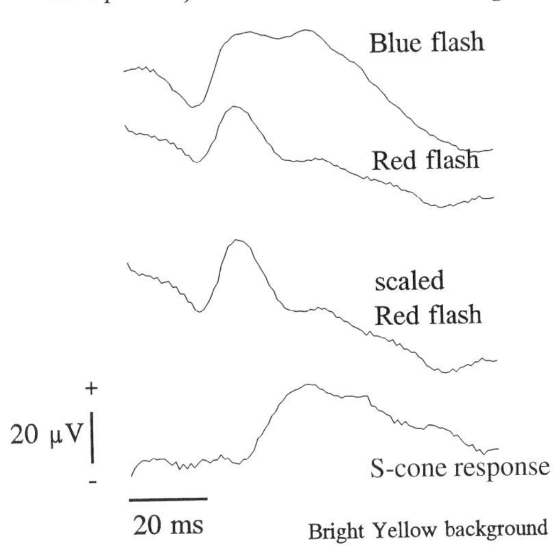

Fig. 8.22 ERG responses to blue and red flashes against a bright yellow background. The blue-flash response is seen to show a positive peak on the trailing edge of the b-wave. This positive peak at about 40 ms is ascribed (Gouras *et al.* 1989) to the short-wavelength sensitive cones (S cones). It can be concluded from the fundamental cone sensitivity curves that the blue flash stimulates the three cone types (L(ong), M(edium), and S(hort) cones). From the same curves it can be concluded that the red flash stimulates the L and M cones only. The response to the red flash does not show the positivity at the trailing edge of the b-wave. The blue and red flashes were chosen such that they were about equal for the L and M cones. This is, however, never exact in terms of the individual subject. So to calculate the S-cone contribution, it was first calculated by linear regression of the first 30 ms of the response, how much L and M cone contribution was present in the blue-flash response. The accordingly scaled red-flash response is the third plotted response and it was subtracted from the blue-flash response yielding the isolated, presumed S-cone contribution (bottom curve).

consider them as a single L and M system. Using a linear regression method from the first 30 ms of the response, we calculated how much of the red-flash response (purely L and M) is present in the blue-flash response, which obviously contains contributions from the three cone systems. This calculated contribution can be subtracted from the original blue-flash response, supposedly yielding a purely S-cone mediated response (lower trace). For the purpose of isolating the S-cone response, a silent substitution method has also been proposed and used by Sawusch *et al.* 1987.

Fig. 8.21 (Top) Responses recorded from the back of the head upon electrical stimulation of the eye with 6-ms pulses of 0.4 mA with alternating polarity to reduce the stimulus artefact. The pulses are delivered by a battery-powered current source through a DTL-electrode (Fig. 8.3) referenced to stick-on electrodes lateral to the outer canthus. Note that the latency of the main positivity (+ve) in these responses increases with 5–10 ms when room light is switched off during testing. This latency increase does not occur for the preceding negative component (–ve) and is not influenced by prolonged dark adaptation (bottom). Since room illumination varies and the density of opacities is generally not known, it is advised that the current pulses are applied in the dark when testing optic pathway integrity. (AVG: group mean; DA RO: darkened room).

8.4.5 Topography of the contributions to the ERG by different parts of the visual field

Ganzfeld stimulation is preferred for the ERG, since local stimulation is hampered severely by stray light within the eye. In 1935 Fry and Bartley were already aware of this problem, but the most elegant proof of this fact was given by Asher in 1951 who demonstrated that stimulation of the optic nerve head produced about the same ERG as stimulation of a similar region of the retina at the other side of the fovea. The stray light produces strong modulation all over the retina, especially in the dark. This is less severe when pattern-reversal (many checks!) stimulation is used, since then the net amount of light in the stimulus remains constant over time. The pattern ERG is, however, small in amplitude since the linear response components will be cancelled by the equal amounts of area that are stimulated in counter phase. In our lab., pattern ERGs are not recorded routinely, since generally there seems to be no need for local evaluation of the retina if the retina can be observed by fundoscopy. If this is not possible, a pattern stimulus can only be imaged on to the retina with a severe reduction of the modulation depth: therefore interpretation of the ERG may not be straightforward. On the other hand, from a research point of view the pattern ERG has received a lot of interest, because the non-linear components comprising this response have been assumed to have their origin in the inner plexiform layer. So the pattern ERG could be complementary to the flash ERG which originates in the outer retina. No consensus has been reached, as may be illustrated by the fact that only guidelines, instead of a standard, for the pattern ERG have been proposed by ISCEV.

A promising technique that has now become available is the technique of local noise-modulated light stimulation published by Sutter and Tran (1992; M-ERG). In this technique many local patches of the stimulus field are modulated with uncorrelated noise. By kernel analysis the signal measured from the cornea can be analysed in terms of the locally produced responses. With this method both the linear and the non-linear response can be extracted, which is not the case with pattern-reversal stimulation. Furthermore, it provides really localized information, whereas the pattern-reversal response is still the algebraic sum of responses at many positions. Since the internal stray light between the stimulus patches plays as important a role as in the pattern ERG, the application of the technique in cloudy media seems impossible. Because of this, it is imperative for interpreting these responses to quantify the amount of stray light produced within the tested eye. A straightforward measurement of the amount of stray light can be obtained using the silent substitution method developed in our lab. (van den Berg and IJspeert, 1991).

8.5 Visually evoked potentials

8.5.1 Averaging

The cortical responses to visual stimulation are of the order of μV in amplitude, and require very high amplification (10^5–10^6 times) in order to be made recordable.

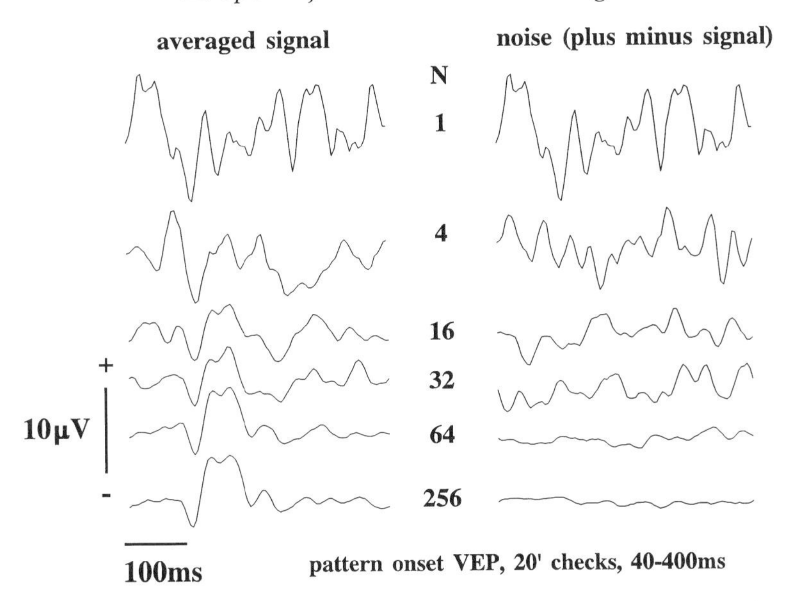

Fig. 8.23 Coherent averaging. Signal segments of 440 ms immediately after checkerboard appearance (pattern on for 40 ms and off for 400 ms) are added and divided by the number of segments, in order to enhance the response against the background noise: those events in the segments that are directly related to the stimulus will appear at the same time in the segment. The noise is different in every segment. An estimation of the remnant noise in the averaged signal is represented at the right, where the segments are added and subtracted alternately ('plus minus signal'). In this way the events that are coherent with the stimulus will be cancelled.

Amplification, however, is not enough, since the electrodes pick up all brain activity, and not only the visually evoked activity. This 'background EEG' can be of the order of 30–50 μV and coherent averaging is required to extract the visual signal from the ongoing 'noise'. In this technique the EEG segments following stimulus presentation are summed. An example is given in Fig. 8.23.

The quality of the recording (left column) improves considerably with the number (N) of averages, since the response is locked ('coherent with') to the stimulus and the ongoing EEG is not. Therefore the response in every signal segment contributes to the summed value, whereas the unlocked noise adds as many positive as negative values to the summed potential, and tends to cancel. An easy and practical way of estimating the remaining noise level in the recording after averaging (right column) can be obtained with the 'addition–subtraction average', in which all even signal segments are subtracted from all odd ones so that the responses are eliminated, whereas the noise is the same as in the averaging procedure. Comparison of the left and right hand traces tells immediately which signal-to-noise ratio has been achieved. By doing this on-line the duration of the recording can be kept at a minimum. Since, as a rule of thumb, the signal-to-noise ratio improves with $N^{0.5}$ it is not very efficient to continue the averaging process for a long time; the more so since the state of the subject might also change by becoming bored or tired of the stimulus.

8.5.2 Special derivations

Another method of enhancing the signal is to make a linear combination of the responses recorded simultaneously from different active sites at the skull. Jeffreys and Axford (1972) were the first to use a bipolar recording to enhance a particular peak (CI) in the pattern-onset response. An example of such a bipolar derivation, consisting of the recording of the potential difference at two 'active' sites in a patient with bitemporal hemianopsia is given in Fig. 8.24.

It can be seen that the responses in the monopolar leads from this patient contain much more background activity (mainly α) than the two bipolar derivations at the bottom. The reason is that in this recording the background activity recorded at the different sites is correlated, since it originates from deeper structures and spreads over a large area of the skull. By being correlated, this background activity is cancelled in the subtraction of the potentials recorded symmetrically across the midline, whereas this is not the case for the evoked activity which originates in one hemisphere only. This enhancement effect can be made even stronger if instead of the signals of two sites the signals of five sites are combined in the so-called Laplacian derivation (Fig. 8.25). This derivation, in which the responses from four surrounding electrodes are weighted subtracted from the central one: Laplacian = $r_2 - \frac{1}{4}(r_1 + r_3 + r_4 + r_5)$, was introduced by Hjorth (1975) and meant to serve the purpose of recording the second-order spatial derivative of the potential distribution across the skull, thus yielding a

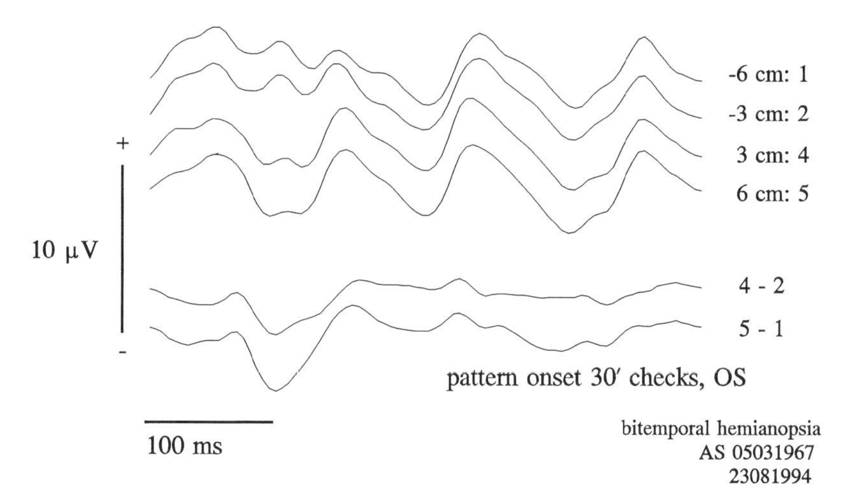

Fig. 8.24 The effect of a linear combination of two derivations in order to obtain a reduction of the coherent noise at different electrode sites: the responses to pattern onset obtained at 6 and 3 cm left and right of the midline (2 cm above the inion), contain a very strong α-component in this subject after averaging, such that any evoked response can hardly be discriminated. However, the bipolar derivation (the response at 3 cm right of the midline minus the response at 3 cm left of the midline) clearly contains a response with a peak at about 100-ms latency. This is also found for the electrodes at 6 cm from the midline (5–1). In this patient with the left (temporal) half-field of the left eye not functioning, only the left hemisphere is active when the full-field is stimulated. Therefore positive activity (CI pattern onset) at this left side produces a negative polarity by the bipolar derivation (see also Figs 8.37–8.39).

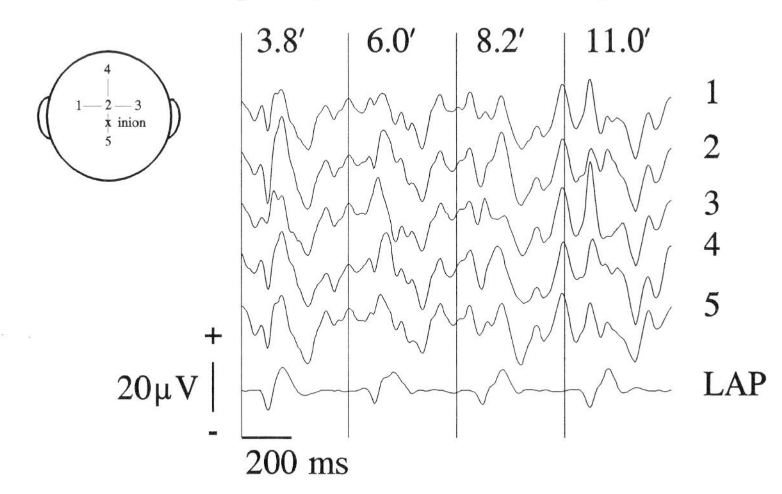

Fig. 8.25 The effect of a linear combination of five derivations in order to obtain a reduction of the coherent noise at different electrode sites: the responses obtained at the central electrode (2) in this calculation are multiplied by 4, and the responses of the surrounding electrodes (1, 3, 4, 5) are successively subtracted. It can be easily deduced that activity that is common in all channels will be cancelled by this procedure, and that activity from sources underneath the central electrode will be enhanced with respect to more peripheral sources. This calculation comprises a first-order approximation of the second spatial derivative of the potential field and is called a Laplacian derivation. The effect of noise reduction can be seen when the calculated Laplacian derivation is compared to the individual channels. The responses given are obtained by only a single presentation of the checkerboards. The reduction of the noise in this case is remarkable. Often less than 10 averages are required to obtain a well-defined response in a Laplacian derivation, which is useful in non-cooperative subjects.

better dipole source localization. Whether this holds true remains a matter of discussion and depends on the orientation of the underlying dipole source with respect to the recording sites. Yet the method can work very efficiently to improve the signal-to-noise ratio as is shown in Fig. 8.25. These recordings were made with an electrode montage which isolates the contributions of the striate cortex to the response, since these sources are assumed to lie directly underneath the central electrode. This is, so to speak, a reverse application of the original method (where it was intended to improve source localization). Once it is known where the different response components originate, one can try to isolate a particular component by appropriately choosing the electrode montage. In this way signal-to-noise ratio can be improved and the behaviour of an isolated response component can be studied better as a function of various stimulus parameters. In the example of Fig. 8.25 this is the dependence of the striate pattern-onset component on the check size.

8.5.3 Latency evaluation

By far the most frequently encountered application of VEP measurement is in the context of optic-pathway, conduction-velocity evaluation. In patients with neurological symptoms that point to multiple sclerosis, a demonstration of the involvement of the visual pathway can be helpful in the final diagnosis. For the evaluation of conduction velocity it is best to use a pattern-reversal stimulus, with fairly large checks (e.g. 60′ of visual angle) and high contrast. Pattern reversal is the stimulus of choice, because the distribution of the latency of the most prominent positive peak at around 100 ms (P100) is narrow: a standard deviation as low as 5 ms has been reported frequently. With such a narrow distribution, there is a fairly sharp distinction between normal and prolonged latency responses. This distinction becomes even sharper when the interocular differences are evaluated in monocular affections: that standard deviation is even less (3 ms). It should be noted that latency prolongation is not specific: many factors can influence latency, e.g. pupil size, refraction, age, and luminance, although by using high contrast and large checks, some of these factors may be overcome. Care should be taken to establish individual laboratory standards.

Figure 8.26 shows the responses of a probable MS patient. It is striking that the response for the affected eye (OD) is prolonged in latency, without a change of the wave shape. This is frequently observed in MS patients. If for instance there is compression of the optic pathways or an active optic neuritis, the VEP changes in other aspects too: the wave shape becomes broader and the response more sluggish, in a manner comparable to when in a normal subject a luminance of three ND units below the usual luminance is used.

It is not possible to evaluate conduction velocity without a recognizable response. Therefore a strategy directed to the aim of recording a recognizable response should be followed. For this purpose steady-state pattern reversals seem indicated when the standard transient pattern reversal does not yield a response. Steady-state reversals reach their maximum amplitude at a rate of about 8 rev s^{-1} (Fig. 8.27).

When stimulating at this high rate, the responses begin to overlap, and at still higher rates pure sine-wave responses are found (left, bottom trace). From the phase characteristics of the fundamental responses an apparent latency can be estimated, which can be scored against normal criteria as well. If the phase ϕ (in radians) increases in proportion to the frequency f (in reversals per second) the apparent latency:

$$\tau = \Delta\phi/(2\pi\Delta f)\ [\mathrm{s}].$$

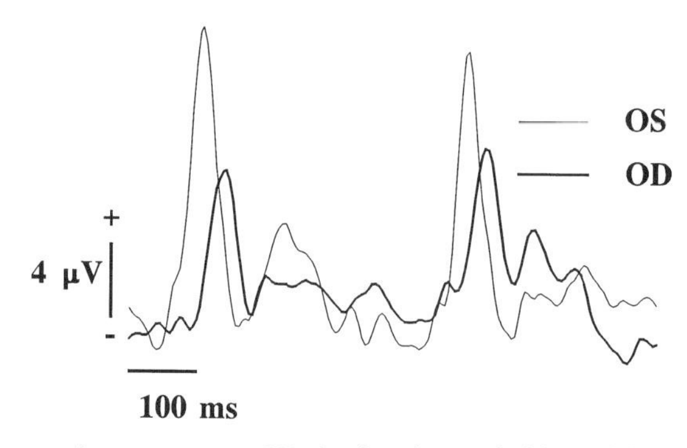

Fig. 8.26 Pattern-reversal responses to 60′ checks of a probable multiple sclerosis patient. The responses of the right (affected) eye show reduced amplitudes with respect to those of the left (normal) eye—with, however, the same wave shape—and are delayed as a whole. The distribution of the P100 latency in normals is very sharp (SD = 3 ms), such that the delay given must be considered pathological.

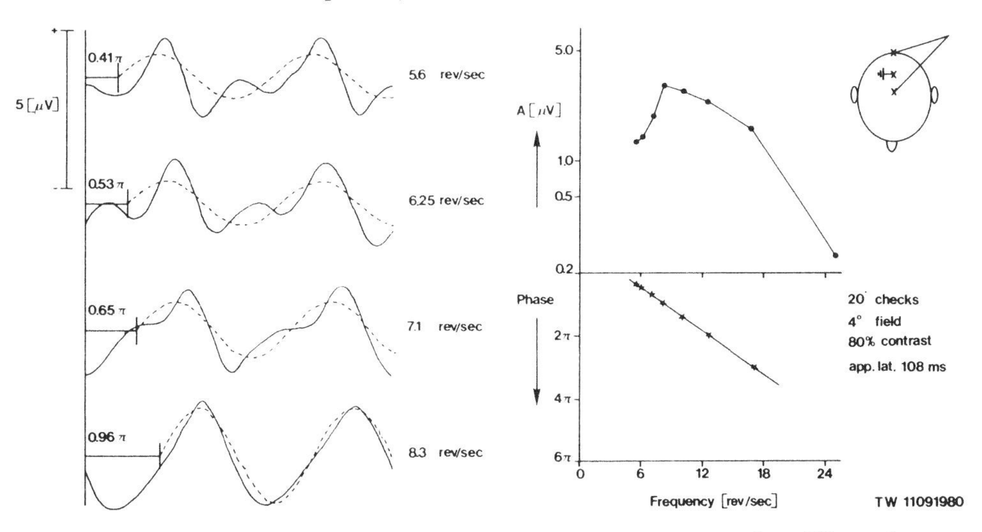

Fig. 8.27 Analysis of steady-state reversals. On the left side, responses to four different frequencies are shown. The dotted lines represent the first harmonic of the responses. The amplitudes and phases of these responses and of those to higher reversal rates are plotted on the right side to give an amplitude and phase characteristic. The phase is seen to vary proportionally to frequency, and from the slope of this phase characteristic an apparent latency of 108 ms can be obtained, which corresponds well with the transient P100.

In this way an estimate can be obtained for the latency when the transient recording does not provide a recognizable response, since apparent latency estimates the peak latency of the transient response. Another appropriate alternative is the use of a pattern-onset stimulus with brief pattern presentations (40 ms every 440 ms for instance). The response obtained is often greater than the pattern reversal one and can be discriminated from the noise more easily. However, its latency is more variable than that of the P100, and the distinction between normal and pathology is less sharp. This is less critical in monocular disorders, because then the actual shape of the response does not matter, but only its relative shift in time, as shown in the example of Fig. 8.28. This figure shows that both normal (for 20′ checks) and prolonged latency pattern-onset responses (for 55′ checks) can be recorded in a definite MS patient who did not produce recognizable reversal responses.

8.5.4 Amplitude evaluation

When visual acuity is to be measured a completely different approach has to be adopted. It is well described that the reversal response is weakly dependent on the image quality, or detail content of the pattern. This is also the reason for the success of the pattern-reversal EP in the clinic: a wide range of sharp or blurred patterns yields the same P100. This is not so for the pattern-onset response. The negative component at 100–140 ms in this response is reduced in amblyopia, and in control subjects when the borders of the checkerboard are outlined with high-contrast (bright or dark) lines. Furthermore, this component seems to be the only one present in the

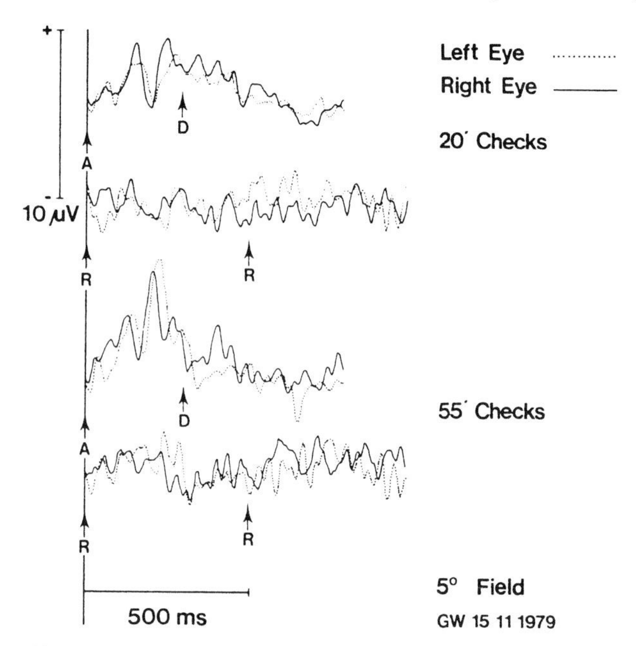

Fig. 8.28 Onset–offset (A–D) and reversal (R) responses obtained from a definite MS patient. In the reversal condition, no significant responses could be obtained for both checks sizes. The onset responses, however, are clearly distinguishable from the background activity. They could be either normal in latency or delayed as can be seen from the responses to 55′ checks.

response if only the edges of the checks are being used as stimulus. In other words, this negativity seems to depend strongly on pattern-detail content, and therefore is the most likely candidate to be used when visual acuity has to be estimated objectively. One way of estimating visual acuity is to use the extrapolated threshold for the negative component to determine the decreasing check size (Fig. 8.29).

For high visual acuities however, this is a difficult task, since one needs to provide very fine pattern sizes. With the TV screens used for this type of stimulation, resolution is limited by the line width. In practice this results in steps of one minute of visual angle, when the TV is far away (4.50 m) from the subject, steps that are too big to enable acuity to be estimated accurately. So, from this method, one can never expect more than a rough estimate of acuity in terms of being 'better than...'. When testing small children with the stimulus TV at a distance of 4.50 m, this almost certainly eliminates their attention, and makes evaluation impossible. In such a situation the latency of the main negativity can be helpful: this latency starts to increase in subjects with good visual acuity (> 1) when the check size is reduced below about 6′ (Fig. 8.30). If similar behaviour is observed when testing a patient, one may assume that visual acuity is at a good level, although it might not have been possible technically to measure a response for checks below 4′.

Another method of estimating visual acuity is to determine, at a number of electrode positions, at which check size no response whatsoever is found—however, sometimes this can be very time-consuming. For this purpose the use of a Laplacian

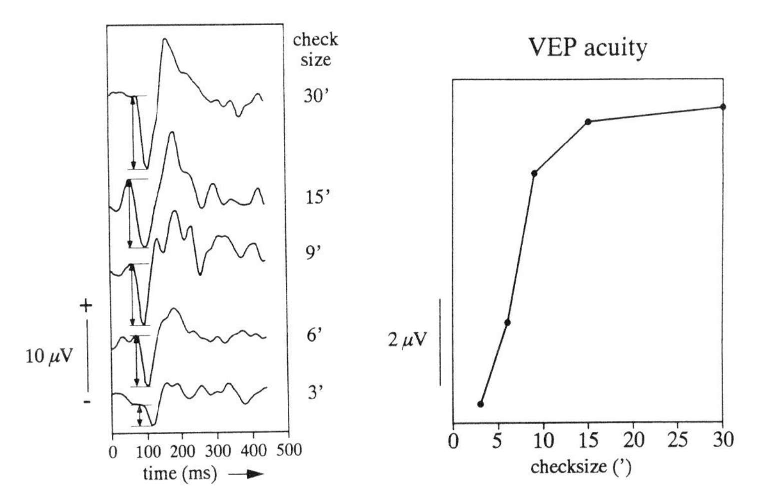

Fig. 8.29 Pattern-onset VEPs to different check sizes (left panel). To the right, the measured amplitudes as indicated by the arrows in the individual responses are plotted against check size. The threshold for onset responses can be seen to be approximately 2′, which is at the border of resolution for the TV screen. It can be readily concluded that resolution of the visual system up to the striate cortex in this subject is better than the reported 1/300.

derivation over the striate cortex may be helpful, since it improves the signal-to-noise ratio in the recordings and thus narrows the region in which the threshold can be pinpointed. This approach, by the way, requires special attention to the calibration of the stimulus. If the onset of the checkerboard is not fully isoluminant, acuity will be overestimated, since this method completely ignores the wave shape encountered; only the presence or absence of a response matters. This criterion also proves helpful in subjects with reduced visual acuity. The example of Fig. 8.31 shows that in those subjects the threshold may also be very sharply defined, suggesting that extrapolation, to zero amplitude check size, may indeed not always be the best solution for estimating the acuity. It should also be noted that the latency of the response near the threshold (10′ in this case) shows a prolonged latency with respect to the normal response for this check size (160 ms vs. 110 ms, see also Fig. 8.30).

An objective evaluation of visual acuity is often needed in subjects who do not cooperate easily, such as psychiatric patients, non-verbal children, patients with hysteric amblyopia, or malingering subjects. It is then important to test the behaviour of the responses in such a manner that the subject is not aware of being tested at threshold levels. This can be done by presenting the stimuli in a series, in which the larger and the smaller checks are presented sequentially and the averaging is carried out over signal segments covering the full series (Fig. 8.32).

When the bigger checks produce normal responses in such a series, one can be sure that the absence of responses to smaller checks is not caused by a lack of attention to or the avoidance of the stimulus by the subject. Testing for every check size separately will always cause ambiguity on interpretation of the data: whether absence of the response for the smaller checks is really indicating reduced acuity or just a lack of attention for the stimulus, either by weariness, boredom, or by deliberately not looking.

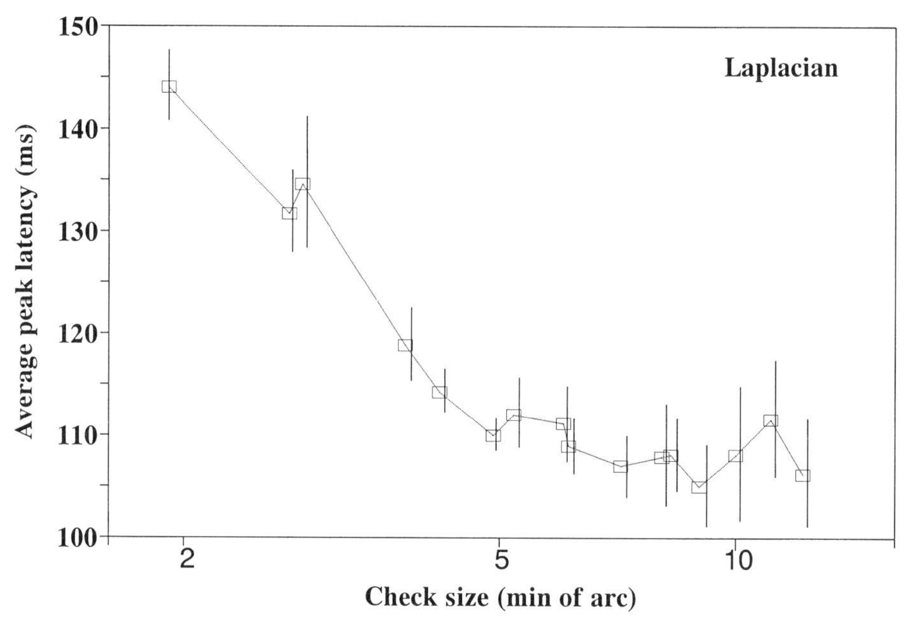

Fig. 8.30 Latency of the negative peak in the Laplacian recording (see Fig. 8.25) as a function of check size in normal subjects. This latency can be seen to be constant for check sizes above 5′, and for lower checks the latency increases with decreasing check size. If a response to bigger checks (like the 10′ response in Fig. 8.31) shows an increased latency, this check size can be concluded to be near the subject's threshold of checkerboard resolution. So when it is not possible to go to lower than 5′ checks because of the resolution of the TV screen (e.g. in small children, when the stimulus monitor cannot be very far away) then from the latency of the response to bigger checks one can conclude whether visual acuity is reduced or not.

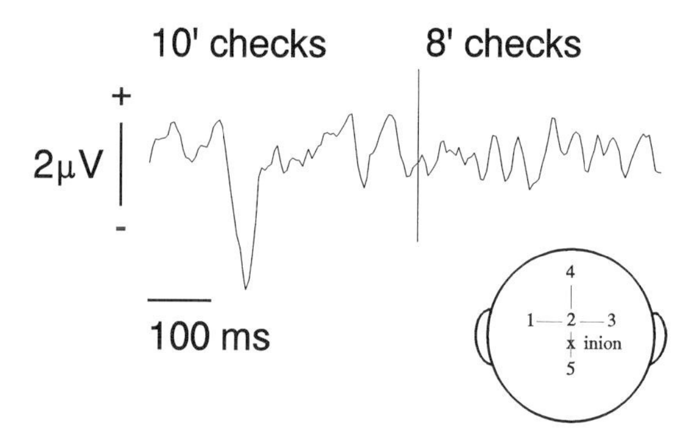

Fig. 8.31 Laplacian responses from a subject with suspicious (from his behaviour—he refused to cooperate in the eye test) reduced visual acuity. The absence of a response to the 8′ checks, with a clearly present response to 10′ checks obtained in a single run where 10′ and 8′ checks were presented alternately and the responses separated by the computer, demonstrated, however, that the subject had indeed reduced acuity. Note that the 10′ response, being so near to the threshold of resolution shows a prolonged latency. One should also note that the threshold between absence and presence of a response can be very sharply defined, even for strongly reduced visual acuities.

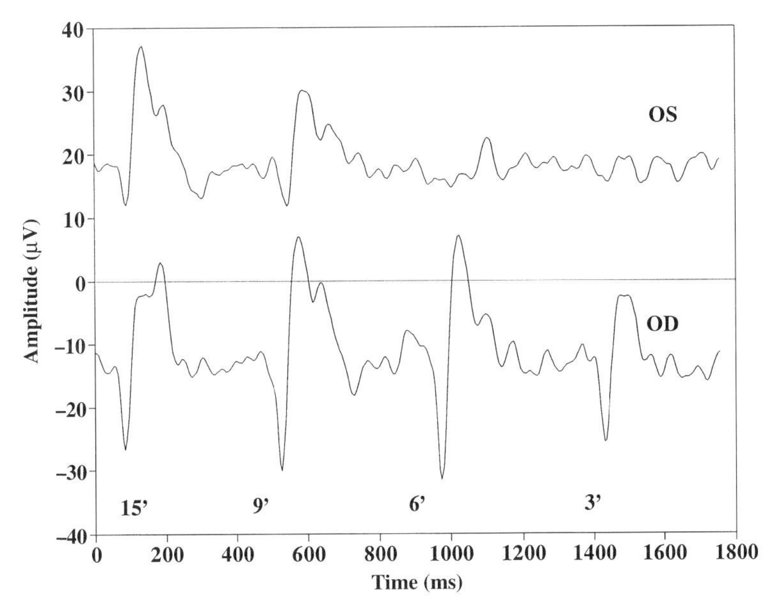

Fig. 8.32 The sequential presentation of different check sizes. The presence in such a run of normal responses guarantees that the subject was looking to the stimulus. Absence of some of the responses in such a run gives information about the visual resolution for checkerboard patterns. In non-verbal children this procedure is very important, because their attention for the stimuli may vary during the recording session, and then this sequential recording provides an internal control. It is also important to note that such a recording requires a four times longer recording time, for which the use of a Laplacian derivation produces an essential speeding up of the recording. For clarity, the responses are shifted arbitrarily along the vertical axis.

When the question at hand concerns deliberate stimulus avoidance, several studies have shown that subjects can actively suppress the response to checkerboards, without the examiner being able to ascertain whether or not such action is being performed even when sitting just next to the subject. This active suppression, for instance by purposely defocusing, is most effective for the smaller checks: exactly those for which the responses are required to prove a possible good visual acuity. This possibility has limited the application of VEP recording in malingering subjects, since interpretation of the absence of responses can be ambiguous. Many malingering subjects claim absence or greatly reduced vision in one eye. In such patients the testing has to be done so that the subject is unaware that the suspected eye is being tested. A completely unaware recording is performed when the spectrum of the ongoing activity is measured with the eyes either opened or closed (Fig. 8.33).

Of course the suppression of the 10-Hz peak (alpha activity) by opening the affected eye only provides evidence that perception is possible through that eye, but it does not indicate the quality of that perception. So once suppression of alpha activity is established, one should go on and try to record pattern responses. One possible way of doing so without the subject being aware that the 'bad' eye is being tested is by making use of a spectacle system in which the two glasses are independently electronically driven Kerr cells that switch from clear vision to oblique vision

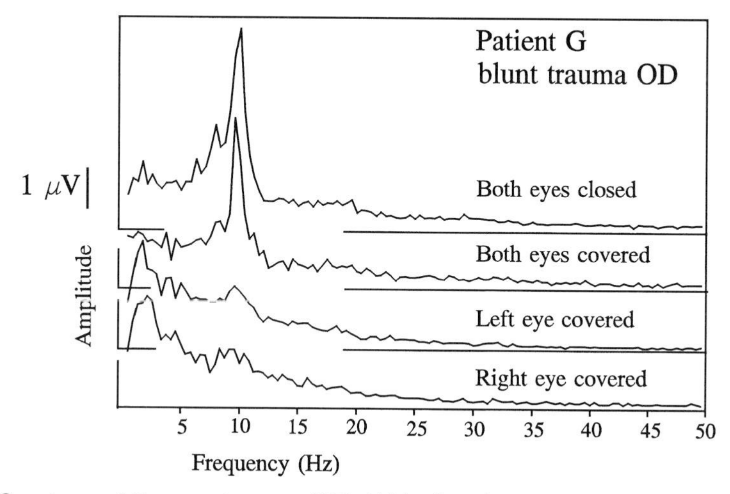

Fig. 8.33 Spectrum of the spontaneous EEG. With closed or covered eyes, the spontaneous ongoing EEG often contains very pronounced activity at around 10 Hz (α-activity), which in the frequency domain is represented by a peak at 10 Hz. Usually opening the eyes produces a strong reduction of the α-peak, as can be seen in the lower trace where only the unaffected eye of this subject was opened. This subject had sustained a blunt trauma to the right eye, and reported no perception at all through that eye, although there was no ophthalmological residue of the trauma. When only the right eye was opened, while keeping the left eye covered, the α-peak was reduced just as much, strongly suggesting that there was perception through that eye. Note that it does not indicate the quality of the perception, and that further pattern tests are indicated to obtain an estimate of acuity, once that the α-test has proven perception. This is a way to test the eye without the subject knowing that the eye is tested.

within milliseconds. When switching between the two eyes is performed randomly, by a computer, the subject is unaware which eye is being tested. Active suppression of the response from the 'affected eye' in this situation is impossible without also suppressing the response from the good eye. Such a frame even allows one to record responses while the subject is observing a static pattern, for example painted on the wall; this can be useful when the subject purposely fixates away from the monitor screen (Fig. 8.34).

Another way to inactivate suppressive action by a subject who claims bad vision in just one eye is to record the response from the good eye while inserting ND filters in front of that eye. It has been reported that the pattern-onset response to lower luminances in normal eyes becomes broader and has an increased peak latency; and, furthermore, that simultaneous stimulation of the other normal eye without ND filters suppresses this response completely. This suppression is so strong that by inserting ND filters in front of the normal eye of an amblyopic subject, the amblyopic eye can suppress the normal eye response thus becoming the dominating eye, although with the characteristic altered wave shape. An example of this procedure is given in Fig. 8.35. Once it has been established that the stimulus produces clear but different responses from the good eye with and without the ND filters, both eyes should be stimulated simultaneously, with the ND filters in front of the good eye.

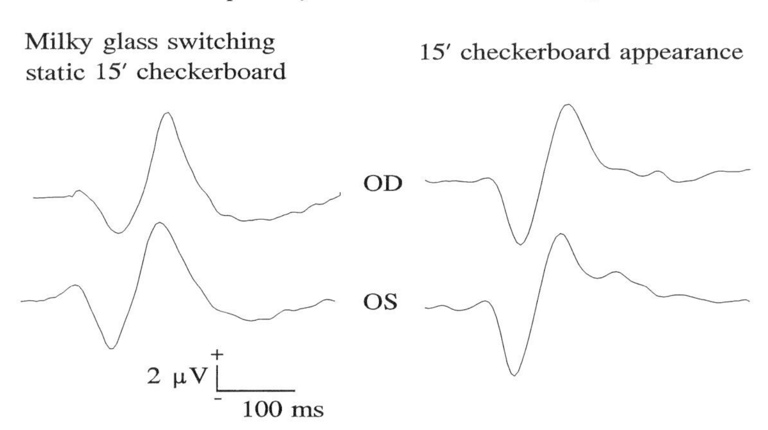

Fig. 8.34 Pattern-appearance responses, obtained by switching the glasses of a Kerr cell frame on and off,: the glasses can go from being milky to translucent within 3 ms by electronic means. By switching between the right and left eye randomly, the subject cannot be aware which eye is being tested. Here the responses of a normal subject obtained with the switching glasses, while he was looking at a static pattern on the TV screen (left responses) are compared with those obtained by making the pattern appear and disappear from the TV screen (right responses). In patients with dubious monocular problems, this technique can be used to force attention for the stimulus, while looking with the bad eye.

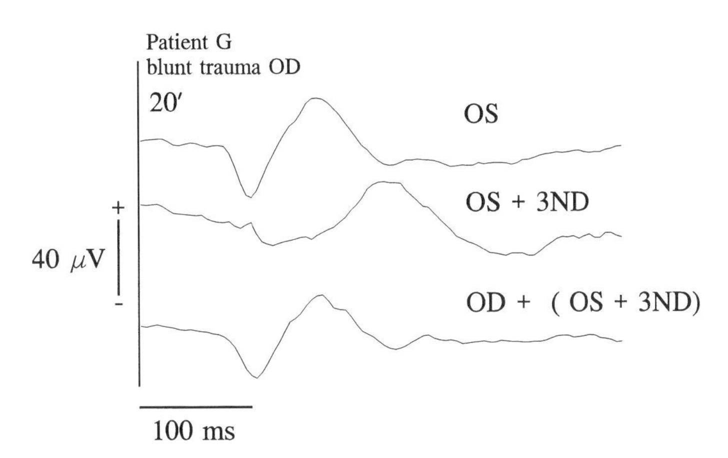

Fig. 8.35 Pattern-appearance responses of the subject in Fig. 8.33. Up to the spectrum recording, no reliable pattern responses were obtained, supposedly because the subject did not look at the pattern attentively. The subject was forced to look with the 'no vision' eye by making him look with the good eye covered with a 3-log ND filter, reducing the luminance of the pattern by a factor 1000. When the good eye was tested alone with the reduced luminance (centre response), a clear response was obtained, but it was delayed in latency and broadened in waveshape. When both eyes were tested with the good eye covered (3 ND), a response (lower response) comparable to the normal good eye response (upper response) was obtained. This could only be generated through the 'no vision' eye (one could end up with no response at all in this condition if the subject still ignores the stimulus, and with a delayed broadened response if the bad eye does not contribute and is really pathological).

Again, the subject will be unable to ignore the stimulus with the affected eye without altering the response from the good eye, and if both eyes yield similar responses in this test, as is the case in Fig. 8.35, then it can be concluded that the 'bad' eye has the same acuity as the good eye, especially so if the similarity holds for small checks. These techniques help to interpret the responses when the cooperation of a subject is not guaranteed. Their application is obvious in the case of a suspicion of malingering, but they can also be used in children, where it may be very difficult to attract their attention and gain their cooperation especially when the child is tested with the good eye covered. The amplitude spectra of the ongoing EEG shown in Fig. 8.36 provide such an example, here a response from the bad eye could not be recorded since the child refused to have the good eye covered— probably because the child could no longer see the cartoon movie used to direct its attention to the monitor screen. The spectra of this 2-year-old child fall into two categories: the ones with the OS closed are similar to the ones with both eyes covered and have a pronounced (α-) peak; the ones with both eyes open are identical to the one with the OD closed and do not show an a-peak. So it had to be concluded, both from the behaviour of the child and from the spectra, that indeed the left eye should be

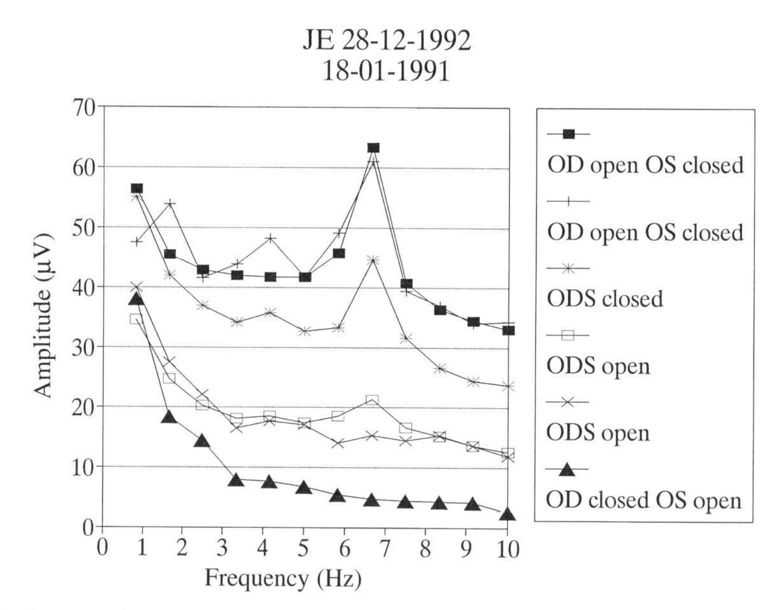

Fig. 8.36 Spectra of the spontaneous EEG of a 2-year-old child suspected of having reduced vision in the OD. Testing the right eye with a checkerboard pattern stimulation was impossible because the child refused to look at the screen with the left (good) eye covered and protested strongly. When both eyes were covered and no task was given, the child quietened down and the spectrum (asterisks) showed a pronounced peak at about 7 Hz. We then uncovered the right eye and measured the spectrum twice (black squares and crosses), again the 7-Hz peak was evident (the spectra were shifted arbitrarily along the amplitude axis for clarity). When both eyes were opened a strong suppression of the 7-Hz peak was observed (open squares and crosses). Subsequently, when the right eye was covered and the left eye opened (black triangles), the 7-Hz peak remained suppressed completely. The absence of suppression with the left eye covered confirmed the absence of perception in the right eye, as was suspected.

considered as performing poorly. Note that the frequency of the α-peak in children progresses from as low as 2 Hz in newborns to 6 Hz in this 2-year-old child, and further to around 10 Hz at later age.

8.5.5 Distribution of potentials across the skull

When the distribution across the skull of the pattern-onset responses to half-field stimulation is studied carefully (that is with multiple (e.g. 24–32) electrodes on the scalp), a clear-cut distinction can be made between those contributions that originate from the striate and those that originate from the extrastriate cortex.

Anatomically, it is known that the left-half visual field of both eyes normally projects to the right hemisphere, and, in contrast, that the right-half visual fields project to the left hemisphere. When these half-field contributions have to be studied it seems unwise to study the striate contributions to the response, since both half-fields are projected into neighbouring parts of the brain (around the calcarine fissure). The activity of those neighbouring parts are spread by the relatively low conductive skull over considerable areas of the scalp. Consequently, the potential distributions of striate activity to both half-field stimulations will show considerable overlap and cannot be separated easily, especially when only a limited number of electrodes is used as in clinical routine examinations. On the other hand, the contributions of the extrastriate cortex are more easily separated, since they originate from parts of the brain that are physically much further apart. Since in the EP to a pattern-onset stimulation with relatively coarse checks (> 50′) the extrastriate activity dominates the striate contribution; the hemispheric contributions to the stimulation of the left- and the right half-field ought to be studied with a limited number of electrodes, placed from left to right across the back of the head, when such a stimulus is used. An example is given in Figure 8.37 which clearly shows that in this subject the positive contributions project contralaterally at around 95 ms after stimulus onset. Note that the distribution for the late positivity with a peak latency of about 160 ms is much more diffuse. Sometimes monopolar recordings may, at first glance, even give an inverted impression. For example, the recordings in Fig. 8.38 seem to indicate that the left half-field projects at the left and the right half-field projects at the right side of the head.

To overcome this problem the use of so-called bipolar derivations, i.e. measuring the difference signals between the two hemispheres, can be very helpful for facilitating interpretation (Jeffreys and Axford, 1972). Since the positive activity (CI) originates only in one of the hemispheres, the difference derivation between the hemispheres shows characteristic behaviour. When for the responses of Figs 8.37 and 8.38 the difference (7 – 2) between the right (electrode 7) and the left (electrode 2) upon left half-field stimulation is calculated, both subjects yield a positive deflection at about 90 ms (Fig. 8.39). For the right half-field the difference derivation is calculated, and inverted by computer to facilitate comparison. Left minus right hemisphere (2 – 7) also yields equal positive activity at about 90 ms. In daily practice the bipolar lead is taken in one direction (e.g. right minus left hemisphere) and should therefore show a 'polarity reversal' when the left and the right half-fields are stimulated. This is another example (like the Laplacian derivation) of how the activity of different sources can be eliminated by astute manipulation of the responses at different positions of the

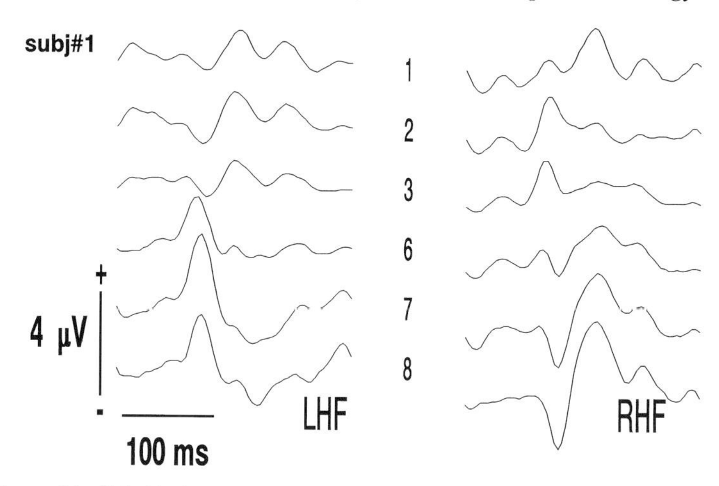

Fig. 8.37 Normal half-field checkerboard onset responses, obtained simultaneously by asynchronous stimulation of the half-fields. The electrodes were positioned in a horizontal row 5 cm above the inion, starting with electrodes 1, 2, and 3 at, respectively, 7.5, 4.5, and 1.5 cm left of the inion, and electrodes 6, 7, and 8 at 1.5, 4.5, and 7.5 cm to the right of the inion. Derivations were referenced against a frontal midline electrode. The positive activity at around 90 ms is clearly present in the right derivations (maximal for el 7) for the left half-field (LHF) stimulation, and in the left derivations (maximal for el 2) for the right half-field (RHF) stimulation, compatible with the topological distribution of the visual half-fields on the left and right hemispheres.

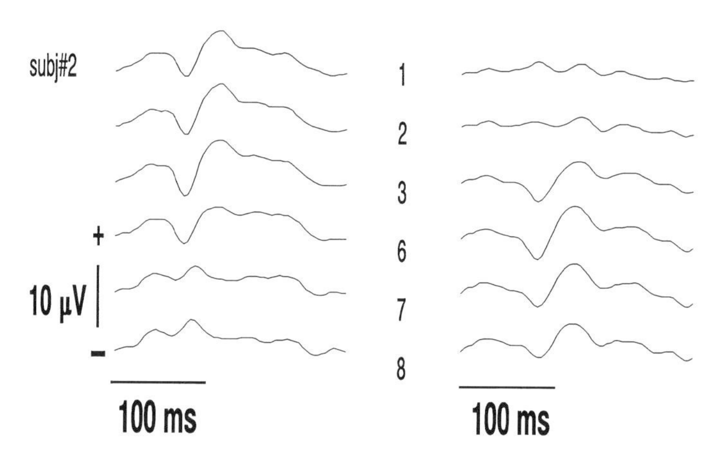

Fig. 8.38 As in Fig. 8.37 in another normal subject. The distribution of positivity is less clear because of interference of striate activity with the extrastriate response, which appears stronger in this subject.

scalp. In this example a strongly present striate response obstructs the interpretation of the extra striate contributions in the monopolar leads. However, since this striate response is almost equally present at both hemispheres (by volume conduction and shielding/spreading by the low conductive skull), it is efficiently eliminated in the difference derivation.

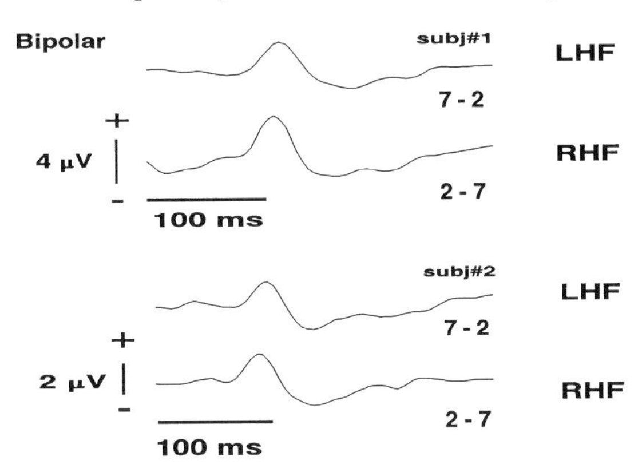

Fig. 8.39 Bipolar recordings for the same subjects in Figs 8.37 and 8.38. For the left half-field, right minus left hemisphere (7 – 2) is taken and for the right half-field, left minus right hemisphere (2 – 7) in order to produce equal polarity of the responses. The positivity at around 90 ms is as conspicuous in both subjects, where there was a big difference in the interpretation of the monopolar data.

This reasoning applies equally to the question of diagnosing tumours at various positions in the brain, and might serve as an indication for a MRI scan. Above that, it may provide information about the functioning of the visual pathway, obstructed by the presence of space-occupying structures. An example is given in Fig. 8.40. Since the full-field responses of this patient with a suprasellar process are completely determined by the half-fields that project to the contralateral hemisphere, it can be concluded that the suprasellar process compresses from the posterior side against the chiasm.

Another application of this method may be found in albinism, where the fibres of the visual pathway to the striate cortex decussate differently from normal. In albinos the full-field of the right eye projects predominantly to the left hemisphere, and the full-field of the left eye to the right. By stimulating the full-field in monocular conditions a distribution across the skull is encountered, as if only one half-field is stimulated. So for the left eye one finds the response mainly at the right side of the occiput, and vice versa. In addition, the bipolar lead provides an extra visualization of the effect by showing polarity reversal when the two eyes are studied separately (Fig. 8.41). Note this polarity reversal is an electrophysiological expression of albino misrouting that is not affected by hemispheric dominance.

Also, in many of the albino subjects, half-field stimulation enhances the visualization of the misprojection: the nasal visual half-fields for both eyes, which normally produce an ipsilateral response, often clearly show a contralateral distribution (Fig. 8.42).

However, this method may be difficult to interpret, especially with subjects like small children who do not fixate well on the centre of the screen. But in subjects who are suspected for misrouting, a clear-cut ipsilateral response proves the opposite: i.e. that there is no misrouting. In children below 4 years of age it is often difficult to draw their attention to the checkerboard stimulus. Therefore, in young infants and

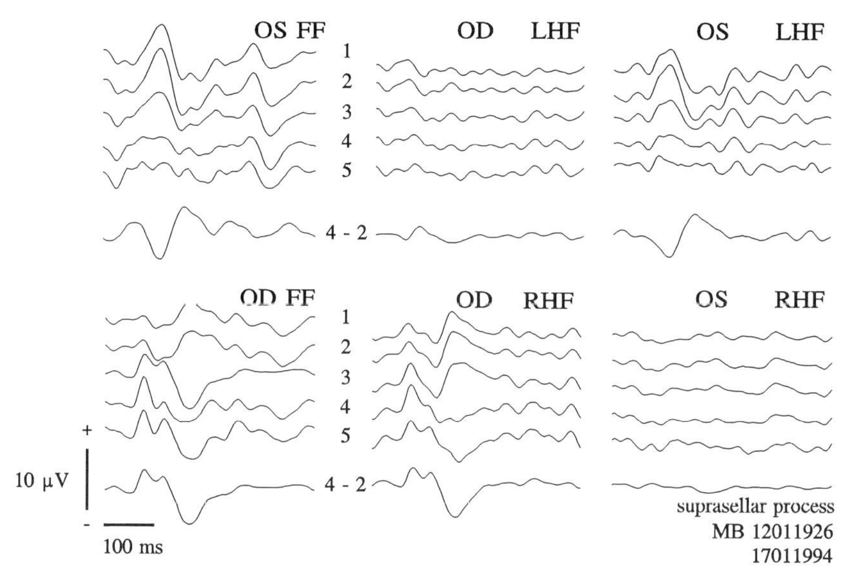

Fig. 8.40 Full-field (FF) and half-field (HF) responses in a patient with a suprasellar space-occupying process. The full-field responses can be seen to be completely determined by the half-fields that project to the contralateral hemisphere. The ipsilateral responses for both eyes are absent, suggesting that the suprasellar process compresses from the posterior side against the chiasm.

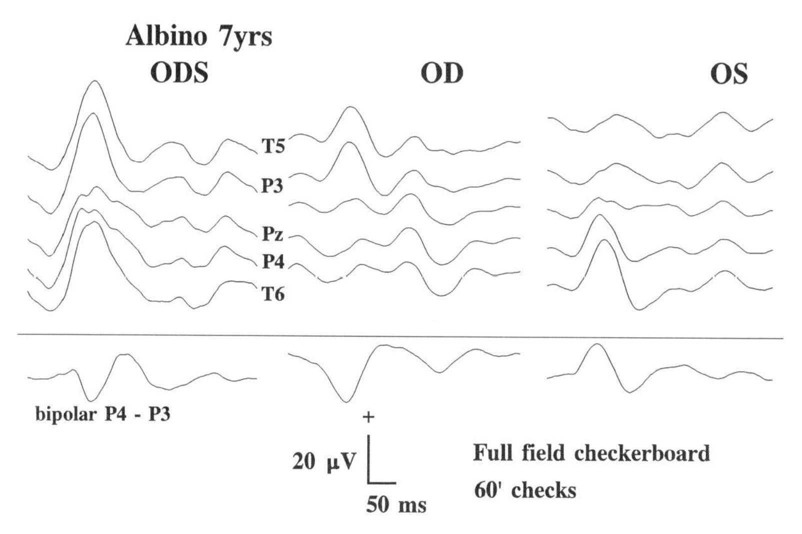

Fig. 8.41 Full-field pattern-onset responses of a 7-year-old normally pigmented boy with reduced visual acuity. From the distribution of the monopolar responses to monocular stimulation it can be readily concluded that projection for both eyes is restricted to the contralateral side, suggesting misrouting. The bipolar responses here produce opposite polarity peaks although full-field stimulation is given. This polarity reversal exemplifies the misrouting explicitly.

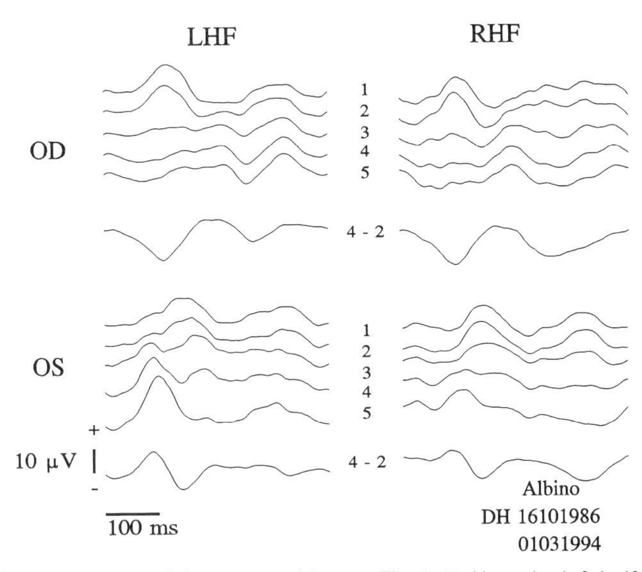

Fig. 8.42 Half-field responses of the same subject as Fig. 8.41. Here the left half-field (LHF) of the right eye can be seen to project to the left, whereas it should project to the right hemisphere. And for the left eye the right half-field (RHF) can be seen to project to the right, whereas it should project to the left hemisphere. Also the bipolar responses can be seen to be of equal polarity for both half-fields of each eye.

toddlers we always use flash stimulation in addition to pattern onset, because flashes seem to attract their attention and calm them. In albino children, the flash response often shows a contralateral distribution, and, just like the pattern-onset response, a polarity reversal in the bipolar derivations (Fig. 8.43). It is obvious that for flash stimulation great care should be taken that no light can enter the patched eye, otherwise the misrouting signature will be lost.

The finding of misrouting in children often provides the reason for reduced vision, evident from either nystagmus or no 'eye contact'. Here it is not the visual acuity measurement that provides the answer to the question at hand, but just the response distribution across the scalp. It gives evidence for a fairly stable deviation of their visual system, with described heritability, and often acceptable outcome of visual function especially when attention is paid to teaching the subjects how to deal with their handicap.

8.5.6 From EPs through MAPS to source localization

Figure 8.28 shows prolonged latency pattern-onset responses recorded in a patient with definite MS. These responses consist of a PNP (positive–negative–positive) complex which is shifted in time without change in waveform: typical for latency increases due to prolonged transportation time along the optic tract. The three peaks in this response are, however, not generated in the same cortical region, as can be seen in Fig. 8.44. This figure shows brain maps, recorded with 32 electrodes, taken at 10-ms intervals after a checkerboard pattern onset in the right visual field. Evoked

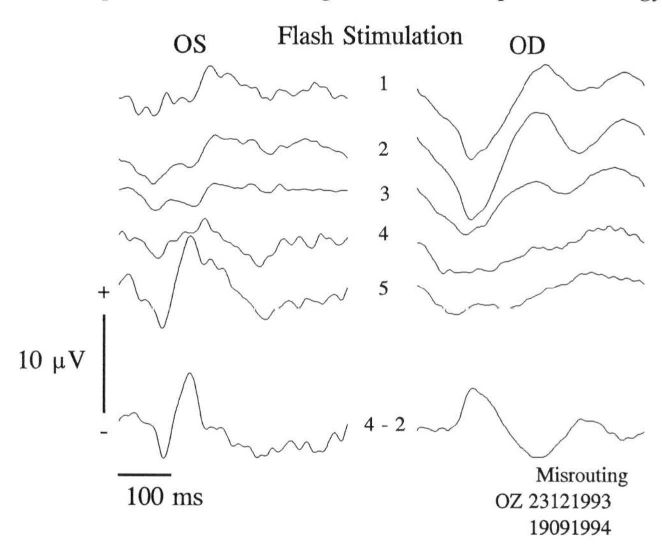

Fig. 8.43 With flash stimulation often crossed representation of the monocular responses can be observed in albino subjects. Especially in very young children this method of stimulation seems more appropriate than pattern stimulation, because the stimulus cannot be escaped even by looking away. It is very important in this case to take care that no light can enter the patched eye.

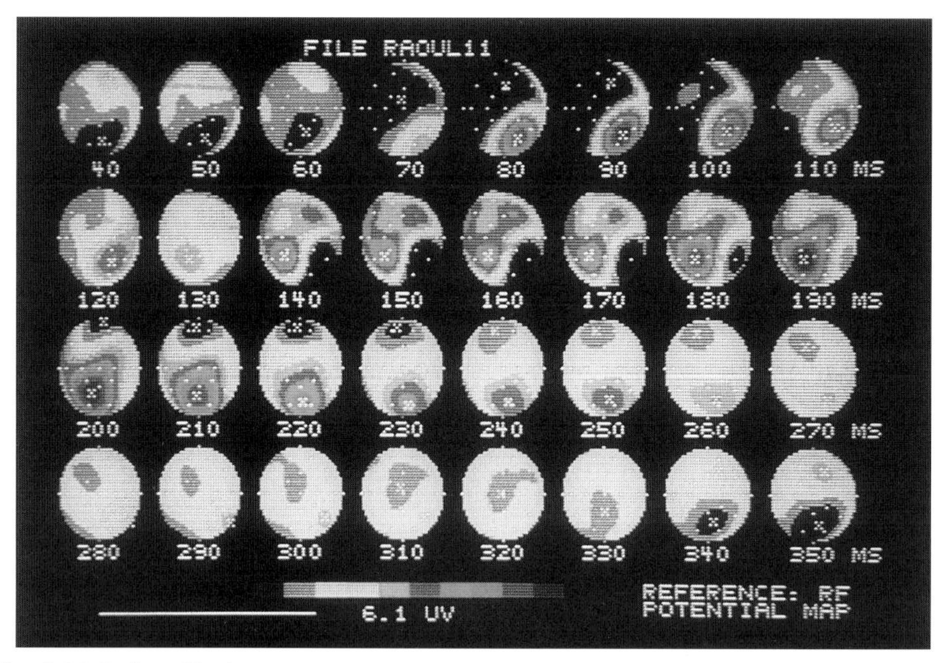

Fig. 8.44 Series of brain maps to pattern-onset stimulation recorded with 32 electrodes whose positions are marked by white dots in the maps. The right visual half-field was stimulated with a checkerboard pattern with 50′ checks for a duration of 300 ms. The extremes in the maps are indicated with a X.

activity can be seen at the back of the head 40 ms after pattern onset. This predominantly radial field remains rather stable for about 20 ms and is then replaced by another, more tangential field for about 40 ms, which at 130 ms after pattern onset is replaced by a third, again tangential field, but of opposite polarity and different localization. About 270 ms after stimulus onset all activity has died out. After 300 ms the pattern disappears. This pattern offset generates activity at the occipital pole 40 ms later. From this sequence of maps it can immediately be concluded that at least three different cortical areas are responsible for the PNP complex of the pattern-onset response. To identify these areas, electrical maps are less suitable than magnetic maps, since the volume conductor properties of the skull and skin smears out electrical but not magnetic maps; the head is magnetically transparent. This can easily be illustrated by comparing the electrical and magnetic responses to high-frequency flicker (Fig. 8.45). Figure 8.45(A) shows the evoked potentials to 40-Hz flicker in the left lower quadrant (2°–4°) of the visual field. This stimulus yields sine-wave responses whose polarity inverts from left to right across the back of the head. Since the electrodes are spaced 3 cm apart, this polarity inversion occurs over a distance of 6 cm. At the same scale, in the centre of the figure, the coils of a 7-channel magnetometer system are depicted; Fig. 8.45(B) gives the magnetic responses picked up by these coils from the back of the head due to the same 40-Hz stimulus but now covering the full lower left quadrant (0°–4°). As can be seen, polarity inversion now occurs from top to bottom and, since the coils are about 2 cm apart, over a distance of 2 cm. If the same generators underlie both fields, then on theoretical grounds two gross differences can be expected between the magnetic and electric responses:

(1) the magnetic and electrical response sets are rotated 90° to each other for sources oriented tangentially with regard to the surface of the head;

(2) the gradient of the magnetic response set is about three times steeper than that of the electric one.

It can therefore be concluded that in this example the same generators underlie both fields, and that preference for a particular type of map is mainly based on practical grounds (e.g. no electrodes have to be glued on for magnetic recording) or costs (low for electrical recording). Figure 8.46 gives the electrical and magnetic contour maps generated by the 40-Hz stimulus for the same subject. Whereas the electrical maps change from purely radial upon foveal stimulation (0°–1°), into mainly tangential upon annular stimulation (2°–4°), stimulation of the full, lower left quadrant only yields a single tangential magnetic map. This is due to the third rule of thumb:

(3) if the part of the head from which the recordings are made is rotationally symmetrical, as is the case for the operculum and the occiput, then the magnetic response set contains contributions of tangentially oriented current sources only.

Therefore magnetic recording seems less suitable to study foveal, striate vision than electrical recording, but it has the advantage that no description of the conductivity of the volume conductor is needed, since in a rotationally symmetrical conductor the depth and location of the source can be estimated directly from the maximum

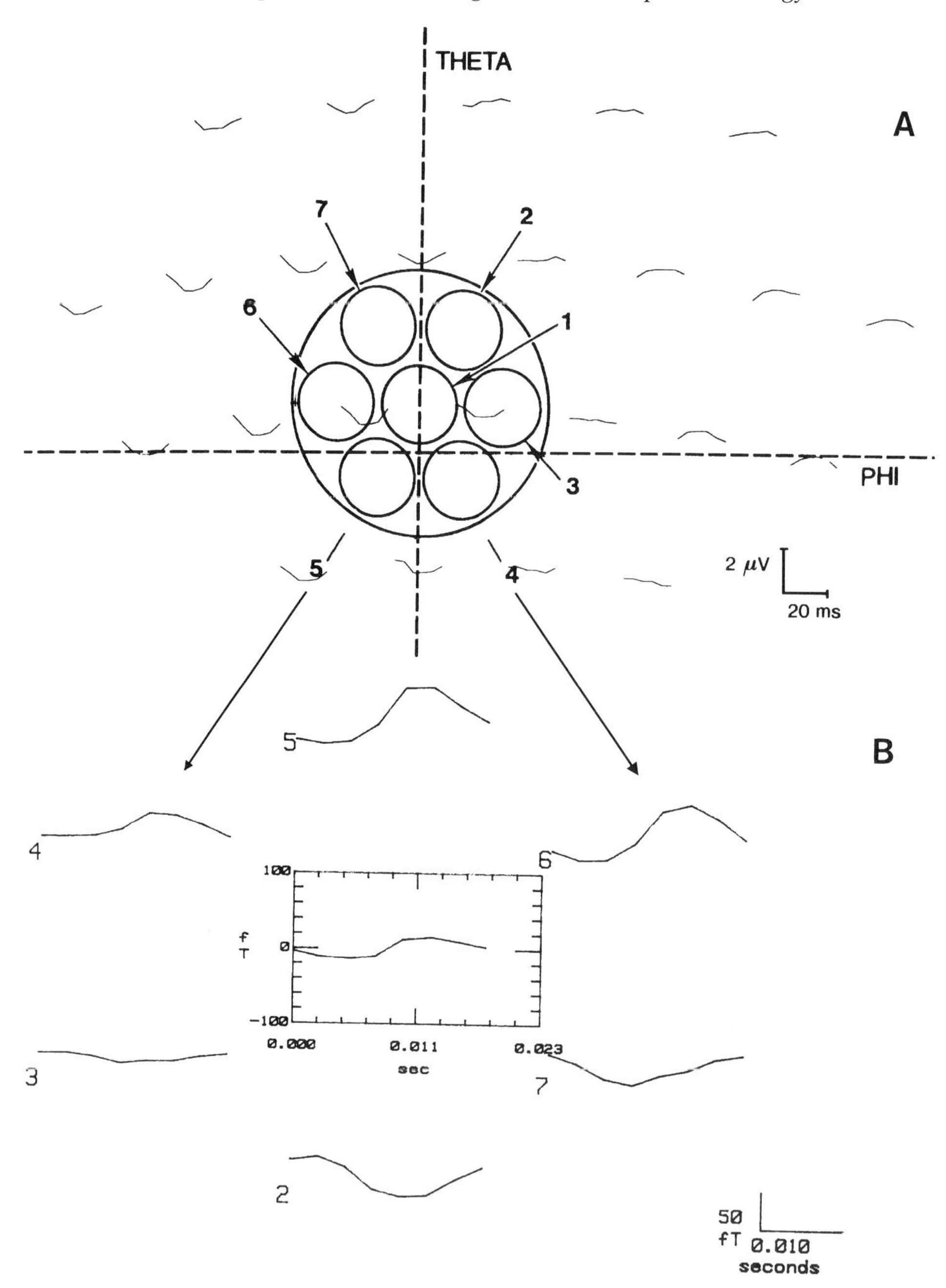

Fig. 8.45 Electric (A) and magnetic (B) responses to 40-Hz flicker. In Fig. 8.45(A) the responses are plotted at the electrode positions on the back of the head from where they were recorded (interelectrode distance is 3 cm). The dashed lines are at 0° azimuth and elevation angles. The inion is approximately at $\phi = 0°$ and $\theta = -10°$. Scaled to this response map and at the recording position, the centre of the figure shows the coils of the 7-channel magnetometer system by which the responses in Fig. 8.45(B) were obtained.

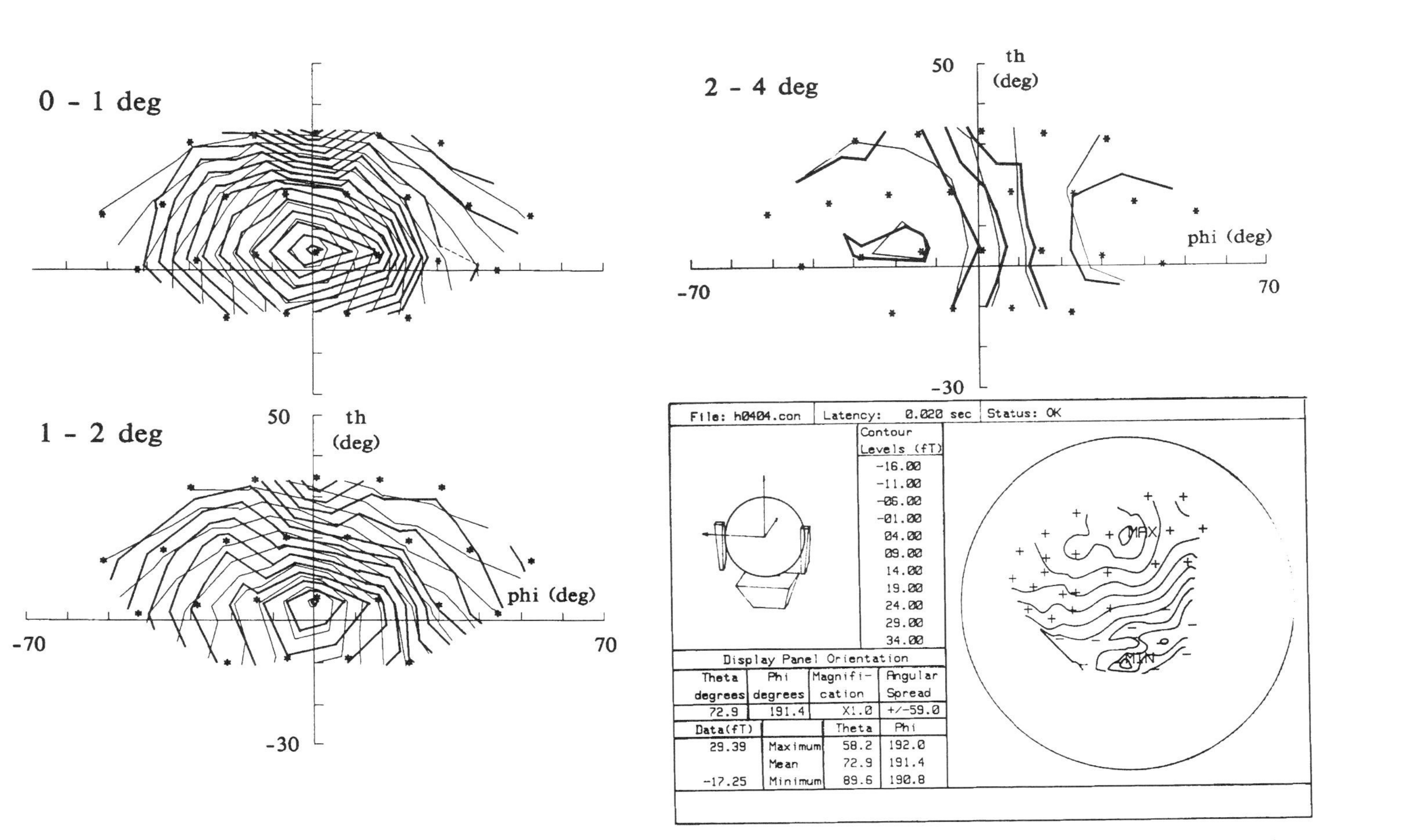

Fig. 8.46 Electric and magnetic isocontour maps generated by 40-Hz luminance modulation (depth 90%) in the lower left quadrant of the visual field. Left: eccentricities 0°–1° (top) and 1°–2° (bottom). Right: 2°–4° (electric, top) and 0°–4° (magnetic, bottom). The electric maps give both the recorded potentials (heavy lines) and the best-fitting dipole distributions (thin lines). The dots mark the electrode locations for the electric and the crosses the centres of the coils for the magnetic maps. In the latter, the extremes are marked by Max and Min.

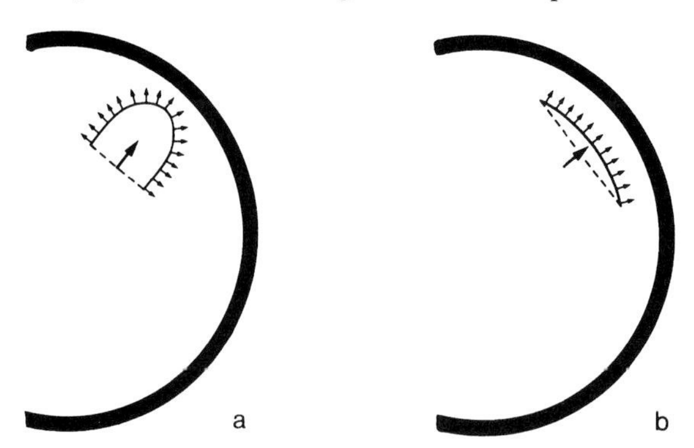

Fig. 8.47 Schematic representation of a patch of activated cortex. In source-localization studies such a dipole layer is replaced by an equivalent dipole (heavy arrow). Since for a homogeneous source layer the contour of the source determines the brain map, this equivalent dipole will be situated deep in the brain for highly curved activated areas (Fig. 8.47(a)) and superficially (Fig. 8.47(b)) if the area is flat.

and minimum of the magnetic map (location halfway at a depth equal to the distance/√2). Both maps have the disadvantage that the amount of cortex activated cannot be estimated, and therefore in most localization studies on the basis of electrical or magnetic recording the dipole layer activated is replaced by an equivalent dipole in the centre of the piece of cortex stimulated. If this patch of cortex is highly curved, as in Fig. 8.47, the equivalent dipole can be strongly mislocated. Given the retinotopic layout of the striate cortex, the localization errors can be minimized by always using half-field stimulation, activating one hemisphere only, and employing the smallest stimulus field still yielding a sufficient signal-to-noise ratio.

The present state of the art allows the localization of no more than two simultaneously active sources with unknown *a priori* positions, since the inverse solutions become unstable with an increasing number of sources. Since the response to the onset of a checkerboard usually consists of three successive peaks (PNP complex) with different distributions over the head (Fig. 8.44), an algorithm that can handle no more than two simultaneously active sources is not applicable. However, as the successive peaks vary in a characteristic way with stimulus parameters, in particular contrast, check size, and monocular vs. binocular presentation, stimuli can always be constructed by which no more than two sources are activated. In this way we were able to demonstrate (Ossenblok *et al.*, 1994) that the pattern-onset maps are due to the simultaneous activity of one striate and two extrastriate regions (Fig. 8.48). All three regions contribute over the entire response interval lasting from about 50 to 250 ms, asking for a high time resolution for source localization. None of the other non-invasive imaging facilities such as PET, SPECT, and functional MRI have such time resolution.

They allow, however, for the absolute localization of the activated cortical patches, and can therefore yield answers that cannot be obtained from the electrical or magnetic recordings. An example is the misrouting in albinos. From their response profile across the back of the head to monocular stimulation (Fig. 8.41) it could be concluded that most of the optic nerve fibres project to the contralateral hemisphere.

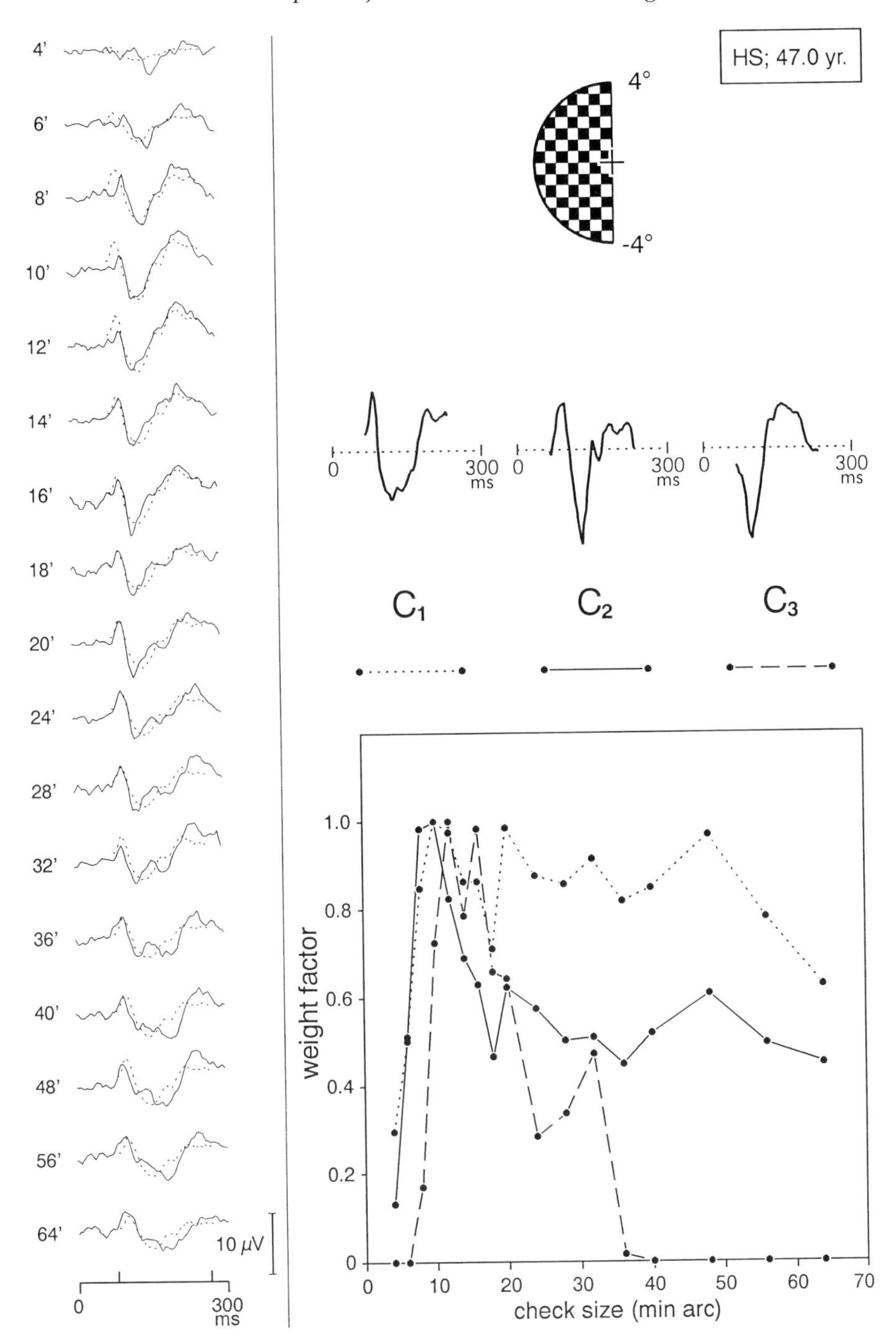

Fig. 8.48 Pattern-onset EPs as a function of check size (left). These responses were least-square fitted with the three components, C_1, C_2, and C_3 (mid figure), representing the dipoles of the three cortical regions contributing to the EPs. For small and large check sizes C_1 and C_2 dominate the responses, for intermediate checks all three regions contribute. (From Ossenblok *et al.*, 1994.)

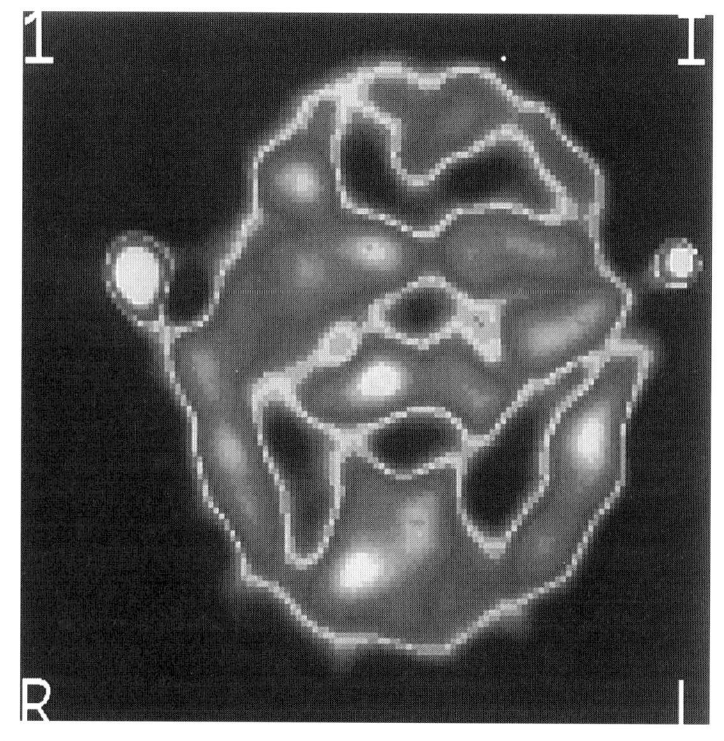

Fig. 8.49 SPECT slice through the head of an albino subject during left eye stimulation with a moving checkerboard (nose top, inion bottom). The white dots are two of the markers used to co-register SPECT hot spots, with MRI and EP coordinates.

However, because of the spread of electric activity by volume conduction, it cannot be concluded whether most means more than 50% or close to 100%. To answer that question, we made SPECT slices through the head of an albino subject. Figure 8.49 shows such a coupe through the lateral geniculate nucleus (LGN) and striate cortex upon left eye stimulation with a moving checkerboard pattern. The figure shows that left eye stimulation activates (hot spots) the right LGN and the right striate cortex, telling one immediately that in this albino nearly 100% of the fibres project to the contralateral hemisphere. This example was also chosen for another purpose. Upon motion onset, responses can be found all over the head with a late negativity far away from the midline (Fig. 8.50). The isocontour plots of this late negativity remain invariant with the visual half-field being stimulated. In 80% of the subjects the late negativity (peak latency about 170 ms) dominates at the right hemisphere, irrespective to which visual half-field the motion stimulus is being presented. It is therefore difficult to decide where the sources for these maps are located. For that purpose we made SPECT coupes during motion onset and stimulation of the same half-field with steady illumination. By using a bite-board to locate the subject at the same position in the SPECT scanner (Fig. 8.51), the coupes in the motion and rest condition could be co-registered and subtracted from each other. This yielded the difference SPECT depicted

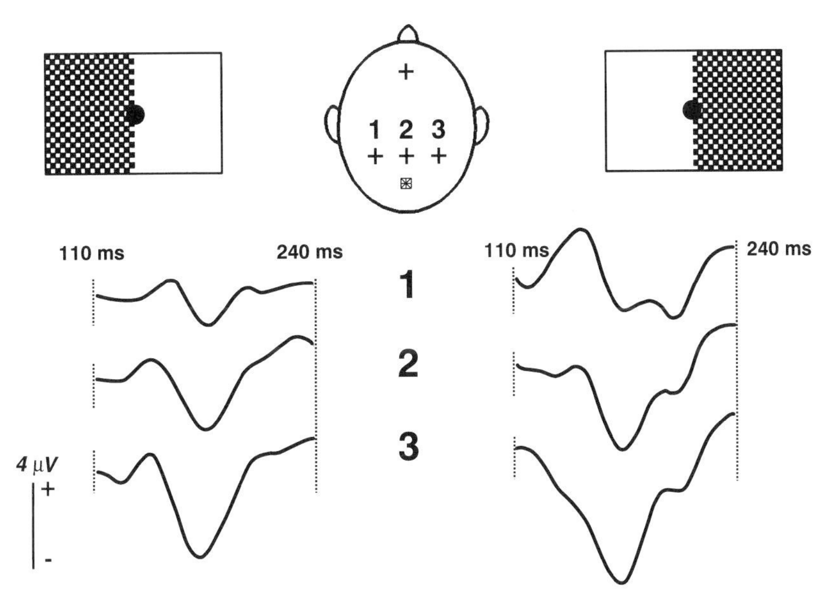

Fig. 8.50 Motion onset responses, derived from three electrode locations spaced 4 cm apart, to left (first column) and right half-field (second column) stimulation. These responses were obtained during SPECT recording.

in the insert of Fig. 8.51 to right half-field motion. In this inset, two hot spots can be seen: one in the left striate cortex and one 5 cm away from the midline in the right hemisphere. These locations were used to fit the motion onset maps, of which Fig. 8.50 shows three recordings upon left and right half-field stimulation, respectively, which were made during the SPECT scan. The EEG amplifiers can be seen in the foreground of Fig. 8.51. The recordings show that the late negativity dominates at electrode 3 (i.e. the right hemisphere) irrespective of the half-field of stimulation, but that the preceding positive wave (peak latency about 120 ms) dominates at electrode 1 for right and at electrode 3 for left half-field stimulation, i.e. flips over to the contralateral hemisphere as expected. Since we used markers in the SPECT study (see the white dots near the ears in Fig. 8.49), and markers at the same locations for MRI slices, the anatomical data could be co-registered with the electrical maps. This resulted in Fig. 8.52, in which a concentric sphere model (bottom) shows the two equivalent dipoles found upon right half-field motion onset: demonstrating that the early positivity originates in the left striate cortex and the late negativity in the right posterior parietal cortex. These locations are transferred into the MRI coupe of this subject (top) showing striate cortex (calcarine fissure) and right posterior parietal cortex activation.

The above examples were chosen to show that by integration of the present-day, non-invasive, brain imaging techniques new and detailed insights can be obtained about the functioning of our brain. The high time-resolution of the electrical and magnetic recordings give insight into how the various brain regions communicate with each other, and by SPECT or PET analysis where they are situated. Combining these four techniques with (functional) MRI and CT offers the possibility of investigating

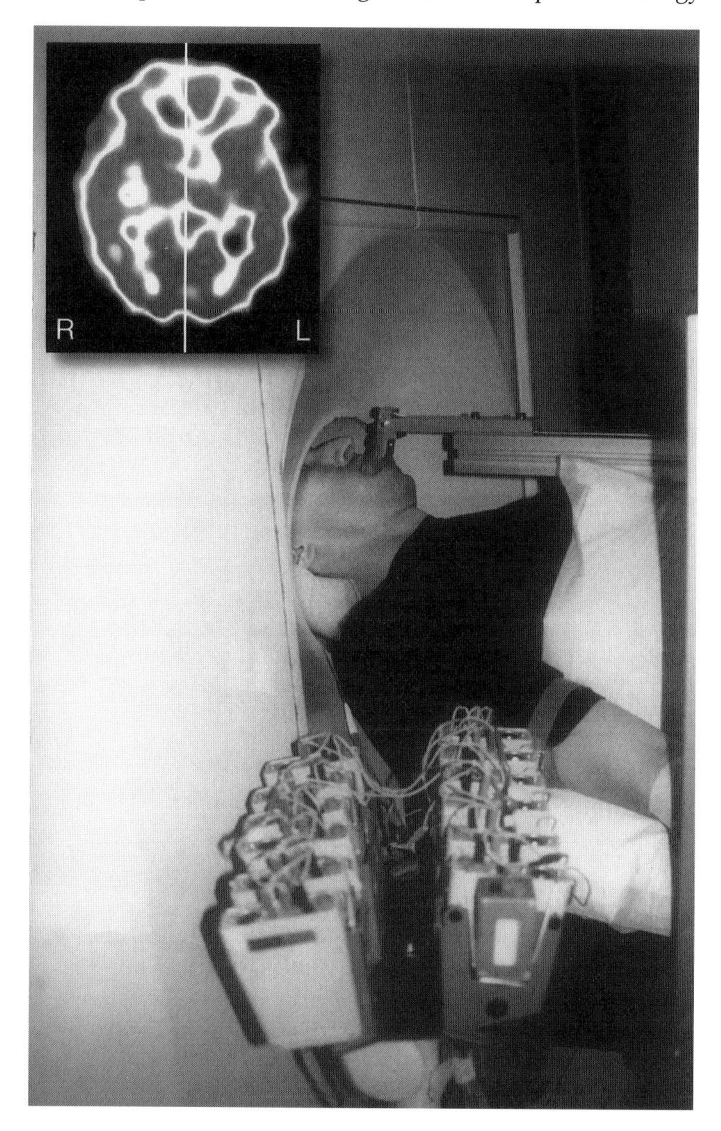

Fig. 8.51 SPECT scanner with the EEG amplifiers used for the EP recordings shown in Fig. 8.50. By means of a bite-board SPECT coupes taken in activated and basal conditions can be co-registered. Such a difference SPECT to motion onset in the right visual half-field is depicted in the top left of the figure. The white line represents the midline through the head. Two regions are activated: the left striate cortex and right posterior parietal cortex. The albino subject of Fig. 8.49 shows to left eye, full-field, motion onset stimulation, activation of the left posterior parietal cortex, which is only the case in 20% of the subjects. In 80% of the subjects the motion negativity is generated in the right posterior parietal cortex, as for the subject in Fig. 8.52.

human functional brain anatomy at millisecond and millimetre intervals. This approach has the additional advantage that the time and space patterns of the functioning brain can directly be related to cognition and behaviour in humans.

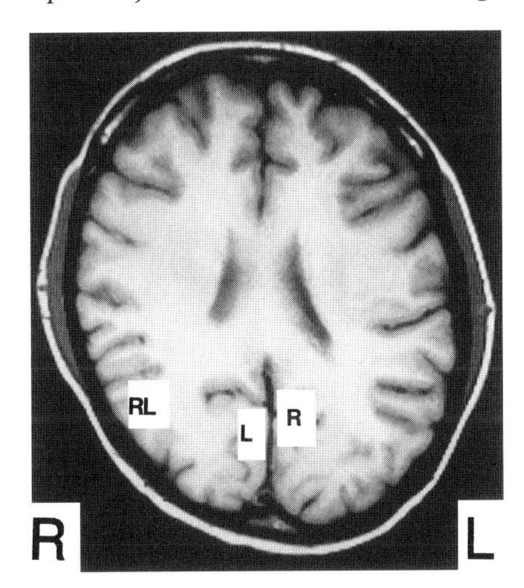

(1) LEFT STRIATE CORTEX

L R

138 MS 207 MS

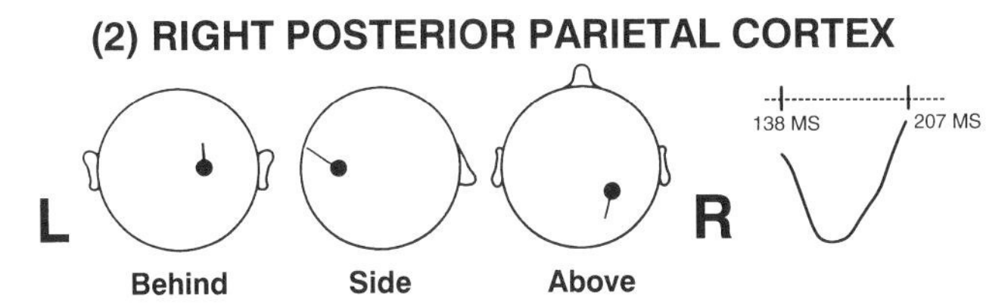

Fig. 8.52 MRI coupe through the head of a subject whose electric sources to motion onset in the right visual half-field are drawn in a sphere model of the back of the head in the bottom half of the figure. The positions of these sources are transferred to the MRI coupe and labelled R. They lie respectively 1 cm left and more than 5 cm right from the midline. The sources to left half-field motion onset are labelled L. Note that the motion onset negativity is generated in the right posterior parietal cortex irrespective of whether the left or right visual half-field is being stimulated.

References

Asher, H. (1951). The electroretinogram of the blind spot. *Journal of Physiology*, **112**, 40P.

Celesia, G. G., Bodis-Wollner, I., Chatrian, G. E., Harding, G. F. A., Sokol, S., and Spekreijse, H. (1993). Recommended standards for electroretinograms and visual evoked potentials. Report of an IFCN Committee. *EEG and Clinical Neurophysiology*, **87**, 421–36.

Estévez, O. and Spekreijse, H. (1982). The 'silent substitution' method in visual search. *Vision Research*, **22**, 681–91.

Fry, G. A. and Bartley, S. H. (1935). The relation of stray light in the eye to the retinal action potential. *American Journal of Physiology*, **111**, 335–40.

Galloway, N. R., Odom, J. V., Anastasi, M., *et al.* (1995). *Visual electrodiagnostics. A guide to procedures*. ISCEV Publications, Nottingham.

Gouras, P. and MacKay, C. J. (1989). Electroretinographic responses of the short-wavelength-sensitive cones. *Investigative Ophthalmology and Visual Science*, **31**, 1203–9.

Harding, G. F. A. (1995). *Standard for visual evoked potentials*. ISCEV Publications, Nottingham.

Hjorth, B. (1975). An online transformation of EEG scalp potentials into orthogonal source derivations. *EEG and Clinical Neurophysiology*, **39**, 526–30.

Jeffreys, D. A. and Axford, J. G. (1972). Source locations of pattern specific components of human visual evoked potentials. I and II. *Experimental Brain Research*, **16**, 1–40.

Marmor, M. F. and Zrenner, E. (1993). Standard for clinical electro-oculography. *Archives of Ophthalmology*, **111**, 601–4.

Marmor, M. F., Arden, G. B., Nilsson, S. E., and Zrenner, E. (1989). Standard for clinical electroretinography. *Archives of Ophthalmology*, **107**, 816–19.

Marmor, M. F., Arden, G. B., Nilsson, S. E., and Zrenner, E. (1994). *Updated standard for clinical electroretinography*. ISCEV Publications, Nottingham.

Ossenblok, P., Reits, D., and Spekreijse, H. (1994). Check size dependency of the sources of the hemifield-onset evoked potential. *Documenta Ophthalmologica*, **88**, 77–88.

Reulen, J. P. H., Marcus, J. T., Kops, D., *et al.* (1988). Precise recording of eye movements: the IRIS technique, Part 1. *Medical and Biological Engineering and Computing*, **26**, 22–6.

Sawusch, M., Pokorny, J., and Smith, V.C. (1987). Clinical electroretinography for short wavelength sensitive cones. *Investigative Ophthalmology and Visual Science*, **28**, 966–74.

Sutter, E. E. and Tran, D. (1992). The field topography of ERG components in man—I. The photopic luminance response. *Vision Research*, **32**, 433–46.

van den Berg, T. J. T. P. and IJspeert, J. K. (1991). Intraocular straylight, studied using the direct compensation technique. *CIE Proceedings*, 22 session, division 1, 83–4.

Further reading

The following are suggested as specialized reading on ERG, EOG, and VEP methods, especially applied in clinical circumstances:

Fishman, G. F. and Sokol, S. (1990). Electrophysiological testing in disorders of the retina, optic nerve and visual pathway. *American Academy of Ophthalmology*, Singapore.

Heckenlively, J. R. (1993). Retina and vitreous. *American Academy of Ophthalmology*, San Francisco.

Heckenlively, J. R. and Arden, G. B. (1991). *Principles and practice of clinical electrophysiology of vision*. Mosby Year Book.

9

Eye movement recording

HAN COLLEWIJN

9.1 Introduction

The art of recording eye movements is about as old as the century; overviews of the early history have been given by Jung and Kornhuber (1964), Jung (1977), Heller (1988), and Westheimer (1989). Dodge and Cline (1901) were among the first to publish (horizontal) eye movement recordings of lasting value, using photography (on a slowly falling photographic glass plate) of a first Purkinje image. The recordings were good enough to document a classification of human eye movements in subtypes that is still used today (Dodge, 1903). Recording techniques have been evolving ever since, and this development is still continuing. Lately, such development is mainly concerned with better technical implementations of existing principles. Sharp reductions in the cost of computational power and advances in video-imaging techniques may, for instance, be leading to affordable instruments that are, once again, based on imaging, but that supply signals complying with modern standards (actually, however, this is much easier said than done!).

The number of fundamental principles used in eye movement recording has always been small and it has not increased for many years, but there is an almost infinite variety in the application of these principles. There are two classes of implementations that can be distinguished in this respect. The first class consists of solutions that have been developed and profitably used by a small group of people, often in a single laboratory and for a very specific purpose, but which have found no general application, probably because they were too difficult to handle. A good example is the 'tracked oscilloscope spot' technique of Rashbass (1960), used effectively in the classical experiments by Rashbass and Westheimer (1961) but which disappeared into oblivion as soon as these investigators developed other interests. The present chapter will deal mainly with the second class: those solutions that have found a wider use. Apart from the intrinsic quality of a method, such widespread use is almost invariably linked to commercial development by an interested industry. The days of 'do-it-yourself' developments in eye movement recording seem to be nearly gone, and the large majority of oculomotor research depends on commercially available equipment. This is a mixed blessing. Industrial development is almost indispensable to guarantee the professional standards required today. On the other hand, the potential market is modest in volume, so that few industries are in a position to invest heavily in further advances of eye movement recording instruments, except as a spin-off of related technology. Nevertheless, various kinds of systems are marketed nowadays that offer practical solutions for most of the needs in basic and clinical research. The choice should depend on the particular requirements of the application, because no system is universally suitable for all applications, and compromises are sometimes unavoidable. Above all, any system requires skilful use and adequate background knowledge. It would be a grave mistake for a newcomer to the field to acquire an apparatus and just assume that whatever signals it produces are a true representation of the eye movements. Appropriate use of any system demands a firm understanding of its principles and the associated requirements, limitations, and pitfalls, plus a good background knowledge of eye movements in general (for recent reviews, see, for example, Kowler, 1990; Carpenter, 1991). When such understanding is lacking, artefacts are more than likely to be confused with genuine behaviour.

To obtain a permanent record of eye movements, the movements of the eyeball have somehow to be transduced into signals that can be stored. The development of storage techniques is at least as important as that of the transduction itself. Not so long ago, researchers were perfectly happy with tracings on paper, obtained by photographic or pen-recorder technology, as the end product of the recording procedures. Such records had to be further analysed by manual, graphical techniques such as measuring the slope of a pursuit movement with a ruler. Nowadays, such techniques have only a minor place, and the essential storage of the signal has to be in a form that lends itself to further processing by digital techniques. These allow presentation of the eye movements in any desired graphical form, and they allow any kind of numerical analysis of the raw data that can be reliably computed with suitable software. Moreover, computing techniques make it possible to elaborate the raw signals of eye movement transducers considerably so as to transform them into proper representations of eye movements. As a simple example: many transducer signals contain non-linearities that can be corrected by computer routines. A much more complex example is the extraction of angular movements from video images of the eye, a technique in which the raw signal yields no direct movement traces at all, and everything depends on computation.

Thus, transduction of the eye movements in motion signals is only the beginning of obtaining veridical representations of ocular angles. It would seem a sound principle that the better the transducer signals reflect the angular eye movements, and the less computation is required, the more reliable the overall results that may be expected. Computation can be reliable only in as far as the relationships between the raw transducer output and the true ocular angles are fully deterministic, stationary, and completely known. We shall see, however, that even the technique with the most direct relationship between eye movements and measured signals (the scleral coil) requires computational corrections to obtain veridical eye angles.

Many reviews of oculomotor recording techniques have been written before (e.g. Shackel, 1967 (especially on EOG); Ditchburn, 1973 (especially on contact lens—optical lever); Young and Sheena, 1975; Carpenter, 1988; see also Gale and Johnson, 1984). These contain extremely valuable information.

9.2 Ideal properties of an eye movement recording system

9.2.1 Veridicality

A good recording system should yield a valid and complete representation of angular eye positions. This implies a number of properties.

9.2.1.1 Spatial considerations

There are two fundamental ways of measuring eye movements. One way is to measure angular eye position relative to the head (*eye-in-head*). Such a signal is obtained by instruments in which the sensors are attached to the head (e.g. EOG electrodes, IR-reflection spectacles, or a head-mounted video system). The second way is to measure eye position relative to the surroundings, or *gaze*. This is done by instruments fixed in the room (e.g. an earth-fixed video camera or magnetic field). When the head is restrained, gaze and eye-in-head are identical, but when the head is free, head movements have to be measured along with eye or gaze movements to calculate one from the other. In a body-centred coordinate system, gaze angle can be simply defined as the sum of head and eye angles. However, in studies addressing the relationship between the line of sight and visual features in the surroundings, one needs true spatial coordinates: in behavioural conditions gaze relates to the intersection of the line(s) of sight with planes or objects in 3-D space. The implication is that in free-head experiments one has to measure head translations along with head rotations in order to define gaze. Even in a partially restrained situation, in which the head is allowed to rotate around some fixed axis, but not allowed to translate, the relationship between head, eye, and target angles is complicated by the fact that the axes of rotation of the two eyes and the head are separated by non-negligible distances, so that a simple addition of angles may be inadequate. (For detailed discussions of these issues see, for example, Hine and Thorn, 1987; Collewijn *et al.*, 1992; Edwards *et al.*, 1994.)

As each eye moves with three degrees of freedom, a full representation requires the measurement of three ocular angles, loosely defined as horizontal, vertical, and torsional orientation. An ideal system provides these angles simultaneously as separate signals, for each of the two eyes, so that binocular eye movements can be studied. The three angular signals should, ideally, show no crosstalk. In practice, it is not always of interest to record in three dimensions or from both eyes, but even for the recording on a single meridian dimensional purity is an important issue. For instance, when vertical eye movements are recorded alone, one wants to be sure that they are not contaminated by horizontal or torsional movements.

An immediately related issue is the definability of the rotational axes of the various components measured. 'Horizontal, vertical, and torsion' as such are mathematically loose terms, until the axes of rotation and their hierarchy (nesting) are defined in a specific coordinate system. Examples are a Fick type gimbal, a Helmholtz type gimbal, or 3-D rotation vectors (see, for example, Carpenter, 1988; Haustein, 1989). Few existing systems allow for such exactness, although advanced study of the control of binocular gaze in 3-D space absolutely requires this.

9.2.1.2 Linearity

Output signals should be linearly related to ocular angles, or the deviation from linearity should be computable and follow a simple function. Irregular nonlinearities require cumbersome calibration procedures and can be corrected only partially.

9.2.1.3 Range

An ideal system should cover the entire oculomotor range, i.e. (for humans) about ±50 deg horizontal and vertical and about ±20 deg in torsion, without saturation or prohibitive non-linearity. Many existing systems operate only within a small (< 20 deg) central part of this range.

9.2.1.4 Immunity to translation

This is a very important point, and an intrinsic weakness of many systems. Consider that, in terms of visual angle, a 1-mm pure *translation* of the whole eye corresponds to as little as arctan 0.001 = 0.057 deg for a target at a distance of 1 m. At the same time, a displacement of the front of the eye by 1 mm as a result of *eccentric rotation* (with an ocular radius of about 12 mm) corresponds to a rotation angle of arctan 1/12 = 4.76 deg. This makes it clear that any system relying on the displacement of features on the surface of the eye is extremely vulnerable to artefacts of translation. Such artefacts (linear displacements masquerading as rotations) may result from displacements between the head and the measuring equipment and also from genuine translations of the eye. Enright (1980, 1984) has documented systematic lateral translations of the eye of about 0.2 mm during changes in vergence between near and far targets. Such a translation would be interpreted as a 1-deg rotation by devices that do not distinguish between translation and rotation. Anteroposterior eye translations associated with horizontal saccades (Enright, 1986) and with various types of fixation (Enright and Hendriks, 1994) have also been described. These systematic translatory eye movements, which form a very interesting but mostly neglected aspect of oculomotor physiology, have obviously negative implications for the validity of measurements based on the orbital location of single ocular landmarks. Such problems can be overcome by using two landmarks, that move differently during rotation (but similarly during translation).

9.2.2 Quality of the signals

9.2.2.1 Signal-to-noise ratio

Apart from the qualities already mentioned above such as linearity and axial veridicality, which together imply *accuracy*, a main requirement is *precision*, i.e. a high signal-to-noise ratio. The noise level determines the minimum size of eye movements that can be reliably discriminated.

Some types of noise (e.g. a single frequency interference related to the power lines) can be cleanly removed by digital filtering techniques without harm to the signal. High-frequency, broad-band noise (> 100 Hz) can usually also be removed effectively by low-pass filtering. Low-frequency, broad-band noise is particularly bothersome as it can not be removed without compromising the signal.

9.2.2.2 Bandwidth

This is the frequency range in which eye movements are reliably represented. It is determined by the dynamics of the recording system, including filtering procedures. Ideally, eye position should be represented fiducially (with a flat gain and phase relationship) between DC (0 Hz) and ≥ 100 Hz, to allow reliable analysis of static position, as well as saccadic dynamics, including differentiation to show peak velocities without clipping. As bandwidth is usually specified within a −3-dB limit; a 150-Hz bandwidth will satisfy virtually all practical demands.

In relation to bandwidth, a word on digital sampling is in order here. Obviously, overall bandwidth should not be diminished by a too low sampling rate. Theoretically, a bandwidth of 100 Hz requires a minimum sampling rate of 200/s. In view of the temporal properties of signals (important, for example, for latency measurements) a sampling rate of 500/s (temporal resolution 2 ms) is desirable for more critical applications. To prevent artefacts of the aliasing of high frequencies to low frequencies (a 'strobing' effect), it is essential that all frequencies higher than those adequately represented digitally be removed *before* digitizing. This should be done with high-quality, steep, active filters (e.g. 4th- or 6th-order Bessel type). Thus, a 500/s sampling rate requires low-pass filtering at 250 Hz. If the sampling rate is lowered, the cut-off frequency of the filters should be lowered commensurately. Furthermore, *all* signals sampled simultaneously should be passed through exactly *identical* filters, to preserve the temporal and frequency relationships between the signals. In the case of equipment that supplies data already in digital form, adherence to these principles is the responsibility of the manufacturer. Disregard of these elementary 'sampling' rules may result in artefacts and poor quality of stored data.

9.2.2.3 Drift

Drift is equivalent to very-low frequency noise (with a DC component if there is a consistent trend in one direction). It is expressed as an unstable value of the output signal for a steady eye position. It is caused by a variety of factors. Thermal instability of electronic components may be an important source; much of this is often eliminated by a suitable warming-up period of the equipment (if a stable thermal state is reached). Other sources are variations in the transduction process, such as fluctuations in the reflectivity of the eye with reflection techniques, or changes in electrode properties and corneoretinal potential, which have driven many a user of electro-oculography to exasperation. Changes in alignment between a subject and the measuring device can also masquerade as drift. The tolerance for drift depends on the application. A drift of 1 deg min^{-1} would be negligible in a study of saccadic dynamics, but disastrous in a study of fixation.

Drift levels are often specified by manufacturers in equivalent deg h^{-1} or similar terms. It should be kept in mind that the figure given usually represents pure system noise, measured, for instance, with an artificial, immobile eye, thus not encompassing sources of drift outside the apparatus.

9.2.2.4 Real-time properties

For many applications it is of little concern whether the output signals emerge from the apparatus simultaneously with the eye movements or after some brief delay. Delays of more than a few milliseconds can be disastrous, however, in experiments that use the eye movement signals in some feedback control loop, e.g. to stabilize retinal images, clamp disparity, etc. Apparatus using video techniques (limited by frame rate) and/or complex computations are the most likely ones to have non-negligible delays in their output signals.

9.2.3 Minimum interference with vision

Vision and visual conditions should be as natural as possible during the measurements. Ideal equipment should leave the visual field and visual acuity of both eyes unimpeded, and introduce no distracting visual stimuli. The method should, ideally, also allow the subject to use his usual spectacle or contact lens corrections. Finally, the method should be insensitive to ambient light conditions, so that measurements can be made under any kind of illumination (including darkness).

9.2.4 Minimum interference with behaviour

Eye movements are an integrated part of behaviour. In the past, researchers have been content with (and have actually favoured) the study of eye movements in isolation, with the exclusion of head movements. This was partly inspired by a reductionistic attitude, but even more dictated by the available equipment. Although there will be a continued need for the further study of aspects of oculomotor control under strictly controlled and isolated conditions, oculomotor research is now clearly evolving to broader questions, in which eye movements are investigated in the context of general visuomotor behaviour, including head, hand, and body movements in natural tasks (see, for example, Steinman *et al.*, 1990). Therefore, there will be an increasing demand for equipment that is capable of recording eye movements under conditions in which subjects are free to move around without restraints. As a corollary, a need will develop for equipment capable of measuring head, body, and hand movements along with eye movements.

9.2.4.1 Head immobilization

Many recording techniques require a critical alignment between the apparatus and the eyes; if the apparatus cannot be carried on the head, this means that the head has to be fixed in space. In principle, this is a nuisance, and the less rigid head immobilization is required, the better. Sometimes, the experiment as such, does not allow head movements (e.g. if a critical alignment of stimuli has to be preserved), or calibration procedures require at least a temporary head fixation. Cooperative subjects can stabilize their heads to a reasonable extent on suitable chin and head supports, but the soft tissues will allow substantial residual movement. Maximum immobility can be achieved only by attachments to the bony skull tissues. In animals, this can be achieved by the permanent implantation of skull fixtures. In human subjects, the only effective solution is the use of an individually fitted bite-board, which supports the maxilla. Even in a free-head experiment a bite-board can be extremely useful, e.g. to carry a coil for the precise recording of head movements, or a head-fixed visual target.

If a bite-board has to be used, one starts with preparing a template (Fig. 9.1) that fits comfortably between the upper and lower dental arcs. The template should be of a rigid and strong material, e.g. a hard type of aluminium (if metal is allowed) or a strong plastic such as polycarbonate. It should be no thicker than 5–6 mm and tapered to a thinner diameter at the posterior end. It should have a frontal extension for fixation and holes drilled through to facilitate anchoring of the impression mater-

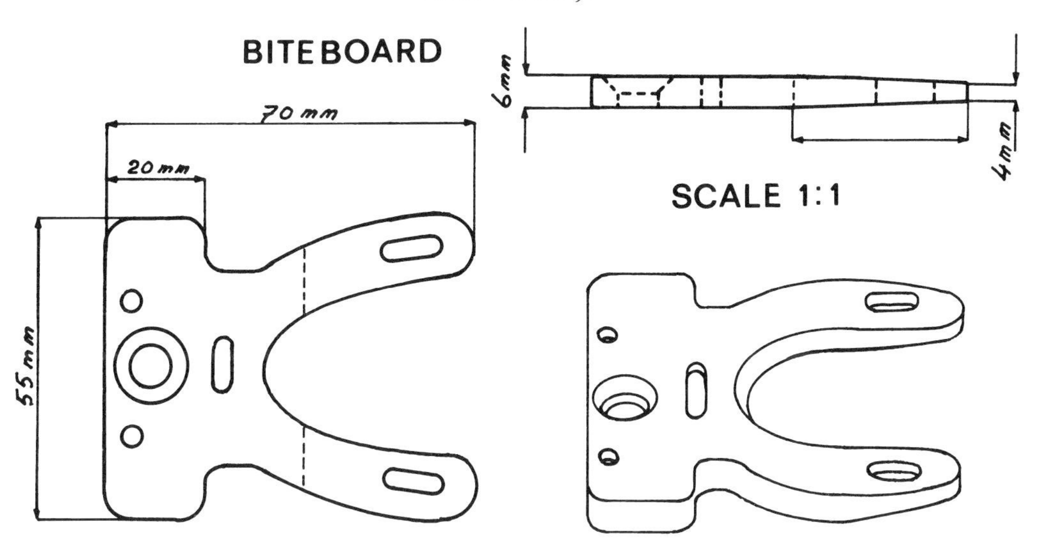

Fig. 9.1 Convenient shape and size of a template for making a dental impression bite-board.

ial, for which one of the modern dental impression substances should be used (poly-siloxan; obtained from any dental supply store). These consist of a malleable, neutrally tasting, silicone base substance, which after mixing with a hardener will solidify to a firm, slightly elastic material in a few minutes. The freshly prepared mixture is kneaded around the template (about finger-thick) and the assembly is placed in the mouth of the subject, who should bite gently through the substance down to the template and then stay immobilized in occlusion for about 4 min, to obtain a permanent impression. The bite-board is then taken out and rinsed under tap water. Any excessive material is trimmed away with a knife. As an alternative, temperature-dependant, reusable, dental impressive compound is available. Inasmuch as the head is supported by the maxilla, efforts should be made to register all of the upper molars clearly to afford the best support for the skull. The lower teeth do not, in themselves, support the head so quite modest registration of the lower teeth does not compromise the effectiveness of the bite-board.

A bite-board should be mounted at a height and angle that is comfortable for the subject, preferably on a stand allowing calibrated 3-D translation to adjust the subject's position in the experimental apparatus. Recording the X, Y, Z settings will allow reproduction of the same head position across sessions. Even an optimally fitting bite-board requires cooperation by the subject, who should rest firmly on the board and clench his teeth around it. The subject should be specifically instructed to avoid 'rocking' movements, which are most likely to occur around the interaural axis. A forehead support can be used along with the bite-board to reduce the amount of change in the pitch of the head. A chair with adjustable height is very helpful, as an uncomfortable head position is difficult to maintain for any long period. Even with the most comfortable settings, few subjects will stay on a bite-board for longer than half an hour. The immobilization is experienced as very unnatural. During large gaze changes, strain in the neck is experienced as a consequence of synergic activation of eye and neck muscles. Also, bite-boards exclude verbal communication.

9.2.4.2 Other attachments to the head

Many lightweight, eye movement recording transducers are built on spectacle frames, inside goggles, or attached to headbands or helmets. Not all of these devices are really comfortable if they are worn for any extended time. They are intended to stay immobile on the head, and therefore often require tightening that may cause pain on pressure points or impaired circulation through the scalp. Discomfort also results from the huge differences in shape between people's skulls and faces, which make universal fitting, even with adjustments, difficult. Velcro tape is one of the more effective and comfortable materials to use in adjustable headbands. Any device resting on the soft tissues of the head will be displaced by changes in tissue volume and activity of facial muscles, so that long-term stability is generally problematic.

9.2.4.3 Attachments to the eyes

An ideal method would not require any object to touch the very sensitive surface of the eye, as this is generally perceived as invasive and uncomfortable. Apart from this, gross mechanical devices are likely to interfere with oculomotor behaviour. For this reason, suction caps and suction lenses have never achieved great popularity. On the other hand, the fact that millions of people routinely wear contact lenses for correction proves that suitably constructed eye-contacting devices are worn without any objection. This point will be taken up again in the treatment of the scleral coil technique.

9.3 Principles and practice in eye movement recording

9.3.1 Purely mechanical techniques

Of historical interest are techniques which established a direct mechanical connection between the eye and some recording device. Although this seems very crude today, such devices, originating from 19th century physiological techniques, have actually been used extensively in this century. The best known example is the 'Hebel-Nystagmographie' used by Ohm (1928), an ophthalmologist practising in Bottrop, Germany (a mining area), who produced an interminable series of articles (much dreaded by journal editors) on miner's nystagmus ('Augenzittern') and almost all other types of nystagmus during the period of about 1914 to 1959 (see Jung, 1977). This earned him the somewhat disdainful nickname of 'der ZitterOhm'. Despite the obvious limitations and artefacts of his crude method, Ohm successfully documented nearly all types of nystagmus known today. There is a useful message for us here: (a) imperfections do not necessarily preclude useful application of a method, as long as they are recognized and understood; (b) we should realize that imperfections can be as plentiful (though more insidious) in today's sophisticated technologies as in those primitive methods that are now only of historical interest.

9.3.2 The optical lever technique

Conceptually, it was only a small step to change from a mechanical to an optical lever, but in terms of quality the step was huge. In its most effective form, the principle was implemented by using a flat mirror, firmly attached to the eye by a suction cap (Yarbus, 1967) or, less traumatically, by a tightly fitting contact lens under negative pressure. A light beam was directed towards the mirror and reflected towards some photosensitive device. Either photographic records or electric analogues of the angle of the reflected beam, which varies by twice the amount of the change in ocular angle, could be obtained by various techniques. The method was first applied nearly a century ago (De La Barre, 1898; Orschansky, 1898), used more extensively in the 1930s (e.g. Adler and Fliegelman, 1934 (in humans); Ter Braak, 1936 (in rabbits)), and brought to great perfection during the 1950s through the 1970s (see Fender and Nye, 1961; Ditchburn, 1973; Steinman *et al.*, 1973). The method could record horizontal, vertical, and even torsional (Fender, 1955) angular eye movements with great sensitivity, precision, accuracy, and bandwidth, and was, in principle, insensitive to translations. It was crucial in the development of our knowledge of fixation and miniature eye movements. The required expertise on contact lenses and optical techniques restricted the method, however, to a small number of specialists. Problems with lens slippage excluded the study of large eye movements, and after about 1980 the progress of scleral coil techniques made the optical lever technique obsolete; to the knowledge of this author it is no longer used in any oculomotor laboratory.

9.3.3 The electro-oculogram (EOG)

The principle of this well-known method was established 60 years ago (Fenn and Hursh, 1934) as rotation of the electrical field derived from the corneoretinal dipole, the cornea being about 1-mV positive relative to the fundus. The field propagates through the tissues and can be picked up from skin electrodes placed around the eye; rotations of the dipole will change the difference in potential measured between an electrode pair. Clinical application soon followed (Jung, 1939) and the EOG has continued to be popular, despite its numerous drawbacks which have been reviewed many times (e.g. Carpenter, 1988; for a general technical treatment see Shackel, 1967; for history see Marg, 1951). In principle, the EOG records eye position relative to the head.

9.3.3.1 Problems of linearity

Although the principle of the EOG—a rotating dipole—seems simple, the actual topography of the field is very complex. The corneoretinal potential probably originates from local potential differences and currents in the photoreceptor and pigment epithelium layers, which thus contain the real sources and sinks. The spatial summation and propagation of these currents and potentials throughout the eye and surrounding tissues—which form an electrically continuous, although inhomogeneous, conducting medium—cause the front of the eye to be positive compared to the back. The actual spatial structure of the electrical field in the moving eye and the stationary surrounding tissues is unknown; in view of the very inhomogeneous composition of the periocular structures it is likely to be very irregular. The concept of a single, symmetrical dipole moving in a homogeneous conducting medium is a strong oversimplification which has actually been shown to be wrong (Berg and Scherg, 1991). The asymmetrical nasotemporal topography of the skull, with an electrode placed near the nasal canthus being much more anterior than its fellow temporal electrode, causes further non-linearities. The limitations of linearity, especially beyond about 20-deg eye deviation, are clearly visible in published records (e.g. Kris, 1964). The situation is complicated further by interference of the field of the fellow eye (after unilateral enucleation, the EOG of the normal eye can be recorded around the artificial eye; see Marg, 1951). This means that really independent EOG recordings of the movements of the two eyes cannot be obtained.

Although the anatomical relationships would seem to provide better opportunities for symmetry in the vertical plane, in practice the situation for recording vertical eye movements with the EOG is much worse than for horizontal eye movements, because of the associated movements of the upper eyelid, which change the conduction of the dipole field enough to preclude representative recording of vertical eye movements with the EOG (see Barry and Melvill Jones, 1965; Yee *et al.*, 1985).

9.3.3.2 Fluctuations of the corneoretinal potential

Another problem in using the EOG is that the corneoretinal potential derives from the photoreceptive retinal activities; thus, the magnitude of this potential—and therefore the calibration of the EOG—will vary markedly with the level of illumination and light-adaptation of the retina, getting smaller in the dark. Measurements need, therefore, to be done under a constant level of illumination, to which the subject should be preadapted for at least 20 min.

9.3.3.3 Noise and drift

These are the worst problems of the EOG. Typically, the signal is in the order of 20 $\mu V\ deg^{-1}$ eye movement; this has to be picked up by small electrodes, attached to the skin, with a transition from electrolytic to metal conduction. Polarization phenomena and a fairly high (and fluctuating) resistance make this transition very noisy. Resistance can be lowered by abrasion of the skin (fine sandpaper is quite effective), and by using suitable electrode jelly between electrodes and skin. The electrode–skin contact has to be stabilized by sealing it with collodion, or disposable, double-stick adhesive collars, or adhesive tape (in decreasing order of reliability). A high input resistance of the amplifier is also required but is standard for any modern amplifier. Autonomic reflexes (the psychogalvanic skin reflex) induce large, erratic changes in skin resistance, causing artefacts. Even electrodes that are well attached to the skin can easily move relative to the eye, causing further artefacts, as facial skin is very mobile and subject to the action of facial muscles; apart from motion such muscular activity produces electromyographic artefacts, a further source of noise. Changes in polarization of the electrode surfaces may cause offsets and drifts far exceeding the signal of interest. Constant offsets can be compensated for by an adjustable balancing voltage in the recording circuit; unfortunately, a sizeable drift component usually remains. Some of these problems can be reduced by using non-polarizable, silver–silver chloride (Ag/AgCl) electrodes that are commercially available in disposable form (Beckman and many other makes). Pairs of such electrodes should be connected to a DC amplifier with high input-impedance and with a possibility for compensation of DC offset. Grounding of the subject through a neutral electrode, placed somewhere in the midline on the forehead, eliminates much noise. AgCl is very unstable and broken down by light and, therefore, Ag/AgCl electrodes are difficult to maintain. For best results, they should be used only once. Equipotentiality of electrode pairs can be further improved by soaking them, with the leads shorted together, in saline for about 30 min after taking them from their package. In practice, most EOG users do not go to this much trouble, and they solve the drift problems by switching from DC to AC recording with a time constant of, for instance, 10 s. Naturally, this eliminates positional accuracy along with drift; the signal will exponentially return to zero from any level. However, saccades and other high-frequency phenomena such as a fast-beating nystagmus are preserved, and these are often of main clinical interest. Anyway, whatever one does, even true DC EOG recording will never yield veridical eye-in-head angles because some drift will always remain. High-frequency noise often necessitates further filtering, limiting the effective bandwidth in practice to a range between about 0.5 and 50 Hz. Even then, the residual noise limits resolution to about 1 deg at best and the recordings continue to be marred by artefacts, while many eye movements even larger than 1 deg remain undiscernible (e.g. Byford, 1963, also reproduced in Carpenter, 1988).

Somewhat better results can be obtained in animals, when miniature Ag/AgCl cups are permanently implanted in the orbital bone, using the procedure described by Bond and Ho (1970). Electrode positioning and stability are improved compared to skin electrodes, but noise, drift, and calibration problems persist (Schlag *et al.*, 1983).

9.3.3.4 When to use the EOG

With all these problems, and other methods available, is there still a place for the use of the EOG? There probably is, for the time being, in situations where it is good enough and in situations where better methods are not practical. It is probably good enough for an—essentially qualitative—evaluation of many clinical features such as nystagmus, vestibular dysfunctions, and gross deficits of saccades or pursuit. Defects should be conspicuous, or be characterized by major asymmetries in the left and right direction, and not require accurate or stable calibrations. For such clinical use, recording of a 'monocular' horizontal EOG (with electrodes placed in the horizontal meridian, near the nasal and temporal canthi) is usually recommended (R. J. Leigh, L. F. Dell'Osso; personal communications), to avoid confusion by binocular averaging. A distinct advantage is that useful data can be obtained over the full horizontal range. The vertical EOG is very unreliable, except in signalling blinks.

Studies of eye movements in infants and very young children may often not allow methods other than the EOG. Calibration is difficult to achieve in these non-instructible subjects. A clever calibration technique (see also Hamada, 1984 for a somewhat similar technique) for the horizontal EOG in 2- and 3-month-old infants was described by Finocchio *et al.* (1991), who observed the centring of the first

Purkinje image of an LED target in the pupil as a characteristic of fixation; by dragging the target—and the eye—to known positions and confirming fixation in this way, an experienced observer was able to obtain a valid calibration in a few minutes. Finocchio *et al.* (1991) also gave a very complete, highly recommended, description of their general EOG procedures. They documented linearity of the horizontal EOG over a range of ± 20 deg with the use of bitemporal, disposable Ag/AgCl electrodes (Beckman type 650950). Such bitemporal recording is probably the best option when conjugate eye movements are the object of interest.

The EOG is not recommended for any research that aims at the quantitative analysis of eye movements where a better method could be reasonably employed.

9.3.4 Techniques using non-contact reflection of light by the eye

9.3.4.1 Types of reflection

Light falling on to the eye can have different fates. Basically, it is either absorbed or reflected. Absorption dominates when light passes through the pupil or falls on strongly pigmented structures, such as a dark iris. Light that is not absorbed is reflected. Reflection is of two different kinds: specular and diffuse. *Specular* reflection requires a smooth, mirror-like surface. Its characteristic is that incident light rays are reflected as unscattered light rays, with the angle of reflection being equal to the angle of incidence. Specular reflection conserves the pattern of the incident light and forms a (real or virtual) image of the source. *Diffuse* reflection scatters the reflected light in all directions, and does not form an image, unless it is captured by an image-forming optical device. Both types of reflection occur in the eye. The front and back surfaces of the cornea and lens constitute four surfaces with specular reflection, which form four 'Purkinje images' (P1–P4) of an external light source. Light falling through the pupil is weakly reflected by the retina and leaves the eye in a direction towards the light source. This means that the pupil will be viewed as a black hole, unless the detection pathway is collinear with the illumination pathway ('bright pupil'). The latter situation is found in an ophthalmoscope, in unsophisticated flash photographs, and in a few oculometers. The iris partially absorbs, partially reflects diffusely; the balance depending on its pigmentation. The sclera mostly reflects diffusely. In reality, the different types of reflection are mixed up. The iris reflects diffusely, but it is covered by the specular cornea. The sclera reflects diffusely, but the overlying conjunctiva, covered by a film of tear fluid, will also act as an (irregular) specular surface. The back surface of the lens forms the (real) 4th Purkinje image by specular reflection, but the slightest bit of cataract will add considerable diffuse reflection. In addition, patterns of reflected light from the eye are easily contaminated by reflection from neighbouring structures such as facial skin, eyelids, and eye lashes. All these factors make the relationship between local densities of reflected light and eye movements much less straightforward than it might seem at first sight. Users of instruments detecting distributions of reflected light should be aware of such pitfalls.

Important improvements in the use of reflected light are: (a) the use of infrared (IR) illumination and detection, to prevent flooding of the eye with visible light and to enhance specificity of detection; (b) the use of chopped instead of continuous illumination, in combination with phase-locked amplification of the transducer signal, to further enhance specificity (diminishing the effect of ambient light). These conditions are now easily realized by using pulsed, solid-state IR emitters (IR LEDs) which are almost universally applied in modern instruments. In the design of such instruments, safety limits for IR irradiation levels should be observed. Some guidelines have been formulated by the American Conference of Governmental Industrial Hygienists (ACGIH), and a good discussion was given by Clarkson (1989). Some important numbers should be kept in mind:

(1) the total IR irradiance (including near- and far-IR) from sunlight in temperate zones is about 1 mW cm^{-2};

(2) chronic exposure of the eyes to up to about 10 mW cm^{-2} is probably harmless;

(3) exposure to the 10–100 mW cm^{-2} range should be brief and occasional, while levels above 100 mW cm^{-2} are definitely inadvisable.

To remain on the safe side, exposure levels of the eyes should be kept below 10 mW cm^{-2}.

9.3.4.2 Instruments using local distributions of diffuse reflection

Characteristically, this type of instrument uses a form of local illumination of the eye and of detection of local light intensity, without real image formation or specific feature detection by image analysis. Thus, the specificity of the method depends on the tightness of the relationship between the illumination of a detector and the position of the eye. A typical, reasonably working arrangement (Stark and Sandberg, 1961; used in many experiments by Stark and his collaborators) is shown in Fig. 9.2(A). Here, two light sources locally illuminate the right and left limbus areas, i.e. the transitions sclera–iris, and two small photodetectors are aimed at these same areas. When the eye moves to the right, more of the iris will be in the viewing zone of the right detector, which will thus receive less diffusely reflected light, and the opposite will be the case for the left detector. The difference between the activation of the right and left detectors is processed as the signal representing horizontal eye position. Obviously, specular reflection disturbs this relationship and should be minimized by suitable placement of the emitters and detectors. In principle, this class of instruments records eye position in the head.

Some fundamental limitations of the schema shown in Fig. 9.2(A) are obvious. For horizontal eye movements in the human, the principle of detection of the sclera–iris border will work well over a range of about ± 15 deg; for larger eye movements linearity will sharply decline until saturation is reached, because one limbus border will be occluded by the lids and the detector on the other side will view only sclera. Furthermore, the schema will not work for vertical movements, as there is no visible sclera in the vertical meridian. In most animals, the limbus is occluded even in the horizontal meridian. Several strategies have been attempted to overcome these limitations.

One approach has been to use highly focused illumination and/or detection; for instance with horizontal slits illuminating the limbus in the horizontal meridian, and vertical slits illuminating the pupil–iris transition as a replacement for the invisible

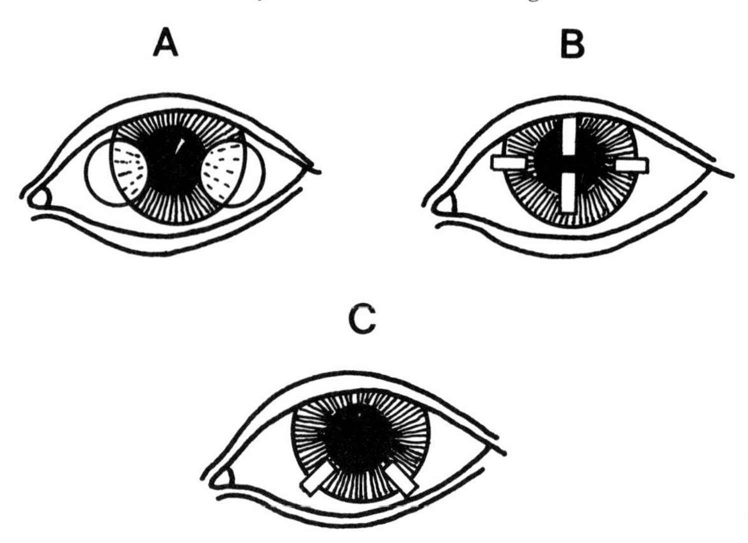

Fig. 9.2 Different schemes for focal illumination of the limbus–pupil area used in IR reflection techniques. A, the basic scheme (e.g. Sandberg and Stark, 1961); B, scheme as used by Wheeless *et al.*, 1966; C, scheme used by Jones, 1973.

limbus (e.g. Wheeless *et al.*, 1966; see Fig. 9.2(B)), or diagonal slits projected on the lower half of the iris (Jones, 1973; see Fig. 9.2(C)). Theoretically, suitable detection of the reflections from each of these projections, with appropriate summation and subtraction, may yield independent signals representing horizontal and vertical eye movements. More specific reflection from the desirable ocular features on the photodetectors, by imaging lenses or strategic positioning of arrays of sensors has also been tried (e.g. Findlay, 1974). None of these strategies has become very popular, because highly focused illumination and/or detection techniques require precise and tedious alignment procedures that are easily upset by slight changes in position, while vertical eye movements can be dissociated from lid movements probably only by retracting at least one of the lids. Yee *et al.* (1985) found that the design of Jones (1973) was inadequate in recording vertical saccades.

The other approach has been to use diffuse illumination, with, for instance, four (local) photodetectors arranged symmetrically around the eye (Nykiel and Torok, 1963; Gauthier and Volle, 1975). Surprisingly, the differences in activation of the two pairs of detectors show a reasonably linear relationship to horizontal and vertical eye position over a limited range. Even more surprisingly, such an arrangement will record (crudely) the movements of a closed eye, probably by differential reflection from the moving corneal bulge protruding through the lids (Comet, 1983). This would seem to imply that also when the eye is open, detection of shifts in the balance of reflection in such an arrangement is global, rather than based on sharply defined ocular features. Notably, the limbus, the pupillary edge, the corneal bulge (which reflects a considerable fraction of IR illumination), and the lid edges are likely to contribute (see Ober, 1994). This makes the relationship between eye movements and output signals somewhat fuzzy; in particular, vertical eye movements will not be discriminated from vertical lid movements.

Many commercial versions of IR diffuse reflection systems have been marketed through the years, with variable success and survival. A complete survey is obviously not practical but some of the better known systems are listed here (addresses are given in the Appendix at the end of this chapter):

1. Applied Science Laboratories (ASL) still market the Model 210 photoelectric tracker as their simplest and cheapest device. This instrument has been fairly widely and successfully used for over a decade. Horizontal (limbus tracking) and vertical (lid tracking) movements are measured over a range of about ±15 deg with an accuracy of 1–2 deg and a precision of about 0.25 deg (vertically somewhat larger). Crosstalk between horizontal and vertical movements is in the order of 10% or less. Notice, however, that although two eyes are measured, for each single eye only one dimension (either horizontal or vertical, set by a switch) is recorded at any time. Horizontal movements as small as a few minutes of arc can be recorded, but this requires effective restraint of the head. The sensor assembly is mounted on either spectacle frames or a headband; in the latter version, a head-mounted scene video camera can be added. The viewing field is unimpeded.

Additional comments: as manufactured, the time constant of the system is 4 ms (effective bandwidth: 40 Hz), but this can been extended to 150 Hz by substituting capacitors; vertical recordings can be improved by taping down the lower eyelids; AC (50 or 60 Hz) lighting interferes and should be minimized (L. F. Dell'Osso, personal communications).

ASL has also prepared an informative '*Eye tracking systems handbook*', dealing not only with their various systems but also with the kind of questions that customers should ask before deciding on any system: candid and recommended literature!

2. Skalar Iris system. This device uses two parallel, linear arrays of photosensitive devices, which leave a viewing field of 90 × 90 deg. Like the ASL-tracker, only one dimension at a time is obtained from each single eye. Bandwidth is DC – 100 Hz (– 3 dB); resolution given as 2 minutes of arc; linearity given as 3% in a horizontal range up to 25 deg and a vertical range up to 15 deg.

3. The Ober 2 System. This computer-controlled, binocular system uses goggles with wide angle illumination and wide angle detection. Horizontal and vertical illumination and photodetection are multiplexed in time to acquire both dimensions simultaneously from both eyes. The diffuse character of the illumination and detection facilitates the set-up, but the penalty for this is a somewhat fuzzy detection with reflections from the whole eye, including the lids, contributing. Lid movement artefacts in vertical saccades are one of the undesired consequences (Ober, 1994), although this is probably true for all IR-reflection systems. The goggles limit the viewing field to about 40 deg horizontally and 30 deg vertically for each eye; in their original design they are fairly uncomfortable due to hard edges. A new type of goggles with a more friendly material is now being introduced.

4. The Microguide Series 1000 System, marketed by Microguide, Inc. This system uses linear arrays of IR LEDs and photodetectors, attached by posts and swivel

joints to a headband and aimed from below and in front of the eyes, to detect horizontal and vertical movements binocularly (up to four signals simultaneously available). Range, linearity, crosstalk, and lid artefacts as specified by the manufacturer are in the range typical for this class of systems (see above).

All these systems have in common that the electro-optical elements are built on spectacle frames, goggles, or headbands which are susceptible to displacement on the head. Thus, long-term stability is difficult to achieve; the most reliable results with IR reflection techniques are obtained with rigorous stabilization of the head and the measuring device on an earth-fixed platform (Dell'Osso and Daroff, 1990).

Distinct advantages of the IR reflection technique in general are a high sensitivity, low noise level, and reasonable bandwidth. Furthermore, simultaneous recording from both eyes is easy. This makes the technique suitable for routine clinical use, especially when precision (low noise) is more important than accuracy. For instance, routine analysis of horizontal nystagmic waveforms or saccades can be done much better with IR reflection than with the EOG, but only within a range of about ± 15 deg. The recording of vertical movements is difficult and probably never ideal; crosstalk between horizontal and vertical is in the order of 10% at best, and the useful vertical range is about half of the horizontal one. The method is also suitable for the study of global gaze control in behavioural studies, and with proper scaling of the hardware it can be adapted for use in young children.

9.3.4.3 Techniques using the position of the first Purkinje image

These techniques use the displacements of the first Purkinje image (P1); the relatively bright, virtual image of a light source, formed by specular reflection by the front of the cornea, and lying about 3.5 mm behind it. The reason that P1 moves with the eye is that the corneal radius of curvature is smaller (about 8 mm) than the distance from the corneal surface to the ocular centre of rotation (about 13.5 mm). The displacements caused by ocular rotation are small (about 1 mm for 10 deg of rotation) but can be resolved very well by suitable imaging or detection instruments, and the relationship between eye rotation and P1 displacement is reasonably linear (except in the case of irregularities of the corneal curvature). The principle works equally well in the vertical as in the horizontal direction. The vulnerable point of methods using the absolute location of P1 alone is their very large sensitivity to translation of the eye or the head (about 10 deg mm^{-1}, which is twice as bad as in limbus reflection systems).

Photographic, eye movement recordings using the location of P1 were obtained by Dodge and Cline (1901) and later, e.g. by Jasper and Walker (1931), in the 1930s. A video technique, allowing the superimposition of the position of P1 (scaled to represent the fixation point) on an image of the viewed scene, was realized in the 1950s by Mackworth and Mackworth (1958). This version used earth-fixed optics and rigorous head stabilization; a few years later (Mackworth and Thomas, 1962) a free-head variant was developed. This used a head-mounted movie camera, viewing the visual scene from a head-centric perspective, in combination with a 'periscope', which projected the P1 of one eye on to this image. Thus, the frame of the image reflected the head direction, and the location of P1 within the frame reflected the eye-in-head, and gaze direction. Although the overall accuracy was no better than a few degrees, this technique was useful in global behavioural studies of gaze control. A limitation was that the output consisted only of images, which were very time-consuming to analyse. A technique to obtain electrical signals representing horizontal and vertical position of P1 (and thus the eye) directly, using an oscilloscope spot-tracking technique akin to Rashbass's (1960) principle, was published by Palmieri *et al.* (1971). Once again, recordings with any fidelity required the most rigorous stabilization of the head; probably this, in combination with the fairly demanding optics and electronics, prevented the method from acquiring popularity, although its signal quality appeared to be quite good.

More recently, new methodological papers based on this principle have been published. Eizenman *et al.* (1984) described an instrument that used a one-dimensional array of 18 phototransistors, in combination with an algorithm as used in gamma-ray cameras, to determine the centre of P1 very precisely. A 2-D version was described by the same group of authors (Frecker *et al.*, 1984). Impressive figures were given on performance (resolution 30-s arc over a bandwidth of 500 Hz; linear range ±18 deg). Obviously, subjects had to hold their head very still; furthermore, linearity in the 2-D version was not very satisfactory (P. E. Hallett, personal communication). It is probably fair to say that this technique has now been superseded by modern video techniques processing the localization of more than one ocular landmark (see below).

9.3.4.4 Detection of the first and fourth Purkinje image

The principles of this technique were described by Cornsweet and Crane (1973). The fourth Purkinje image (P4) has almost the same size, and is formed in almost exactly the same plane, as the first Purkinje image (P1), although it is fainter by about a factor of 100. During translations of the eye, both images move through the same distance and in the same direction as the eye, while during rotations the separation between the two images changes proportionally with the sine of the angle of rotation. Detection of the relative position of P1 and P4, disregarding their absolute position, will therefore yield the angular eye position in a form uncontaminated by translations. Although the geometrical principle (Fig. 9.3) is fairly straightforward, its implementation in a reliably working instrument is obviously no trivial matter. It has been realized and further developed in five generations of the SRI Double Purkinje Image Tracker by a group at the Stanford Research Institute (see Clark, 1975; Crane and Steele, 1978, 1985). Crucial elements are the specific detection of the P1 and P4 images of an IR source, the accurate measurement of their horizontal and vertical distances, rejection of translations, and fast recovery from blinks (which occlude the images). The positions of the Purkinje images are continuously tracked by servo-controlled mirrors that keep images of P1 and P4 stabilized on quadrant photodetectors; the horizontal and vertical output signals, representing eye positions, are in fact derived from the angles of these mirrors. Thus, the dynamics of the servo-tracking mechanisms determine the overall bandwidth of the system.

The combination of dedicated development with effective marketing has made the SRI tracker into a fairly widely used, though expensive instrument. It has found its

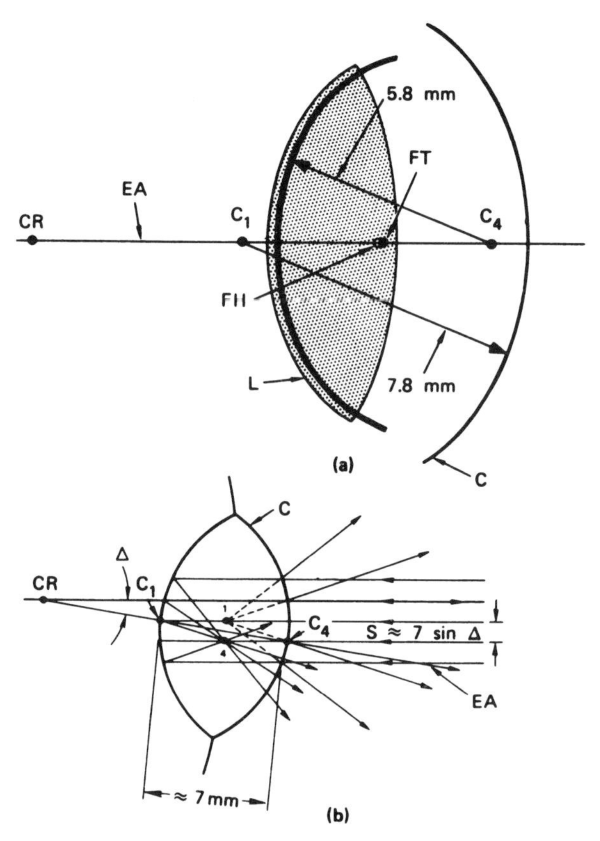

Fig. 9.3 Diagram to illustrate the position of the first and fourth Purkinje images, and their mutual distance *S*, being about equal to 7 times the sine of the angle of eye deviation, Δ. (From Cornsweet and Crane, 1973.)

application mainly in basic oculomotor research requiring precise and accurate recording in combination with demanding psychophysical tasks that would be compromised by any ocular irritation or reduction in acuity. The merits of the SRI tracker were extensively discussed by Steinman and Levinson (1990). A major strength of the system is that it allows very precise (noise-level about 20-s arc), no-contact measurements of horizontal and vertical eye position with adequate bandwidth (> 100 Hz). Measurement of both eyes is possible, but essentially requires duplication of the whole instrument. Its range is limited by the occlusion of P4 by the iris during eye deviations larger than about 15 deg from the centre (or less with small pupils); thus, the instrument is fundamentally unsuitable for the study of large movements. Optimal, constant performance requires rigorous head stabilization: changes in the distance between eye and instrument, for instance, will defocus the Purkinje images and degrade performance. Auto-focusing and auto-staging systems have been built into generations IV and V of the tracker to correct for this; unfortunately, these operations affect the calibration of the instrument, a problem that has made it necessary to disable such auto-correction devices in studies requiring reproducible calibration with precision as good as the stability of the fixating human eye (see, for example, He and Kowler, 1992). Furthermore, homogeneity of calibration throughout the range depends on the ideal shape of two reflecting surfaces (the front of the cornea and the back of the lens). Any irregularities in these surfaces will introduce local non-linearities of calibration. Other optical aberrations (such as local opacities in the lens) may cause spurious reflections that can be confused with P4. In some subjects the P4 image is too faint to allow the tracker to do its work. There is a further problem with dynamics, deriving from the use of P4. The high accelerations associated with saccades will cause movements of the lens within the eye: because of the elasticity of the zonula, the lens will lag somewhat in the beginning of a saccade and overshoot at the end. As a result, saccades measured with the tracker will show a small 'backshoot' in the beginning and a considerable overshoot at the end. Deubel and Bridgeman (1995) recently found that, as a result of these distortions, peak velocities of saccades, measured with the SRI tracker, are about twice as large as peak velocities measured simultaneously with the search coil (Fig. 9.4). These distortions are not due to bandwidth limitations (fifth generation trackers can follow the accelerations of 15-deg saccades without ever losing the Purkinje images) but to the mechanics of the eye.

Importantly, these lens oscillations (with possibly also some sloshing of the lens in its capsule) will not only corrupt saccadic dynamics as measured in the SRI tracker, but also cause real instability of the retinal image during all saccades; see Deubel and Bridgeman, 1994. The somewhat sad implication is that *no* recording of ocular rotation, however accurate, will be a veridical representation of retinal image position.

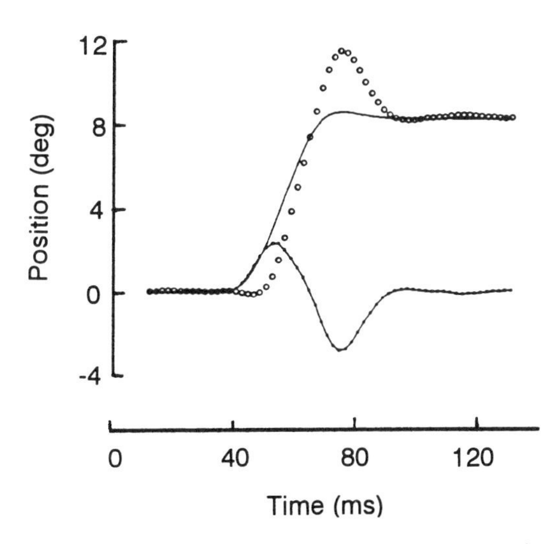

Fig. 9.4 Horizontal saccade of about 8 deg, recorded simultaneously (from the same eye) with the search coil (continuous line) and the SRI tracker (generation V; open circles). The *difference* (coil – tracker; line with dots) trace shows that the initial error of the tracker is a lag + backshoot, and the late error is a large overshoot. Also, the slope of the tracker trace, representing peak saccadic velocity, is substantially larger than in the coil trace. (From Deubel and Bridgeman, 1995.)

The size of the linear range depends on the subject, and is seldom found to exceed 5 or 6 deg when complete and careful calibrations are made and repeated frequently (R. M. Steinman, personal communication).

Finally, corrective spectacles or contact lenses cannot be worn in the tracker, and if subjects need these they have to be substituted by bulky lenses that are positioned far from the eye (and thus change image size).

In summary: the SRI tracker should be used with prudence. Subjects should be selected for good performance and extensively calibrated. Stimuli should fit within the restricted range of the instrument; usually they will be imaged on a CRT. This makes the tracker especially suitable for experiments requiring high visual and oculomotor resolution over a small range with repeated sessions and many trials in relatively few subjects with sufficient experience. Not surprisingly, a prime application of the instrument has been the study of stabilized images (Kelly, 1979; but see Steinman and Levinson, 1990). The tracker is not suitable for addressing the dynamics of saccades; neither is it suitable for one-time experiments in arbitrary, inexperienced subjects such as patients, or experiments requiring head movements or natural visual surrounds.

9.3.4.5 Techniques using the first Purkinje image and the pupil

By using the location of P1 relative to the (centre of) the pupil, these techniques are able to dissociate ocular rotations from translations. In principle, the distance between the centre of the pupil and P1 varies proportionally with the sine of the change in gaze angle (Fig. 9.5); the centre of the pupil travels about twice as fast as P1. The principle is well known in clinical orthoptic practice, where it is applied in the 'Hirschberg test' to evaluate binocular alignment of the eyes (see Miller *et al.*, 1993). It is also applied in the EOG calibration procedure of Finocchio *et al.* (1990). All instruments based on this principle use video-imaging in combination with digital data-processing. In many cases, they will provide the pupillary diameter along with horizontal, vertical, and (sometimes) torsional eye movements.

An early version of such an instrument was the 'Honeywell oculometer', described by Merchant *et al.* (1974). It used the 'bright-pupil' technique (collimated IR illumination from a frontal source being reflected towards a camera in effectively the same direction), in combination with the first Purkinje image of the same IR source. It was a monocular instrument. Thresholding techniques were used to extract the spatial coordinates of the pupil and P1 from the video raster. Horizontal and vertical eye positions were computed from these and provided as electric signals (an essential improvement compared to the earlier Mackworth video recording technique). It was quite an advanced instrument. It could be used in a remote position (up to about 2 m from the subject) and allowed considerable freedom of the head: the instantaneous working area of the camera, within which the tracked eye had to be, was a cubic inch (about 16 cm^3); additional servo-mirror and auto-focusing systems extended the effective working space to a cubic foot (about $2.8 \times 10^4 cm^3$ (with temporary loss of signals during fast head movements). Digital processing included dealing with borderline conditions (such as partial obstruction of the pupil), calibration, and linearization procedures. The overall accuracy and precision of the resulting output signals was in the order of 1 deg, with temporal resolution

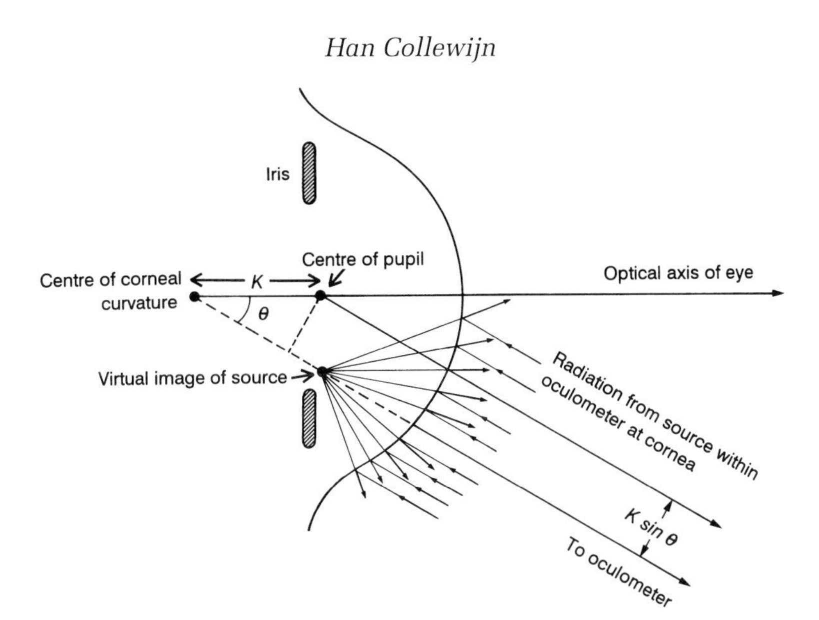

Fig. 9.5 Principle of oculometers using the first Purkinje image and the centre of the pupil. The distance between these features is a factor K times the sine of the angle of eye deviation, θ. K is drawn as the distance between the centre of ocular curvature and the centre of the pupil. Figure adapted from Merchant *et al.*, 1974. Actually, this scheme is not quite correct, because the oculometer views the entrance pupil, i.e. the pupil seen through the cornea. As a consequence, K is slightly larger than indicated here; its value is about 4.75 mm (see Gale, 1982).

restricted by the video frame rate and additional filtering (time constant 0.1 s). The working range was 30 deg to the right and left, 30 deg upward and 10 deg downward. The pupil diameter had to be 3 mm or greater; therefore, ambient light had to be at a modest level.

More recently, there has been a surge in the development of modern versions of this technique. Important improvements have become possible due to: (a) the availability of miniature TV cameras that can be easily worn on the head; (b) the revolutionary development of computers, with cheap PC systems now having the computing power found only in main-frames decades ago. Most of the current systems use the dark-pupil technique. Several such systems are now marketed; clearly they are all still being further developed, and undoubtedly others will come on the market. With the very rapid changes in suppliers, capabilities, and prizes, potential buyers are advised to obtain the most recent information and, if at all possible, a hands-on try-out before acquisition. Some of the current systems have many options to fit different applications. Some systems that the present author is aware of are described below (addresses are given in the Appendix to this chapter).

1. The Applied Science Laboratories Series 4000 systems. There are two versions of this system. One has the optical package remote, and includes an option in which an extra mirror tracks the head over ranges of ±15 cm. The principles and performance of this version show a striking resemblance to the 'Honeywell oculometer' in almost every detail. In another version, the optical package is mounted on a helmet, so that

subjects are entirely free to move or even walk around. The head-mounted version includes a scene camera. A further option is a magnetic head-tracking system, providing head position and orientation, which can be used in subjects who are seated or whose head will remain within about a 2-cubic foot volume of space (about $5.7 \times 10^4 cm^3$). Under these restrictions, it is possible to compute gaze with respect to the environment with the helmet-mounted version. The magnetic head-tracking system can also be used to enhance the speed of the head-tracking of the remote optical package in the cubic foot version, enabling, in addition, auto-focusing capability. All versions are monocular, bright pupil systems. This has the disadvantage that the contrast between pupil and iris decreases with miosis, because a smaller pupil reduces the light reaching the retina, and thus the intensity of the reflected light (minimum pupil diameter is specified as 3 mm; ambient light has to be at low level). Horizontal and vertical gaze (of one eye) is provided with an overall accuracy of 1–2 deg. The update (sampling) frequency is 60 Hz (European versions: 50 Hz); output delay equals three video frames (50 or 60 ms). These systems are especially suited for behavioural and ergonomic types of studies, where global gaze behaviour in relation to the environment (e.g. lay-out of cockpit or automobile instruments) is the main concern; the remote versions allow gaze studies in unencumbered subjects. Obviously, spatial and temporal resolution are limited and measurements of binocular movements or torsion are not included.

2. The 'EL-MAR' System 2020. This binocular system uses two miniature video cameras, mounted on a spectacles–helmet assembly; the cameras view IR images of the two eyes by way of 'hot' mirrors that reflect IR light but leave the visual field largely free. The system tracks the P1 images of two to three sources (rather than one, to increase range) and the pupil, of which the centre is computed. The following details were kindly provided by two users (Drs R. J. Leigh and G. D. Paige, personal communications; see also DiScenna *et al.*, 1994). The video sampling rate is 120 Hz (bandwidth 60 Hz). The system has a set-up procedure to customize the software for the geometry of individual subject's eyes; together with the subsequent calibration procedure this preparatory stage takes about 2 min. The system provides horizontal and vertical positions of both eyes, in almost real-time (processing delay 2 samples = 16.7 ms). With an artificial eye, the noise level (SD) is 0.04 deg, and linearity is acceptable over a range of about ±40-deg horizontal and ±30-deg vertical. Insensitivity for translation is quite good: about 0.3-deg artefactual rotation was measured for a 5-mm displacement between the system and an artificial eye; all these values appear to be slightly less favourable during the tracking of a real eye. In a comparative test (DiScenna *et al.*, 1995) this video system approached the quality of the search coil (although with higher noise, especially during motion, and with overestimation of saccadic peak velocities) and was much superior to the EOG (Fig. 9.6). Present drawbacks are limited bandwidth (borderline for obtaining velocities), pixel noise, and the unavailability of torsion.

3. ISCAN systems. This supplier offers a range of eye movement monitors based on similar principles as outlined above, but using the dark pupil technique. This makes the systems suitable for any ambient light condition. The basic parameters measured are horizontal and vertical positions and pupillary diameters of both eyes. Maximum

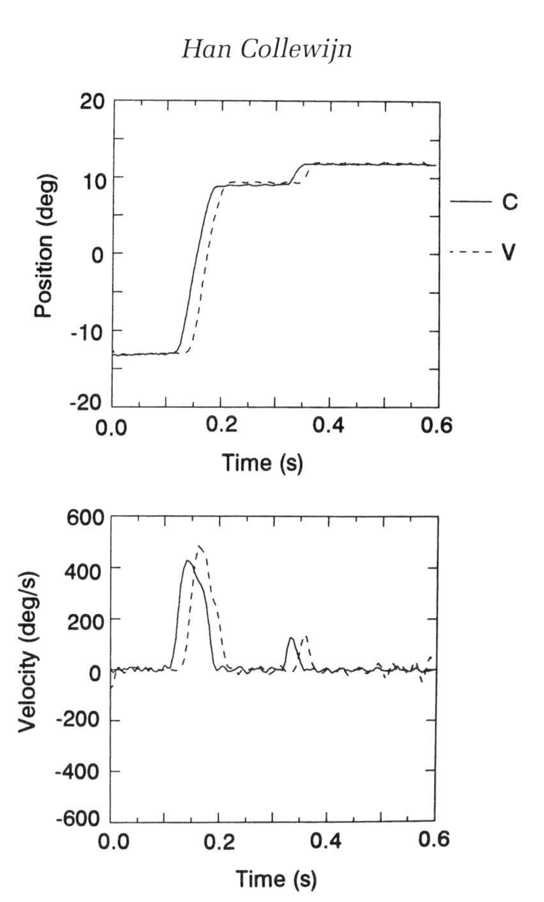

Fig. 9.6 Comparison of vertical saccades as recorded (from the same eye) with the search coil (continuous line) and the EL-MAR video system (interrupted line). Upper panel, position; lower panel, velocity. Notice the time delay (2 frames = 16.7 ms) of the video system, and the slight overestimate of peak velocity. (From DiScenna *et al.*, 1995.)

sample rates are specified as 240 Hz (standard is 60 Hz). Options include remote optics and head-mounted versions (headband, video mask, or helmet). The helmet-mounted version can be fitted with a scene camera and a magnetic head-tracking system that allows complete gaze analysis with a freely moving head, with software to map the line of sight to user-defined planar surfaces. For the latter option, an accuracy of 1 deg in a range of ±25 deg at 60 Hz sampling rate is specified. Probably, higher accuracy and precision can be achieved in conditions with the head fixed.

Recently, a connection has been formed between ISCAN (USA) and SensoMotoric Instruments (SMI) in Europe (see Appendix to this chapter for the address).This European group has been actively developing 'video-oculography' systems for several years (Clarke *et al.*, 1991; Scherer *et al.*, 1991); the special attraction of their system is that it includes the measurement of torsion, however, only off-line, not in real-time!

Several laboratories are committed to further developments of video procedures for the measurement and display of 3-D eye movements, particularly torsion: notably the ENT group in Sydney, Australia (see Moore *et al.*, 1991; Curthoys *et al.*, 1994; temporal frequency at present limited to 6 Hz) and the ENT group at the University of Limburg, Maastricht, The Netherlands (Kingma, 1994).

In summary: video techniques, combined with computer processing, using the positions of first Purkinje image(s) and pupillary centre, appear to be a promising development that could well become the technique of first choice for many types of research in the coming years. The technique will work under most experimental conditions, except with closed eyes, and except that it will be compromised, at least to some extent, by corrective glasses or contact lenses worn by the subject. Main aspects to consider critically now are temporal and spatial resolution, range, and veridicality of coordinates. For instance, not all systems specify the hierarchy of rotational axes, and especially the measurement of torsion may become unreliable at horizontal or vertical eccentricities of more than 15 deg (H. Kingma, personal communication). The successful measurement of torsion with video techniques will probably also remain subject-dependent, as the amount of contrast and the width of the iris vary widely. When performance is optimized, the video techniques are likely to become the method of first choice. In particular, they should be expected to replace the EOG and the limbus reflection techniques in most applications.

9.3.4.6 More simple techniques using video images

For completeness sake, three video-based techniques are briefly mentioned which are much less sophisticated but that have been used in recent years with success in specific applications.

The important oculomotor research by J. T. Enright has been wholly done by off-line, manual frame-by-frame analysis of close-up video recordings of the eyes of subjects who were firmly stabilized on a bite-board. By the strategic placing of two cameras under various angles, Enright has been able to analyse diverse types of eye movements, including vergence, torsion and, most uniquely, translational movements of the eyes (see Enright, 1984; Enright and Hendriks, 1994). Naturally, the analysis is extremely tedious and time-consuming, and unsuitable for the detailed analysis of fast phenomena such as saccadic velocities. Nevertheless, Enright made several important discoveries with this simple method.

A second example is the recent work of Land (1992), who studied eye head coordination during driving with a helmet-mounted TV camera, which recorded the (enlarged) eye as well as the road scene. Off-line, frame-by-frame analysis was achieved by superimposing a cursor generated by a computer program, which modelled the outline of the iris of an eye rotating about its centre. The angular coordinates to generate the fitting cursor were taken as the angles of the line of sight. They were stored and also used to generate a spot in the image of the scene, representing the gaze direction. Head movements were inferred from changes in the position of distant objects relative to the video frame (notice the resemblance to the old Mackworth technique!). The method was apparently satisfactory for the analysis of large gaze changes during brief sequences, e.g. negotiating a road junction. Obviously the method could not pretend any analysis exceeding the global characteristics of gaze behaviour.

This reminds us that even direct observation tells us a lot about the eye movements of our fellow people, not only in clinical examinations but in everyday life. Just by looking at somebody we get quite a good idea of what that person is looking at, and in many cases such observations have strong connotations with behaviour, emotions, and interpersonal relationships. Along the same line, static photography is often still very useful in documenting oculomotor conditions in the clinic (Buquet and Charlier, 1994). For completeness sake, subjective after-image techniques may be mentioned here as well: although they are technically not a recording method, they may occasionally be useful in informal, qualitative observations or—with prudence—in problems involving retinal image position. For some examples see Verheijen (1961) and De Bie (1985).

Finally, Ott and Eckmiller (1989) obtained video recordings of fundus images in monkeys and human subjects using a scanning laser ophthalmoscope; changes in torsion were analysed in off-line, frame-by-frame measurements of the coordinates of the fovea and of a conspicuous vascular landmark. The technique is reminiscent of Cornsweet's (1958) fundus-scanning technique, which never caught on. With the modern developments in video techniques described above, there is probably no future for fundus-scanning techniques for the measurement of eye movements. However, fundus images produced by scanning laser ophthalmoscopes are extremely useful, even unique, in determining the retinal location of preferential fixation. Such determinations are essential in clinical studies of eccentric, parafoveal fixation (see, for example, Guez *et al.*, 1993). Another example of an oculomotor application of fundus photography is the finding of a permanent change in torsional eye position after unilateral section of the vestibular nerve (Curthoys *et al.*, 1991).

9.3.5 Search-coil techniques

9.3.5.1 Principles of search-coil measurements

A good 30 years ago, D. A. Robinson (1963) published his classical paper on the scleral search-coil technique. Entering oculomotor research from an industrial background revolving around applications of magnetic principles, Robinson, who needed a good recording system for his planned oculomotor work, saw immediately how to use magnetic induction for this purpose. The principle is as follows. An alternating (AC) magnetic field is created by passing an AC current through suitable field coils. When a sensor coil is placed in this field, an AC voltage is induced in it with the same frequency as the field and a magnitude proportional to the surface area and number of turns of the sensor coil, and the number of field lines encompassed by the coil. This number depends on the strength of the field and the orientation of the coil. When the sensor coil is parallel to the field lines (i.e. roughly orthogonal to the field coils) the induced voltage is zero (Fig. 9.7(A)). Rotation of the coil increases this voltage (Fig. 9.7(B)) to a maximum at 90 deg, following a sine function (Fig. 9.7). The magnitude of the induced signal is also proportional to the field frequency, because the induction depends on the *velocity of change* in the magnetic flux (Faraday's law; transformer principle). The sinusoidal relationship between induced voltage and orientation is almost linear for angles up to 20 deg; it has opposite polarities for the two directions of rotation from the zero position and it can be unambiguously linearized for

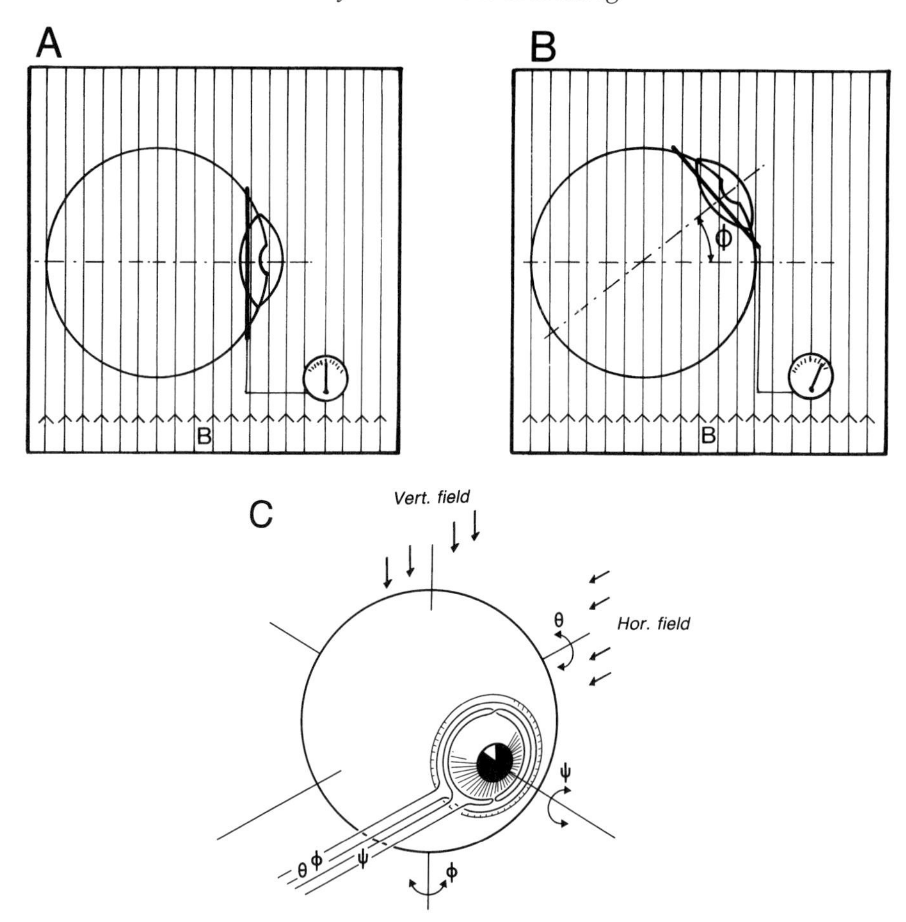

Fig. 9.7 Arrangement of AC magnetic fields and search coils according to Robinson (1963). A, B: relationships for a single dimension. Voltage induced in the search coil is zero (A) when the coil is parallel to magnetic field lines. Rotation of the eye and coil through an angle ϕ results in the induction of an AC voltage that is proportional to *sine* ϕ. C: three-dimensional arrangement with two AC fields in spatial and phase quadrature, and two search coils. Notice that the coil signals as such do not strictly follow angular coordinates of conventional systems (such as Fick's ϕ, θ, and ψ) and that pure coordinates require the appropriate computational corrections (see text).

larger angles, so that eye movements throughout the oculomotor range (with the head stationary) can be measured. The magnitude (and polarity) of the induced signal is easily measured with a lock-in amplifier. Under the control of a reference signal (derived from the field) such an amplifier selectively detects signals that have the frequency of, and a specified phase relationship to, the reference signal. Signals at other frequencies, and signals at the proper frequency but 90-deg phase-shifted relative to the reference, are rejected. This property was used by Robinson to make the system two-dimensional. By creating two AC fields at the same frequency, but orthogonal in space (horizontal and vertical) and phase (90 deg shifted in phase), two orthogonal (90 deg out of phase) signals are induced in a single sensor coil. One component represented horizontal orientation, the other vertical orientation; the components are perfectly separated by a dual-phase lock-in amplifier. (An alternative to separation by phase-orthogonality is the use of two sufficiently different frequencies for the horizontal and vertical field.)

Robinson (1963) positioned two pairs of field coils (diameter 2 ft (about 61 cm)) in an open, cubic arrangement, with the coils generating the horizontal field in parasagittal planes, and the coils for the vertical field in horizontal planes. The subject was positioned with his eye near the centre of this cube. To attach the sensor coil to the eye, Robinson mounted the coil on a large, scleral suction lens of the type used at that time in optical lever studies. In the chosen field configuration, the induced voltages were minimal when the subject looked straight ahead. Robinson completed the capability of his system by winding a second coil on the lens which, following a 'figure of 8' pattern in its turns, effectively became a sagittally oriented coil. In combination with the vertical field and a third lock-in amplifier, this coil yielded the torsional position of the eye (Fig. 9.7(C)).

Thus, 3-D magnetic measurement of eye movements was born. It was immediately applied in fundamental work on the mechanics of human eye movements (Robinson, 1964, 1965). A procedure for the permanent implantation of (single) search coils in monkeys was soon developed by Fuchs and Robinson (1966); an improved procedure was later published by Judge *et al.* (1980). Use of the search coil in humans was later facilitated by replacing the suction contact lens by self-adhering, silicone annuli: first for single coil, 2-D detection (Collewijn *et al.*, 1975) and later for dual-coil, 3-D detection (Ferman *et al.*, 1987). Robinson (1963) had created a superior method that in many ways became the 'golden standard' of oculomotor recording. Many factors contribute to the excellent properties of the scleral search coil technique:

1. The signals from the coil are a direct representation of the orientation of the coil and thus—provided attachment to the globe is solid—of the eye. There are no indirect links such as optical correlates of eye rotation. Within the homogeneous region of the field, the coil is totally insensitive to translation.
2. The technique can provide 3-D positions of both eyes in real-time without any restriction of the visual field and under any lighting conditions; it works even when the eyes are closed.
3. Accuracy, precision, range, bandwidth, and linearity can be easily made to surpass any realistic requirement for oculomotor recordings. Apart from configurations of metal that distort the field, there are virtually no physical factors that interfere with fiducial recording.

The only disadvantages of the use of the coil in human subjects are its mildly invasive character, which causes slight discomfort and limits wearing time to about 30 min, and the expense caused by the limited life-time of the coils.

The use of the principle is not limited to eye movements. The recording of head rotations with a head-fixed coil is a logical extension; a less obvious application is the reliable recording of upper eyelid movements with a miniature search coil attached to the lid (Guitton *et al.*, 1991).

9.3.5.2 Procedures for coil insertion

Humans For humans, the only commercially available coil is the silicone annulus construction manufactured by Skalar, Delft (address given in the Appendix). (An alternative construction, using sandwiching of a coil between two soft lenses, has been described (Kenyon, 1985) but the procedure appears rather awkward and it is not known if it is actually used). The Skalar coil is supplied in two versions: single coil (for horizontal and vertical) and dual coil (for horizontal, vertical, and torsion;

Some practical advice

1. It is easiest to store the coils with the leads stretched out, in long boxes, to prevent entangling. Before use, the coil should be sterilized by immersion in 3% hydrogen peroxide (H_2O_2) for about 10 min, followed by immersion in 70% ethanol and drying, or thorough rinsing in sterile saline (0.9% NaCl). Hydrogen peroxide and ethanol are reasonably effective against viruses and bacteria, and evaporate without leaving any aggressive material. However, true sterility is not achieved by methods of this kind. As the coils do not tolerate more aggressive sterilization techniques, it is best to discard a coil after any use involving a realistic risk of bacterial or viral contamination (especially in a clinical setting). When repeated sessions are run with the same subjects, it is good practice to assign each individual his/her own coils. Coils should not be left immersed in fluid for extended time, as this may damage the wire and insulation.

2. Immediately before insertion, the eye(s) should be topically anaesthetized with two successive drops of an ophthalmic anaesthetic, such as oxybuprocaine–HCl 0.4%. Obviously, the anaesthetic should meet the usual clinical criteria of sterility and expiration time; these are best guaranteed by the now available single-use packages ('minims'). The effect of the anaesthetic is almost immediate and wears off in about 20 min; repeated administration is not recommended.

3. After confirmation of integrity (resistance check), the annulus is briefly immersed in saline, carried on a fingertip and picked up with the suction tool (which is connected to a vacuum pump). Suction is in effect as long as the air vent is covered by a finger. Slight pressure of the annulus against the suction ring may be needed to stop air leakage and obtain firm adherence.

4. Insertion requires the subject's cooperation, who should open the eye widely, assisted by gentle manual retraction of the lids by the subject or the experimenter (Fig. 9.8(B)). The coil should be positioned immediately in the correct position, concentric with the cornea, and with the lead wires towards the nasal canthus. As the upper lid is the most mobile one, it is a good strategy to slip the annulus first of all in position under the upper lid, then angle the tool and coil down to achieve full insertion. The annulus is sealed to the eye by slight pressure on the tool, after which it is released by removing the finger from the vent hole. Strong pressure should be avoided.

5. Inspection should confirm that the coil is in the correct position, and comfortable. If the position is unsatisfactory, it is best to take the annulus out and reinsert. Pushing the annulus around on the eye is often unproductive and potentially traumatic.

6. The lead should be taped somewhere near the top of the nose, leaving a smooth loop to the annulus, allowing maximum eye deviations. Intrusion of the wire into lids or eyelashes should be minimized.

Fig. 9.8(A)). The silicone annulus fits around the corneal bulge and is shaped such that it adheres firmly to the eye when fluid and air between annulus and conjunctiva are evacuated by mild pressure (Collewijn *et al.*, 1975). It is important that the coil is inserted atraumatically and immediately in the correct position (i.e. concentric with the cornea) and, when a torsion coil is used, with the plane of symmetry in the vertical meridian (as in Fig. 9.8(C)). In practice, the suction tool as described by Collewijn *et al.* (1975), and supplied by Skalar, is the most effective way to handle and place the coils (Fig. 9.8(B)).

7. As soon as measurements are completed, the coil should be taken out. One method is to slip the blunt (!) tip of a pair of fine, curved forceps under the annulus to grab it and take it out. This is facilitated by having the subject look away from the forceps. An alternative method (slightly more dangerous to the coil) is to simply pull at the wire. With both methods of removal, the subject should try to keep the eye widely open to prevent damage to either eye and coil.

8. Immediately after use, the annulus should be cleaned by thorough rinsing under lukewarm tap water, dried with tissue paper (Kleenex) and inspected for damage. With gentle handling and luck, coils will last, on average, for about five sessions. Breakage of a lead wire near the point of entry in the annulus is the most common failure. Tears in the silicone annulus, exposing the coil, are another reason to discontinue use in view of possible trauma.

9. While the procedures above work well, some users find it easier to slip an annulus on to an eye without any tools; with sufficient dexterity and confidence this is certainly an alternative, although alignment, especially of a torsion coil, can be less well controlled.

10. The single coil annulus will fit almost any adult subject except in cases of strongly abnormal ocular shape, such as in extreme myopia. In most subjects, adherence is tight and repeated calibration will confirm stability. Such confirmation is recommended when stability of absolute position is crucial. The dual coil is slightly less flexible, and therefore less tolerant to differences in ocular shape; accordingly, it will not adhere well to some eyes. Even with good adherence, slip around the torsional axis is intrinsically more likely than around other axes due to the mechanical relationships. The corneal bulge offers no resistance against torsional slip of the annulus on the conjunctiva or of the conjunctiva on the sclera. Empirically, torsional slip appears to occur mainly during large eye movements and forceful blinks. In evaluating torsional slip, it should be realized that the eye is intrinsically less stable in torsion than in the horizontal and vertical directions (Ferman *et al.*, 1987). The fact that in binocular recordings torsional movement is largely conjugate (Collewijn *et al.*, 1985; Van Rijn *et al.*, 1994) argues against any continuous slip of the torsional coil.

11. With proper handling, the annuli are tolerated well by a majority of subjects, including patients. Wearing time should normally not exceed 30 min to prevent corneal drying and possible build-up of ocular pressure. Subjects should be encouraged to blink or close the eyes between measurements. Some subjects benefit from a tear-substitute after wearing the coil. Slight, transient blurring of visual acuity (which wears off within 30 min after removing the annulus) occurs occasionally; this makes the annulus less suitable for experiments in which maximum acuity cannot even be marginally compromised. The annulus can be easily worn together with hard contact lenses (which should be simply left in the eye) and, of course, with most spectacles (heavy metal frames should be checked for interference with the field). Experience for over 20 years in basic and clinical research with the annulus has been positive; research groups that give the coil technique a serious try will usually stick with it forever because of the unsurpassed quality of the signals, and, as one colleague recently remarked, 'It is much easier to slip a coil into an eye than to prepare a subject for EOG recording.'

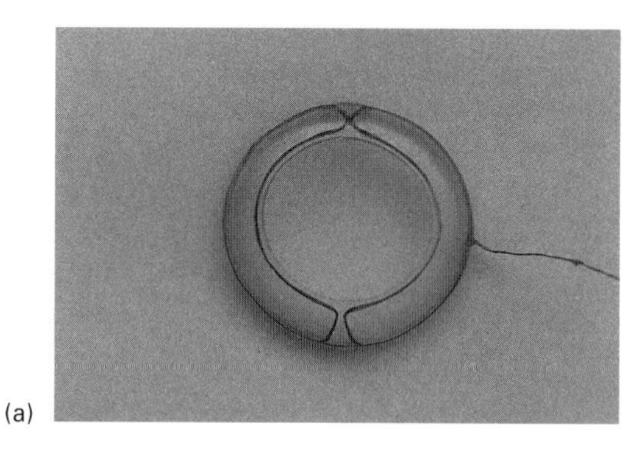

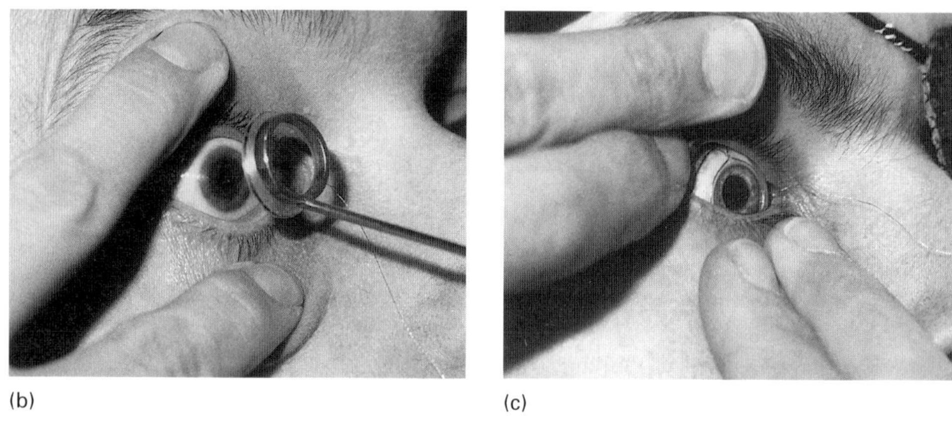

Fig. 9.8 The Skalar search coil and its insertion. A, dual coil, embedded in silicone annulus; B, coil about to be inserted with the suction tool— eyelids are gently retracted by experimenter or subject; C, view of a properly inserted dual coil.

Animals In animal oculomotor research, coil techniques are now standard in all major laboratories. Permanent implants following techniques pioneered by Fuchs and Robinson (1966) and Judge *et al.* (1980) are used in animals with relatively large eyes, such as monkeys, cats, and rabbits. The appropriate wire for implanting is multistranded stainless steel, Teflon-insulated (type AS632, Cooner Wire, price about $1.14/ft for 500 ft). In smaller animals, implants are not practical, but tiny search coils can be glued (with tissue-glue) on to the eye for the duration of a session. Such coils (80 turns of double-insulated copper wire (diameter 0.02 mm), coil diameter 1.8 mm, weight 1 mg) are commercially available from Sokymat S. A.; see Kasper *et al.*, 1987; Kasper and Hess, 1991). For 3-D recording in animals, the Robinson 'figure of 8' design is not practical, but two, approximately orthogonal, coils can be implanted. As an alternative, Hess (1990) has developed a prefabricated dual-coil in which two miniature coils (measuring torsion) are mounted upon a regular coil.

9.3.5.3 Coordinate systems: corrections and calibrations

Robinson's 1963 paper did more than document a technique to obtain reliable eye position signals by using the electromagnetic induction principle and a lock-in amplifier. It also addressed the kinematics of ocular rotation about several axes, a subject that has its roots in the 19th century, when Fick, Helmholtz, and Listing thought about appropriate coordinate systems for describing eye movements. They proposed systems with various hierarchies of nested rotation axes in gimbals, thus defining 3-D coordinate systems in which it was exactly clear what an angle of 'horizontal' or 'vertical' rotation really meant. Obviously, such subtleties were of little practical interest as long as the EOG or IR reflection were the technical standards in oculomotor recording, because there was no way to distinguish between axial systems with such techniques. They became relevant with the introduction of the coil, and quite appropriately Robinson (1963) gave the exact relationships between the three coil signals and a specified 3-D rotational coordinate system, for which he chose a Fick system. Specifying the coordinate system was necessary, because for rotations about more than a single axis the relationships between the raw coil signals and the ocular rotation angles depend on the exact definition of the axes. In addition, the geometry of the coil and its projection on the magnetic fields had to be taken into account. The main message here is, that the raw output signals from a coil system are *not* equivalent to the angular coordinates of any specific system; the signals need appropriate mathematical treatment. A very recent survey of the mathematical background of 3-D rotations of the eye is given by Haslwanter (1995). A treatment of the various coordinate systems that are in use and the necessary transformations between raw coil signals and specific systems is beyond the scope of this chapter and only a few references to the pertinent literature can be given here.

1. Robinson's (1963) treatment, specifying Fick angles, is correct, but does not take into account the effects of misalignment of the coil-zero position with regard to the line of sight. Such misalignment originates from both the variability in the positioning of the coil on the eye and from the non-coincidence of the foveal line of sight with the optical axis of the eye. This type of error, which can amount to several degrees, causes impurity (cross-couplings) of axes when not corrected. A correction procedure, also including linearization, was described by Ferman *et al.* (1987) for Fick angles. Unfortunately, the polarity of the angles was not clearly specified in that paper; the given relationships are correct when *positive* signs are attributed to vertical (θ) upward, horizontal (ϕ) leftward, and torsion (ψ) with the upper pole rotating to the right side of the subject.

2. A further complication is caused by spurious inductions from the field into cables and amplifiers when the coil is in the zero position. These cause offsets on the signals that corrupt the relationships given above. Every precaution should be taken to minimize such false induction. Appropriate measures are:

(i) electrical grounding of the subject with a large, skin electrode connected to the system's earth;

(ii) suitable electrostatic shielding of the field coils;

(iii) tight twisting of the lead wires—this works best with *thin* copper wire (e.g. 0.1 mm) which is tightly twisted (use a drill);

(iv) connectors are notorious sites of false pick-up. Mu-metal shielding helps, but is often impractical. A very ingenious, effective solution (invented by Mr Glen Davis, University of Maryland) is to use connectors with four contacts in a

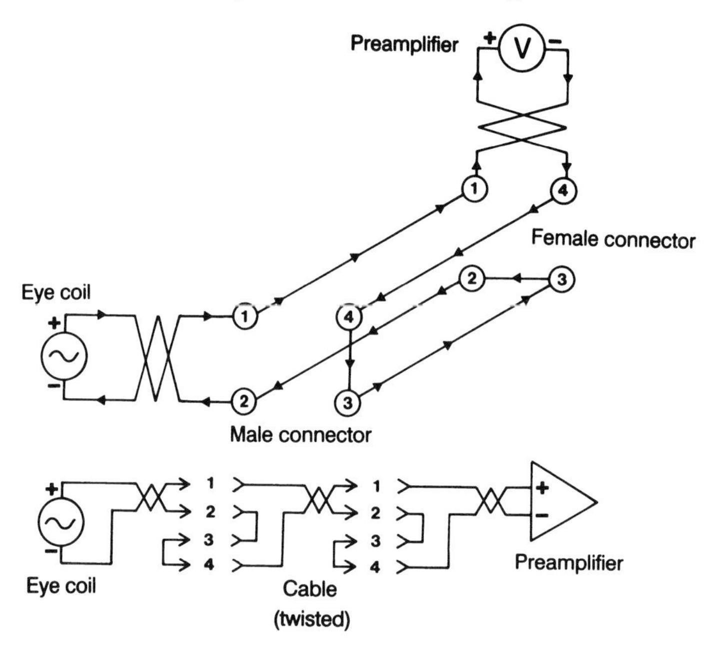

Fig. 9.9 Reduction of false pick-up of field in connectors by using 4-contact instead of 2-connectors. The contacts should be arranged in a square configuration, and in the male and female connectors one particular pair of contacts should be shorted with a jumper. As a result, two opposite loops are formed, with mutual cancelling of any false pick-up. This scheme was designed by Mr Glen Davis, University of Maryland, for use in the Maryland Revolving Magnetic Field Monitor (Edwards *et al.*, 1994).

square configuration instead of connectors with two contacts; the extra contacts are used to build (with jumpers) mutually cancelling loops, as shown in Fig. 9.9.

In our experience, such measures usually reduce false induction to insignificant levels. However, situations may occur where problems remain. For such cases, Hess *et al.* (1992) have described (complex!) mathematical procedures for correcting offsets in the calibration of a dual-coil, two-field system.

Many alternatives to a Fick coordinate system are used nowadays; each system has its own advantages. For binocular gaze control, the Helmholtz coordinate system is attractive, because for any binocularly fixated target the vertical angle (elevation, λ) is equal for the two eyes (the two lines of sight being in a single plane of regard), and the angle of horizontal vergence between the lines of sight is always equal to the difference between the azimuths (μ) of the two eyes.

Transformation between Fick and Helmholtz angles (Lemij, 1990):

Fick to Helmholtz:

$$\lambda = \arctan(\tan\theta / \cos\phi) \qquad \mu = \arcsin(\sin\phi \times \cos\theta);$$

Helmholtz to Fick:

$$\phi = \arctan(\tan\mu / \cos\lambda) \qquad \theta = \arcsin(\sin\lambda \times \cos\mu).$$

For other studies, notably those involving Listing's law, rotation vectors (e.g. Haustein, 1989) and quaternion representations (e.g. Tweed *et al.*, 1990) have been found very appropriate.

A returning problem is how to calibrate the measured signals. As with any recording system, whenever possible it is wise to build a subjective calibration into each experimental protocol. This provides an internal check on magnitude and polarity of signals. The safest procedure is to record steady fixation of single targets at known angular positions for one eye at a time (covering the other eye) for a duration of a few seconds. It is better not to use targets that jump around, to avoid presaccadic anticipatory smooth eye movements, saccadic anticipations, and postsaccadic drifts. Obviously, this procedure will not work for torsion or for subjects with unsteady gaze. Fortunately, the search-coil technique has the additional unique advantage that it can be objectively calibrated. Prior to use, a coil can be mounted on a protractor device in the position where the eye will be, and rotated through calibrated angles around any desired axis. Experience in human subjects has shown that such 'dry' calibrations remain valid for the measurements '*in vivo*'.

9.3.5.4 Alternative configurations

The original Robinson configuration (two orthogonal fields generated by two pairs of coils in a cube; phase-locked amplitude detection) has remained the most popular one. Naturally, it is easy to scale the field coil size down (for use in small animals, or reduced mass for use on rotating devices) or up (for experiments requiring ample space and easy access). A disadvantage of this type of field is the fairly small size of the central, homogeneous area (diameter roughly 10% of the field coil diameter). A much better homogeneity is reached with, for each field direction, five square coils in series, spaced by one-quarter of the diameter to make a cube again; the number of turns in the successive coils should be proportional to 19:4:10:4:19. This configuration, designed by Rubens (1945) and applied by Collewijn (1977) to eye movements, has a central homogeneous area (variations $< 1\%$) with a diameter of 50% of the diameter of the field coils. An intermediate solution is the use of true Helmholtz coil pairs, which are spaced by *half* the diameter of the coils. The diameter of the homogeneous area in this configuration is about 25% of the coil diameter. Apart from homogeneity, correct magnitude and phase quadrature of the fields are important conditions for good performance of the Robinson system; these can be easily compromised by slight imperfections in the circuits or metals in and around the field. Optican *et al.* (1982) have described a regulated magnetic-field generator to prevent such problems.

Another limitation of Robinson's configuration is the sinusoidal relationship between eye deviation and output signal; obviously the relationship will saturate at 90 deg and lose quality well below that value. Thus, the scheme will not work in a free motion condition. This problem was essentially solved by Hartmann and Klinke (1976). They used Robinson's field configuration, turned around to have all coils in vertical planes, to create a field with a horizontal vector, rotating 360 deg around during each AC cycle. A (vertically oriented) sensor coil in this field always receives the maximum inductive AC voltage, but its *phase* will vary linearly over 360 deg with the orientation of the sensor coil about a vertical axis. Thus, phase instead of amplitude detection will provide a signal representing search-coil orientation over the full range of 360 deg. This principle was applied to freely moving rabbits by Collewijn (1977), using two orthogonal Rubens (1945) coil configurations for better homogeneity. It is technically difficult to implement phase detection in more than one dimension, but this has been achieved for three dimensions in one unique instrument in the laboratory of R. M. Steinman at the University of Maryland (see, for example, Collewijn *et al.*, 1988 and Edwards *et al.*, 1994 for the theory of the instrument and the treatment of the data it generates). Unfortunately, details of the hardware are not yet in print and a commercial version of this elaborate instrument is not available at this time. Becker and Renner (1985) described a non-linear filtering technique for separating signals induced by horizontal and vertical magnetic fields with different frequencies, without changing the signal's phases. Also, a 2-D apparatus, suitable for small animals, has been developed in Zürich (Kasper *et al.*, 1987; Kasper and Hess, 1991). A special advantage of phase-detection systems is that the magnitude of the induced voltage is immaterial; therefore, variations in coil sensitivity do not affect the output and these systems have an *absolute* calibration. Obviously, this is of great potential benefit in animal and patient studies. Combinations of phase and amplitude detection to reach 3-D capability have also been applied (Van der Steen and Collewijn, 1984).

A further twist is the use of three, instead of two, orthogonal fields, in combination with two, not necessarily orthogonal, sensor coils (Tweed *et al.*, 1990). This method offers a solution for obtaining reliable 3-D coordinates (e.g. expressed in quaternions) when the exact position and orientation of implanted coils (in this case in the monkey) can not be known with certainty.

A final variant is the 'double magnetic induction' technique (Reulen and Bakker, 1982; Bour *et al.*, 1984), in which a lead-less, short-circuited coil in the form of a metal annulus (shaped like the silicone annulus) is placed on the eye. The current in this annulus obviously varies with orientation as in Robinson's design, generating a secondary magnetic field, which is picked up in a sensor coil placed in a frontal plane, concentric with the annulus. The signals from this sensor coil are treated as in Robinson's technique. Though ingenious in avoiding broken leads and being more gentle on the eye, the method is difficult to use: the sensor coil has to be aligned with extreme precision in order not to pick up the powerful primary field, and the input–output relationships are inherently non-linear, thus requiring an elaborate calibration and linearization procedure.

Several commercial versions of the search-coil system, essentially following the Robinson configuration, are marketed. Well-known manufacturers and comments on the products are given below (for addresses see the Appendix at the end of this chapter):

1. Skalar offers the S 3000 system with up to six output channels (3-D from each eye), and field coils of 70- or 31-cm diameter. Skalar is also the only manufacturer of the silicone search-coil annuli. This is a proven and very reliable system.
2. C-N-C Engineering offers another widely used, proven system. A special feature is the large (6 feet diameter, about 183 cm) field coil which offers easy access and a

large uncluttered visual field, but its area of homogeneity is not very large (about 10% of 6 feet = 0.6 ft). Thus, care must be exercised with respect to where the head is located in the magnetic field and in keeping it in this relatively small region.

3. Remmel Labs offer the EM3 monitor; this is an inexpensive design, the principles of which were published by Remmel (1984). This is not a bad choice when economy is an overriding factor; an additional advantage is that the (single-turn) field coils can be supplied (or made later) in any size.

9.4 Processing eye movement signals

Apart from the specific transformations required in several of the monitor types discussed above in obtaining signals that represent veridical eye rotations, most investigations will require quantitative treatment to extract the desired kind of data from the raw recordings. Digital storage and subsequent analysis is almost universally practised now. Obviously, the possibilities are almost unlimited in the present computerized era, as long as the limitations inherent in each specific recording technique are duly respected. Computers are no cure for poor recording techniques, and they require intelligent use. There is no substitute for looking first of all critically at the 'raw' signals, and subsequently scrutinizing and verifying every step in their analysis. This requires a firm understanding of the software operations, their limitations and assumptions, and a keen alertness for anomalous or inconsistent results. This chapter is not the place for a treatment of signal analysis. For completeness sake, however, the following references are provided to papers that discuss important operations such as the recognition of saccades, digital filtering, and the computation of velocities: Tole and Young (1981), Juhola *et al.* (1985), Ebisawa *et al.* (1988), Arzi and Magnin (1989), Sauter *et al.* (1991), and Carpenter (1994).

Finally, in concluding this chapter, it is useful to mention the existence of discussion groups on the Internet that may be a rich source of practical advice; e.g. on commercial devices, a group worth attending to is eyemov-l@spcvxa.spc.edu.

Acknowledgements

Thanks are due to Mr B. L. F. Weijer for drawing several of the figures, and to Drs L. F. Dell'Osso, R. J. Leigh, G. D. Paige, and R. M. Steinman for sharing their experience.

References

Adler, F. H. and Fliegelman, M. (1934). Influence of fixation on the visual acuity. *Archives of Ophthalmology*, **12**, 475–83.

Arzi, M. and Magnin, M. (1989). A fuzzy set theoretical approach to automatic analysis of optokinetic records. *IEEE Transactions in Biomedical Engineering*, **36**, 954–63.

Barry, W. and Melvill Jones, G. (1965). Influence of eye lid movement upon electro-oculographic recording of vertical eye movements. *Aerospace Medicine*, **36**, 855–8.
Becker, W. and Renner, A. (1985). Measuring eye movements with a search coil: non-linear filter allows simultaneous recording of horizontal and vertical eye position by means of the phase modulation method. *Vision Research*, **25**, 1755–8.
Berg, P. and Scherg, M. (1991). Dipole models of eye movements and blinks. *Electroencephalography and Clinical Neurophysiology*, **79**, 36–44.
Bond, H. W. and Ho, P. (1970). Solid miniature silver–silver chloride electrodes for chronic implantation. *Electroencephalography and Clinical Neurophysiology*, **28**, 206–8.
Bour, L. J., Van Gisbergen, J. A. M., Bruijns, J., and Ottes, F. O. (1984). The double magnetic induction method for measuring eye movement—results in monkey and man. *IEEE Transactions on Biomedical Engineering*, **BME-31**, 419–27.
Buquet, C. and Charlier, J. R. (1994). Quantitative assessment of the static properties of the oculomotor system by the photo-oculographic technique. *Medical Electronics and Biological Engineering*, **32**, 197–204.
Byford, G. H. (1963). Non-linear relations between the corneo-retinal potential and horizontal eye movements. *Journal of Physiology*, **168**, 14P-15P.
Carpenter, R. H. S. (1988). *Movements of the eyes* (2nd edn). Pion, London.
Carpenter, R. H. S. (ed.) (1991). Eye movements. In *Vision and visual dysfunction*, Vol. 8 (general ed. J. R. Cronly-Dillon). Macmillan, London.
Carpenter, R. H. S. (1994). SPIC: a PC-based system for rapid measurement of saccadic responses. *Journal of Physiology*, **480**, 4P.
Clark, M. J. (1975). A two-dimensional Purkinje eye tracker. *Behavior Research Methods and Instrumentation*, **7**, 215–19.
Clarke, A. H., Teiwes, W., and Scherer, H. (1991). Video-oculography—an alternative method for measurement of three-dimensional eye movements. In *Oculomotor control and cognitive processes* (ed. R. Schmid and D. Zambarbieri), pp. 431–43. Elsevier, Amsterdam.
Clarkson, T. (1989). Safety aspects in the use of infra-red detection systems. *International Journal of Electronics*, **66**, 929–32.
Collewijn, H. (1977). Eye- and head movements in freely moving rabbits. *Journal of Physiology*, **266**, 471–98.
Collewijn, H., Van der Mark, F., and Jansen, T. C. (1975). Precise recording of human eye movements. *Vision Research*, **15**, 447–50.
Collewijn, H., Van der Steen, J., Ferman, L., and Jansen, T. C. (1985). Human ocular counterroll: assessment of static and dynamic properties from electromagnetic scleral coil recordings. *Experimental Brain Research*, **59**, 185–96.
Collewijn, H., Erkelens, C. J., and Steinman, R. M. (1988). Binocular coordination of human horizontal saccadic eye movements. *Journal of Physiology*, **404**, 157–82.
Collewijn, H., Steinman, R. M., Erkelens, C. E., Pizlo, Z., Kowler, E., and Van der Steen, J. (1992). Binocular gaze control under free-head conditions. In *Vestibular and brain stem control of eye, head and body movements* (ed. H. Shimazu and Y. Shinoda), pp. 203–20. Karger, Basel.
Comet, B. (1983). An eye movement recording method operating with a closed eye. *Medical Electronics and Biological Engineering*, **21**, 628–31.
Cornsweet, T. N. (1958). A new technique for the measurement of small eye movements. *Journal of the Optical Society of America*, **48**, 808–11.
Cornsweet, T. N. and Crane, H. D. (1973). Accurate two-dimensional eye tracker using first and fourth Purkinje images. *Journal of the Optical Society of America*, **63**, 921–8.
Crane, H. D. and Steele, C. M. (1978). Accurate three-dimensional eye tracker. *Applied Optics*, **17**, 691–705.

Crane, H. D. and Steele, C. M. (1985). Generation V dual-Purkinje-image eye-tracker. *Applied Optics*, **24**, 527–37.

Curthoys, I. S., Dai, M. J., and Halmagyi, G. M. (1991). Human ocular torsional position before and after unilateral vestibular neurectomy. *Experimental Brain Research*, **85**, 218–25.

Curthoys, I. S., Moore, S. T., Haslwanter, T., Black, R. A., and Smith, S. T. (1994). Video procedures for the measurement and display of the three dimensions of eye movements. In *Eye movements in reading* (ed. J. Ygge and G. Lennerstrand), pp. 39–50. Elsevier/Pergamon, Oxford.

De Bie, J. (1985). An afterimage vernier acuity method for assessing the precision of eye movement monitors: results for the scleral coil technique. *Vision Research*, **25**, 1341–3.

De la Barre, E. B. (1898). A method of recording eye movements. *American Journal of Psychology*, **9**, 572–4.

Dell'Osso, L. F. and Daroff, R. B. (1990). Eye movement characteristics and recording techniques. In *Neuro-ophthalmology* (ed. J. S. Glaser), pp. 279–97. Harper and Row, Hagerstown, MD.

Deubel, H. and Bridgeman, B. (1995). Fourth Purkinje image signals reveal eye-lens deviations and retinal image distortions during saccades. *Vision Research*, **35**, 529–38.

DiScenna, A. O., Das, V., Zivotofsky, A. Z., Seidman, S. H., and Leigh, R. J. (1995). Evaluation of a video tracking device for measurement of horizontal and vertical eye rotations during locomotion. *Journal of Neuroscience Methods*, **58**, 89–94.

Ditchburn, R. W. (1973). *Eye-movements and visual perception*. Clarendon Press, Oxford.

Dodge, R. (1903). Five types of eye movement in the horizontal meridian plane of the field of regard. *American Journal of Physiology*, **8**, 307–29.

Dodge, R. and Cline, T. S. (1901). The angle velocity of eye movements. *Psychological Review*, **8**, 145–57.

Ebisawa, Y., Minamitani, H., Mori, Y. and Takase, M. (1988). New method for removing saccades in analyses of smooth pursuit eye movement. *Biological Cybernetics*, **60**, 111–19.

Edwards, M., Pizlo, Z., Erkelens, C. J., *et al.* (1994). The Maryland revolving-field monitor: Theory of the instrument and processing its data. Technical Report CAR-TR-711, Center for Automation Research, University of Maryland (College Park, MD).

Eizenman, M., Frecker, R. C., and Hallett, P. E. (1984). Precise non-contacting measurement of eye movements using the corneal reflex. *Vision Research*, **24**, 167–74.

Enright, J. T. (1980). Ocular translation and cyclotorsion due to changes in fixation distance. *Vision Research*, **20**, 595–601.

Enright, J. T. (1984). Changes in vergence mediated by saccades. *Journal of Physiology*, **350**, 9–31.

Enright, J. T. (1986). The aftermath of horizontal saccades: saccadic retraction and cyclotorsion. *Vision Research*, **26**, 1807–14.

Enright, J. T. and Hendriks, A. (1994). To stare or to scrutinize: 'grasping' the eye for better vision. *Vision Research*, **34**, 2039–42.

Fender, D. H. (1955). Torsional motions of the eyeball. *British Journal of Ophthalmology*, **39**, 65–72.

Fender, D. H. and Nye, P. W. (1961). An investigation of the mechanisms of eye movement control. *Kybernetik*, **1**, 81–8.

Fenn, W. O. and Hursh, J. B. (1934). Movements of the eyes when the lids are closed. *American Journal of Physiology*, **118**, 8–14.

Ferman, L., Collewijn, H., Jansen, T. C., and Van den Berg, A. V. (1987). Human gaze stability in horizontal, vertical and torsional direction during voluntary head movements, evaluated with a three-dimensional scleral induction coil technique. *Vision Research*, **27**, 811–28.

Findlay, J. M. (1974). A simple apparatus for recording microsaccades during visual fixation. *Quarterly Journal of Experimental Psychology*, **26**, 167–70.

Finocchio, D. V., Preston, K. L., and Fuchs, A. F. (1991). Infant eye movements—quantification of the vestibulo-ocular reflex and visual-vestibular interactions. *Vision Research*, **31**, 1717–30.

Frecker, R. C., Eizenman, M., and Hallett, P. E. (1984). High-precision real-time measurement of eye position using the first Purkinje image. In *Theoretical and applied aspects of eye movement research* (ed. A. G. Gale and F. Johnson), pp. 13–20. Elsevier, Amsterdam.

Fuchs, A. F. and Robinson, D. A. (1966). A method for measuring horizontal and vertical eye movement chronically in the monkey. *Journal of Applied Physiology*, **21**, 1068–70.

Gale, A. G. (1982). A note on the remote oculometer technique for recording eye movements. *Vision Research*, **22**, 201–2.

Gale, A. G. and Johnson, F. (ed.) (1984). *Theoretical and applied aspects of eye movement research*. Elsevier, Amsterdam.

Gauthier, G. M. and Volle, M. (1975). Two-dimensional eye movement monitor for clinical and laboratory recordings. *Electroencephalography and Clinical Neurophysiology*, **39**, 285–91.

Guez, J., Le Gargasson, J., Rigaudiere, F., and O'Regan, J. K. (1993). Is there a systematic location for the pseudo-fovea in patients with central scotoma? *Vision Research*, **33**, 1271–9.

Guitton, D., Simard, R., and Codere, F. (1991). Upper eyelid movements measured with a search coil during blinks and vertical saccades. *Investigative Ophthalmology and Visual Science*, **32**, 3298–305.

Hamada, T. (1984). A method for calibrating the gain of the electro-oculogram (EOG) using the optical properties of the eye. *Journal of Neuroscience Methods*, **10**, 259–65.

Hartmann, R. and Klinke, R. (1976). A method for measuring the angle of rotation (movements of body, head, eye in human subjects and experimental animals). *Pflügers Archiv* (Suppl.), **362**, R52.

Haslwanter, T. (1995). The mathematics of 3-dimensional rotations of the eye. *Vision Research*, **35**, 1727–39.

Haustein, W. (1989). Considerations of Listing's law and the primary position by means of a matrix description of eye position control. *Biological Cybernetics*, **60**, 411–20.

He, P. Y. and Kowler, E. (1992). The role of saccades in the perception of texture patterns. *Vision Research*, **32**, 2151–63.

Heller, D. (1988). On the history of eye movement recording. In *Eye movement research: physiological and psychological aspects* (ed. G. Lüer, U. Lass, and J. Shallo-Hoffmann). C. J. Hogrebe, Toronto, pp. 37–51.

Hess, B. J. M. (1990). Dual-search coil for measuring 3-dimensional eye movements in experimental animals. *Vision Research*, **30**, 597–602.

Hess, B. J. M., Van Opstal, A. J., Straumann, D., and Hepp, K. (1992). Calibration of 3-dimensional eye position using search coil signals in the rhesus monkey. *Vision Research*, **32**, 1647–54.

Hine, T. and Thorn, F. (1987). Compensatory eye movements during active head rotation for near targets: effects of imagination, rapid head oscillation and vergence. *Vision Research*, **27**, 1639–57.

Jasper, H. H. and Walker, R. Y. (1931). The Iowa eye-movement camera. *Science*, **74**, 291–4.

Jones, R. (1973). Two dimensional eye movement recording using a photo-electric matrix method. *Vision Research*, **13**, 425–31.

Judge, S. J., Richmond, B. J., and Chu, F. C. (1980). Implantation of magnetic search coils for measurement of eye position: an improved method. *Vision Research*, **20**, 535–8.

Juhola, M., Jantti, V., and Pykko, I. (1985). Effect of sampling frequencies on computation of the maximum velocity of saccadic eye movements. *Biological Cybernetics*, **53**, 67–72.

Jung, R. (1939). Eine elektrische Methode zur mehrfachen Registrierung van Augenbewegungen und Nystagmus. *Klinische Wochenschrift*, **18**, 21–4.

Jung, R. (1977). An appreciation of early work on gaze control in man and of visuo-vestibular research before 1940. In *Control of gaze in brain stem neurons* (ed. R. Baker and A. Berthoz), pp. 2–10. Elsevier, Amsterdam.

Jung, R. and Kornhuber, H. H. (1964). Results of electronystagmography in man: the value of optokinetic, vestibular, and spontaneous nystagmus for neurologic diagnosis and research. In *The oculomotor system* (ed. M. B. Bender), pp. 428–88. Harper and Row, New York.

Kasper, H. and Hess, B. J. M. (1991). Magnetic search coil system for linear detection of 3-dimensional angular movements. *IEEE Transactions on Biomedical Engineering*, **38**, 466–75.

Kasper, H. J., Hess, B. J., and Dieringer, N. (1987). A precise and inexpensive magnetic field search coil system for measuring eye and head movements in small laboratory animals. *Journal of Neuroscience Methods*, **19**, 115–24.

Kelly, D. H. (1979). Motion and vision. I. Stabilized images of stationary gratings. *Journal of the Optical Society of America*, **69**, 1266–74.

Kenyon, R. V. (1985). A soft contact lens search coil for measuring eye movements. *Vision Research*, **25**, 1629–33.

Kingma, H., Gullikers, H., and De Jong, I. (1995). Real time binocular simultaneous detection of horizontal, vertical and torsional eye movements by a video-eye tracker. *Acta Otolaryngologica (Stockholm)*, Suppl. **520**, 9–15.

Kowler, E. (ed.) (1990). Eye movements and their role in visual and cognitive processes. In *Reviews of oculomotor research*, Vol. 4. Elsevier, Amsterdam.

Kris, E. C. (1964). Discussion. In *The oculomotor system* (ed. M. B. Bender), pp. 355–61. Harper and Row, New York.

Land, M. F. (1992). Predictable eye–head coordination during driving. *Nature*, **359**, 318–20.

Lemij, H. G. (1990). *Asymmetrical adaptation of human saccades to anisometropic spectacles*. Doctoral Thesis, Erasmus University, Rotterdam (The Netherlands).

Mackworth, J. F. and Mackworth, N. H. (1958). Eye fixations recorded on changing visual scenes by the television eye-marker. *Journal of the Optical Society of America*, **48**, 439–45.

Mackworth, N. H. and Thomas, E. L. (1962). Head-mounted eye-marker camera. *Journal of the Optical Society of America*, **52**, 713–16.

Marg, E. (1951). Development of electro-oculography. *American Medical Association Archives of Ophthalmology*, **45**, 169–85.

Merchant, J., Morrissette, R., and Porterfield, J. L. (1974). Remote measurement of eye direction allowing subject motion over one cubic foot of space. *IEEE Transactions on Biomedical Engineering*, **BME-21**, 309–17.

Miller, J. M., Mellinger, M., Greivenkemp, J., and Simons, K. (1993). Videographic Hirschberg measurement of simulated strabismic deviations. *Investigative Ophthalmology and Visual Science*, **34**, 3220–9.

Moore, S. T., Curthoys, I. S., and McCoy, S. G. (1991). VTM—an image-processing system for measuring ocular torsion. *Computer Methods and Programs in Biomedicine*, **35**, 219–30.

Nykiel, F. and Torok, N. (1963). A simplified nystagmograph. *Annals of Otology, Rhinology and Laryngology*, **72**, 647–54.

Ober, J. (1994). Infra red reflection technique. In *Eye movements in reading* (ed. J. Ygge and G. Lennerstrand), pp. 9–20. Elsevier/Pergamon, Oxford.

Ohm, J. (1928). Die Hebelnystagmographie. *Albrecht von Graefes Archiv für Ophthalmologie*, **120**, 235–52.

Optican, L. M., Frank, D. E., Smith, B. M., and Colburn, T. R. (1982). An amplitude and phase regulating magnetic field generator for an eye movement monitor. *IEEE Transactions of Biomedical Engineering*, **BME-29**, 206–9.

Orschansky, J. (1898). Eine Methode die Augenbewegungen direkt zu untersuchen (Ophthalmografie). *Zentralblatt fur Physiologie*, **12**, 785–90.

Ott, D. and Eckmiller, R. (1989). Ocular torsion measured by TV-ophthalmoscopy and scanning laser ophthalmoscopy during horizontal pursuit in humans and monkeys. *Investigative Ophthalmology and Visual Science*, **30**, 2512–20.

Palmieri, G., Oliva, G. A., and Scotto, M. (1971). C.R.T. spot-follower device for eye-movement measurements. *Kybernetik*, **8**, 23–30.

Rashbass, C. (1960). New method for recording eye movements. *Journal of the Optical Society of America*, **50**, 642–4.

Rashbass, C. and Westheimer, G. (1961). Disjunctive eye movements. *Journal of Physiology*, **159**, 339–60.

Remmel, R. S. (1984). An inexpensive eye movement monitor using the scleral search coil technique. *IEEE Transactions of Biomedical Engineering*, **BME-31**, 388–9.

Reulen, J. P. and Bakker, L. (1982). The measurement of eye movement using double magnetic induction. *IEEE Transactions of Biomedical Engineering* **29**, 740–4.

Robinson, D. A. (1963). A method of measuring eye movement using a scleral search coil in a magnetic field. *IEEE Transactions of Biomedical Electronics*, **BME-10**, 137–45.

Robinson, D. A. (1964). The mechanics of human saccadic eye movement. *Journal of Physiology*, **174**, 245–64.

Robinson, D. A. (1965). The mechanics of human smooth pursuit eye movement. *Journal of Physiology*, **180**, 569–91.

Rubens, S. M. (1945). Cube-surface coil for producing a uniform magnetic field. *The Review of Scientific Instruments*, **16**, 243–45.

Sauter, D., Martin, B. J., Renzo N. D., and Uomschied, C. (1991). Analysis of eye tracking movements using innovations generated by a Kalman filter. *Medical Electronics and Biological Engineering*, **29**, 63–9.

Scherer, H., Teiwes, W., and Clarke, A. H. (1991). Measuring three dimensions of eye movement in dynamic situations by means of videooculography. *Acta Otolaryngologica*, **111**, 182–7.

Schlag, J., Merker, B., and Schlag-Rey, M. (1983). Comparison of EOG and search coil techniques in long-term measurements of eye position in alert monkey and cat. *Vision Research*, **23**, 1025–30.

Shackel, B. (1967). Eye movement recording by electro-oculography. In *Manual of psychophysical methods* (ed. P. H. Venables and I. Martin), pp. 229–334. North-Holland, Amsterdam.

Stark, L. and Sandberg, A. (1961). A simple instrument for measuring eye movements. *MIT Electronics Research Laboratory Quarterly Report*, **62**, 268–70.

Steinman, R. M. and Levinson, J. Z. (1990). The role of eye movement in the detection of contrast and spatial detail. In *Eye movements and their role in visual and cognitive processes. Reviews of oculomotor research*, Vol. 4 (ed. E. Kowler), pp. 115–212. Elsevier, Amsterdam.

Steinman, R. M., Haddad, G. M., Skavenski, A. A., and Wyman, D. (1973). Miniature eye movements. *Science*, **181**, 810–19.

Steinman, R. M., Kowler, E., and Collewijn, H. (1990). New directions for oculomotor research. *Vision Research*, **30**, 1845–64.

Ter Braak, J. W. G. (1936). Untersuchungen über optokinetischen Nystagmus. *Archives Néerlandaises de Physiologie de l'homme et des animaux*, **21**, 309–76.

Tole, J. R. and Young, L. R. (1981). Digital filters for saccade and fixation detection. In *Eye movements: cognition and visual perception* (ed. D. F. Fisher, R. A. Monty, and J. W. Senders), pp. 247–56. Erlbaum, New Jersey.

Tweed, D., Cadera, W., and Vilis, T. (1990). Computing 3-dimensional eye position quaternions and eye velocity from search coil signals. *Vision Research*, **30**, 97–110.

Van der Steen, J. and Collewijn, H. (1984). Ocular stability in the horizontal, frontal and sagittal planes in the rabbit. *Experimental Brain Research*, **56**, 263–74.
Van Rijn, L. J., Van der Steen, J., and Collewijn, H. (1994). Instability of ocular torsion during fixation: cyclovergence is more stable than cycloversion. *Vision Research*, **34**, 1077–87.
Verheijen, F. J. (1961). A simple after-image method demonstrating the involuntary multi-directional eye movements during fixation. *Optica Acta*, **6**, 309–11.
Westheimer, G. (1989). History and methodology. In *The neurobiology of saccadic eye movements. Reviews of oculomotor research*, Vol. 3 (ed. R. H. Wurtz and M. E. Goldberg), pp. 3–12. Elsevier, Amsterdam.
Wheeless, L. L., Boynton, R. M., and Cohen, G. H. (1966). Eye-movement responses to step and pulse-step stimuli. *Journal of the Optical Society of America*, **56**, 956–60.
Yarbus, A. L. (1967). *Eye movements and vision*. Plenum, New York.
Yee, R. D., Schiller, V. L., Lim, V., Baloh, F. G., Baloh, R. W., and Honrubia, V. (1985). Velocities of vertical saccades with different eye movement recording methods. *Investigative Ophthalmology and Visual Science*, **26**, 938–44.
Young, L. R. and Sheena, D. (1975). Survey of eye movement recording methods. *Behavior Research Methods and Instrumentation*, **7**, 397–429.

Appendix: Addresses of suppliers

Applied Science Laboratories (ASL), 335 Bear Hill Road, Waltham, MA 02154, USA (Tel.: 617–890 5100; Fax: 617–890 7966; e-mail: asl@world.std.com).
C-N-C Engineering (Mr Charles Chase), PO Box 75567, Seattle, WA 98125, USA (Tel.: 206–742 7160; Fax: 206–742 7206).
Cooner Wire, 9265 Owensmouth, Chatsworth, CA 91311, USA (Tel.: 818–882 8311; Fax: 818–709 8281).
EL-MAR, Downsview, Ontario, Canada; scientific contact person: Dr Moshe Eizenman, University of Toronto (Tel.: 416–978 5523; Fax: 416–978 4317).
ISCAN Inc., 125 Cambridgepark Drive, Cambridge, MA 02140, USA (Tel.: 617–868 5353; Fax 617–868 9231). (http://www/iscaninc.com)
Microguide, Inc., 14846 Shepard Drive, Dolton, IL 60419, USA (Tel.:708–964 3335). (e-mail: gkro1@uic.edu)
Permobil Meditech AB, Box 120, S-861 00 Timrå, Sweden (Tel.: +46–60–572606, Fax: +46–60–575250). (http://www.tiac.net/users/permobil/)
Remmel labs (Dr Ron Remmel), 26 Bay Colony Drive, Ashland, MA 01721–1840, USA (Tel.: 508–881 5086).
SensoMotoric Instruments (SMI) GmbH, Warthestraβe 21, 14513 Teltow Berlin, Germany (Tel: +49–3328/39–55–20; Fax: +49–3328/39–55–99; scientific contact person: Dr W. Teiwes). (http://www.smi.de)
Skalar Medical, PO Box 233, 2600 AE Delft, The Netherlands (Tel.: +31–15–619494; Fax: +31–15–560401). (http://www.wirehub.ul/~skalar/eye.htm)
Sokymat SA, CH-1614 Granges, Switzerland.
Stanford Research Institute Menlo Park, CA 94025, USA.

10

Techniques for the dynamic measurement of accommodation and pupil size

STUART JUDGE

10.1 Accommodation

10.1.1 Introduction

10.1.1.1 Definitions

Ocular accommodation is the ability to change the optical power of the eye. In primates this is achieved almost entirely by altering the curvature of the anterior surface of the lens. If the plane conjugate with the retina is at a distance of x metres from the principal plane of the eye, accommodation is $D = 1/x$ dioptres.

A dynamic optometer seeks to measure the change in position of the plane conjugate with the retina (reckoned in terms of dioptres, not metres) as accommodation changes; if the eye is astigmatic, the orientation of the stimulus whose conjugate plane one wishes to track will have to be specified. To determine the conjugate plane, a bright source is used to form an image on the retina. Because the retina is to some extent reflective, some light is re-emitted from the eye, and forms an image in the plane conjugate with the retina. It should be emphasized that very little light is re-emitted from the eye and this secondary image is faint. The estimated distance of this image from the eye can be used to determine the degree of accommodation. Most optometers do not use the elegant 'null' method of directly tracking the plane conjugate with the retina, but instead rely on one or other methods of estimating the position of this plane.

There are various ways of estimating the position of the plane conjugate with the retina; I shall concentrate on two methods that have been particularly widely used. The Scheiner principle relies on the fact that (thinking in the spatial rather than the spatial frequency domain) the fundamental characteristic of defocus is that light rays from a given point in the object which pass through different parts of the pupil end up at different points in the retinal image. Retinoscopy relies on a related, but less obvious, characteristic of defocus—that light rays re-emitted from a bright point in the retinal image and reaching a particular point in the object plane are non-uniformly distributed across the pupil.

Occasionally (O'Neill *et al.*, 1968) accommodation dynamics have been inferred indirectly, by tracking changes in the curvature of the anterior surface of the lens, but clearly more direct methods are to be preferred.

There is an excellent general survey of the principles on which optometers work by Henson (1983). This is a very useful source, but note that the instruments he describes are generally designed to determine refraction rather than follow accommodation dynamically. In this chapter I shall use the term optometer to apply only to instruments that give a dynamic output.

10.1.1.2 Design parameters

Some parameters to consider when selecting or designing an optometer for a particular use are:

- range of accommodation to be measured;
- bandwidth and precision of measurement required;
- extent of eye and head movements to be tolerated;
- tolerance of pupil constriction.

Another less obvious parameter that can be important is:

- field of view required.

Optometers tend to be fairly bulky instruments and care is needed if the instrument itself is not to restrict the field of view unduly. If the field of view is to be large, the optometer must be placed well out of the field of view, either to one side or above or below the subject, and must view the eye via a dichroic mirror that transmits visible light but reflects infrared light. The mirror itself must take up some space if its edges are to be eccentric in the field of view, and in practice this sort of consideration leads to the working distance of the optometer (i.e. the distance between its first lens and the eye) having to be at least 100 mm—200 or even 300 mm is often useful). The Canon Autoref R1 (no longer produced but perhaps available second-hand) was a particularly useful instrument in this regard because it was mounted below the subject's line of sight and viewed via a large dichroic mirror so that the field of view was unobstructed. The instrument was marketed as an autorefractor rather than a dynamic optometer, but Pugh and Winn (1988, 1989) have described how it can be modified to work as a dynamic optometer. It is good to ask whether a large field of view is needed. If one is studying accommodation itself, then a large field of view should not be necessary, because it is said that accommodation relies exclusively on foveal and near-foveal stimulation.

10.1.1.3 Important issues

Some issues that turn out to be important are:

- choice of wavelength;
- preventing artefacts associated with eye and head movements, or with changes in pupil size;
- preventing artefacts associated with corneal reflections;
- safety.

Some of the issues to be considered in designing or selecting a dynamic optometer follow directly from the nature of the task.

Wavelength Because a very small fraction of the incident light is re-emitted from the eye, a bright source must be used and this will distract the subject if it is visible. A near-infrared source should therefore be used. The spectral transmission curve of the human eye is shown in Fig. 10.1. The notch near the 1000-nm wavelength is caused by absorption by water. In practice, one wants to use a wavelength that is as close as possible to the visible range, not hazardous, and well-transmitted by the eye. A wavelength of between 850 and 900 nm is therefore ideal. Generally, two kinds of sources have been used:

- an incandescent source filtered to block visible light (e.g. with a Kodak Wratten no. 88A filter);
- a solid-state source (generally a gallium arsenide infrared-emitting diode).

Movement artefacts To get an adequate amount of IR radiation both into the eye and back into the optometer most optometers focus the input beam in the plane of the pupil. The retinal illumination is therefore constant, and the focus of the retinal image does not vary with the state of accommodation. This makes it simpler to work out whether the retinal illumination is safe under all conditions; and also, because it is only the output beam that is affected by accommodation, one has only to consider the effects of accommodation on the output beam, rather than the compounded effects of two passes through the optics. The price to pay for this 'Maxwellian' focus, as it is sometimes called (Westheimer, 1966), is that it is necessary either to keep the pupil in a well-defined position or to have a system for automatically tracking the movements of the pupil caused by eye or head movements. If the optometer is to avoid the complication of incorporating a system for tracking the pupil, it is simplest

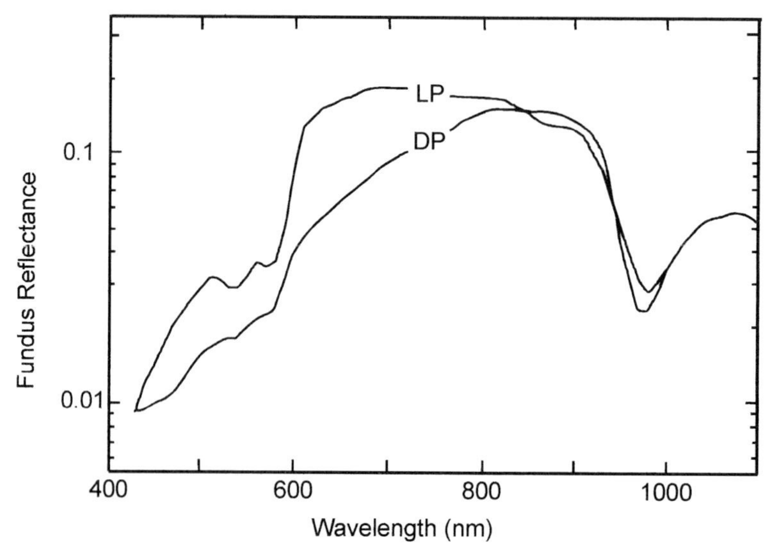

Fig. 10.1 *In vivo*, double-pass fundal reflectance as a function of wavelength, for two subjects: one (DP) darkly pigmented and the other (LP) lightly pigmented. (From Elsner *et al.* (1996) with permission.)

to have the subject's head fixed with a deep dental bite-bar and headrest so that residual head movements are less than ±0.5 mm or so. What really matters is that the pupil does not move, and thus it is also necessary to avoid larger eye movements. If we assume that the centre of rotation of the eye is approximately 10 mm behind the entrance pupil, then the line of sight can deviate by up to 3 deg or so without causing a pupil displacement of more than 0.5 mm.

To make optometers less sensitive to fluctuations in pupil size it has often been necessary to dilate the pupil (e.g. Cumming and Judge, 1986; Gamlin *et al.*, 1989). The mydriatic most commonly used is phenylephrine hydrochloride. A 2.5%, rather than a 10% solution, is adequate for many purposes. Pupil dilation is not ideal, first because one cannot exclude the possibility that even the most selective mydriatics have a small effect on the ciliary muscle, and second because dilating the pupil will alter the degree of blur. In principle, one could ameliorate the latter problem by incorporating an artificial pupil in the input optics which is dichroic, so that the pupil is large in the IR but small in visible light.

Corneal reflection The corneal reflection is very bright compared with the images of the retina, and all optometers need to include some means of avoiding artefacts from this. The simplest method is to align the optometer somewhat off-axis so that the corneal reflection falls outside the optometer. Geometrically, having the optometer axis parallel to but slightly displaced from the eye's optic axis is a more effective way of excluding the corneal reflection than having the optometer axis at an angle to the optic axis of the eye, but in practice excluding the reflex may require the optometer to be off-axis in both senses. The penalty for having the optometer off-axis is that it will not then be measuring the on-axis accommodation of the eye, and various optical aberrations of the eye will affect the measured value. Displacement of the beam by 2 or 3 mm may be associated with a dioptre or so of spherical aberration (Fig. 6 in Cornsweet and Crane, 1970), and off-axis astigmatism will amount to about 1 dioptre at an eccentricity of 30 deg (Millodot and Lamont, 1974).

It can sometimes be arranged that the corneal reflex falls on a blocking stop (see Section 10.1.3). Another possibility is to plane-polarize the incident light so that reflection is minimized (see Section 10.1.5 below).

Safety Because the sources used are intense, it is wise to explore safety issues. It should be realized that even ordinary ophthalmological equipment can deliver dangerous levels of illumination under certain circumstances (e.g. Calkins *et al.*, 1980).

Safety considerations would be particularly critical if one were to incorporate laser diodes in an optometer. I have given sample safety calculations in Section 10.1.2.8.

10.1.2 A Scheiner-principle optometer

10.1.2.1 The Scheiner principle

If a point target is viewed (in a modest illumination so that one's pupil is not too small) through a pair of pin-holes about 1 mm in diameter whose centres are separated by about 2.5 mm, then altering one's focus by accommodating D dioptres (relative to the target) causes the image to double, with the separation of the two images being proportional to D. If one or other pin-hole is occluded, then as one pin-hole is closed and the other opened, the target appears to move from one position to another, with the direction and sense of motion being proportional to the direction and sense of accommodation relative to the position of the image of the target seen by the subject.

10.1.2.2 Virtual pin-holes and the input optics

The Scheiner principle has been incorporated into a number of optometers. A real pin-hole is of course impractical, so an optical arrangement is used that has the effect of viewing through pin-holes: the pin-holes are projected into the subject's pupil (Fig. 10.2).

Note that from the point of view of geometrical optics the image of target ST_1 will be in focus on the retina regardless of the state of accommodation of the eye, because if the object ST_1 is in a collimated beam, only one ray of light goes from each point in the object to each point in the retina. Diffraction will limit the extent to which this is exactly true. If we arrange that S_1 and S_2 can be operated independently, then we have a means of opening one or other effective pin-hole. I have shown the two sources as physically separate (e.g. IR diodes), but the same effect can be achieved by using a single source, and a bi-prism, separated by a 'chopper' (i.e. rotating sector disc that alternately blocks the path between the source and the two halves of the prism.

10.1.2.3 The output optics: usefulness of the optical relay

To turn this instrument into a practical optometer we need a means of peering into the eye to observe the displacement of the image on the retina as S_1 and S_2 are alternately illuminated. The usual way to do this is to have a beam-splitter between ST_1 and L_2 and a separate output channel of the optometer. Pellicle beam-splitters (semi-silvered nitrocellulose films) are useful, though fragile.

It is often convenient to incorporate an 'optical relay' (see Crane and Steele, 1978, p. 698): in its simplest form this is two lenses of equal power separated by twice their focal length; lenses of focal lengths f_1 and f_2 separated by distance $f_1 + f_2$ will also function as a relay, provided that a change in magnification is of no significance. Planes P_1 and P_2 which are equal distances to the right of the respective outer focal planes of L_1 and L_2 are not only conjugate (i.e. rays leaving any point on one will converge on one point on the other) but have a stronger relationship that I call

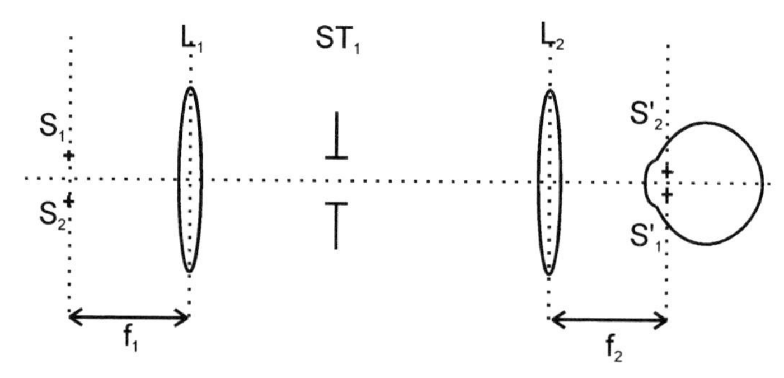

Fig. 10.2 Imaging two sources, S1 and S2, in the plane of the pupil, by means of a collimator lens, L_1, with focal length f_1, and a Maxwellian lens, L_2, with focal length f_2.

equivalent, in which a ray crossing plane P_1 at eccentricity h with angle α (to the optical axis) will cross plane P_2 at the same eccentricity and angle α. By using such a relay one can remove items to a physically remote plane for which there may be little space in the original plane. In Fig. 10.3 lenses L_3 and L_4 form a relay so that the detector, D, is remote. An additional advantage of this particular relay is that it can be placed so that L_2 and L_3 also form a relay, so that there is a plane equivalent (not merely conjugate) to the pupil, where a stop ST_1 can be placed which has exactly the effect on the output beam of a stop that is actually in the pupil.

The operation of a complete Scheiner-principle optometer (Cumming and Judge, 1986) can now be described in some detail.

10.1.2.4 Theory of operation of the Cumming and Judge optometer

In this optometer the Scheiner principle is realized by alternately illuminating (at a frequency of 400 Hz) two infrared-emitting diodes, that are imaged in the pupil.

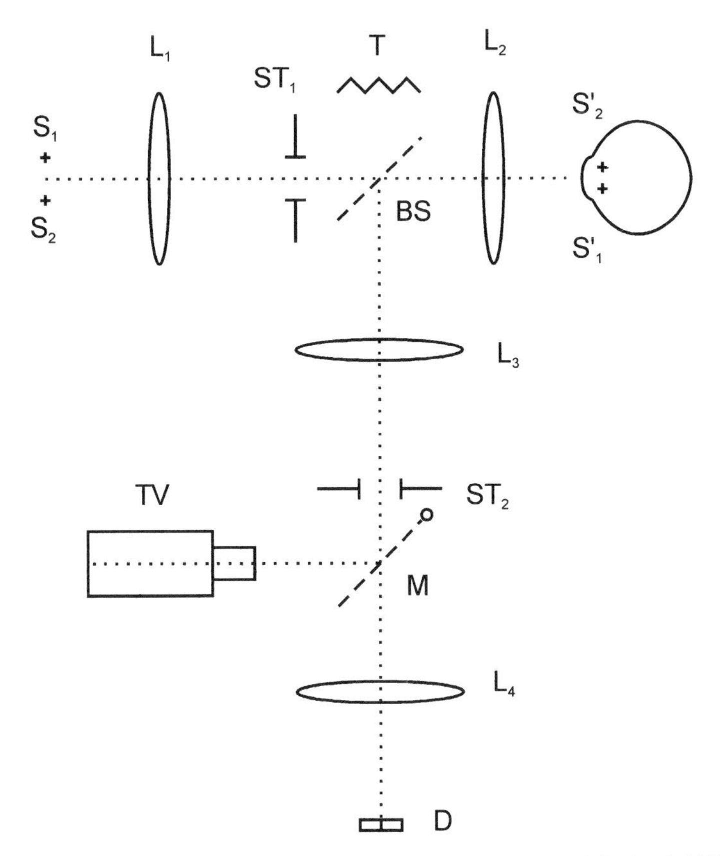

Fig. 10.3 Layout of the Scheiner-principle optometer of Cumming and Judge (1986). Sources S_1 and S_2 are imaged in the plane of the pupil. A pellicle beam-splitter, BS, allows the radiation re-emitted from the eye to be detected. Lenses L_3 and L_4 form an optical relay to allow the detector, D, to be effectively in the position of the target aperture ST_1. Stop ST_2 is an artificial pupil. Mirror M allows the real pupil to be viewed with a TV camera when aligning the optometer.

Infrared radiation reflected from the retina falls on to a detector that provides a measure of the magnitude of the movement of the image on the retina, as described below.

The optical layout of the instrument is shown in Fig. 10.3. The lenses L_1, L_2, L_3 and L_4 are all achromats. Two infrared sources, S_1 and S_2, in the focal plane of lens L_1 (focal length approximately 0.08 m) illuminate a rectangular aperture, ST_1. A second lens, L_2 (focal length approximately 0.11 m), is placed with its far focal point in the plane of the pupil, so that the images of the sources are formed (at a modest magnification) in the pupil, and an image of the aperture ST_1 is formed on the retina. This image is in focus regardless of the refractive state of the eye, since the infrared radiation from ST_1 is focused in the plane of the pupil. Infrared radiation returning from the retina is reflected by a pellicle beam-splitter, BS, into the detection limb of the instrument. The reflected radiation is relayed via a pair of lenses L_3 and L_4, on to a four-quadrant planar silicon diode detector, D. Because the detector is optically coincident with ST_1, when the eye is accommodated at a distance equivalent to the plane of ST_1 (-2 dioptres) the two images of the aperture are superimposed on the detector. If the eye is accommodated at any other distance, the images are not superimposed on the retina or on the detector. Light trap, T, absorbs stray light reflected from the back of the beam-splitter, BS.

There are two advantages to using the relay formed by lenses L_3 and L_4. First, it produces a plane between the two lenses optically conjugate with the pupil, so that by placing an aperture stop (ST_2, an iris diaphragm) in this plane it is possible to ensure that changes in the size of the actual pupil do not alter the quantity of light returning from the retina to the detector (provided the real pupil is always larger than the aperture stop). Second, the relay provides space for a mirror, M, to be moved into the path of the reflected radiation, so that an image of the pupil can be obtained on a TV camera suitably sensitive to infrared radiation (i.e. an ultricon tube or an appropriate CCD camera). The TV camera is used to align the optometer so that no IR radiation is reflected from the cornea into the instrument, and that the source images are not vignetted by the iris of the eye.

As described above, images of ST_1 formed on the retina are reflected onto the detector in the optometer. This detector measures the displacement of these images from the optical axis of the instrument. It is possible to calculate the magnitude of the image displacement, and thus the performance of the optometer, in the following way. If the eye is accommodated D dioptres, then the image of the retina formed by the reflected IR radiation is at a distance $1/D$ from the principal plane of the eye. Since L_2 is positioned so that its near focal plane coincides with the principal plane of the eye, applying the Newtonian lens formula to lens L_2 gives the distance, y, of the image formed by lens L_2 from its far focal plane as:

$$y = Df_2^2. \tag{10.1}$$

The displacement of this image from the optical axis can be calculated with the help of the construction show in Fig. 10.4. A ray is shown leaving one IR source (A) and traversing the path BCEFGH to the retina. Note that the ray passes through the centre (C) of the aperture ST_1. Because the source is in the focal plane of L_1, all rays

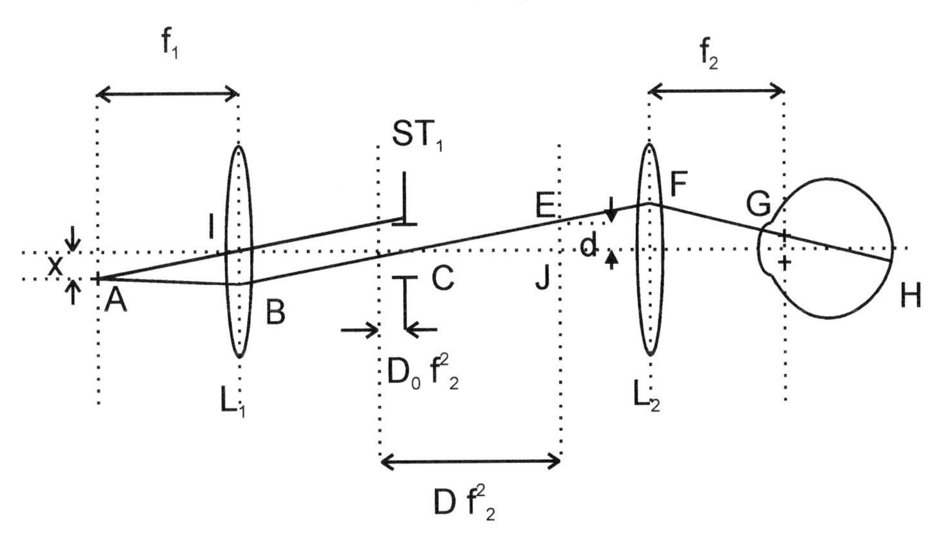

Fig. 10.4 Construction to simplify the calculation of the optical gain of the Scheiner-principle optometer.

leaving L_1 must be parallel. Hence angle α between BCEF and the optical axis is the same as the angle between ray AI and the axis, i.e.:

$$\tan \alpha = x/f_1 \tag{10.2}$$

Applying the principle of reversibility, one ray *leaving* the eye must traverse the path HGFECB, and because the eye is focused in plane EJ, *all* rays leaving the eye must pass through E. Hence to calculate the displacement, d, of the image from the optical axis, we need only to note that in triangle CEJ:

$$\tan \alpha = d/((D - D_0)\, f_2^2). \tag{10.3}$$

Substituting from eqn 10.2:

$$d = (D - D_0)\, f_2^2\, x/f_1; \tag{10.4}$$

i.e. the displacement d is a linear function of D, the accommodation of the eye.

Because the detector is conjugate with ST_1, not to the plane (EJ) of the image, it is necessary to calculate the displacement of the (blurred) image in the plane of ST_1. This is in fact practically identical to d because the beam passing out of the eye and through L_2 runs essentially parallel to the axis. This can be seen from consideration of the ray defining the centre of the beam—that which passes through the nodal point of the eye. The nodal point of the eye is close to the focal plane of lens L_2 because the distance between the nodal point and principal plane of the eye is small compared with the focal length of L_2, so this ray emerges from L_2 nearly parallel to the axis.

The optometer output signal consists of that component of the difference between the signal from the right and left halves of the detector which alternates with the alternation of the sources. This is achieved by phase-locked amplification of the signal from the two halves of the detector: with one source 'ON', the difference between the left and right half-detectors is passed to the amplifier and with the other source 'ON', the inverted signal is passed to the amplifier. The amplified signal is smoothed to produce a voltage proportional to the displacement of the image which occurs as one source is switched 'ON' and the other 'OFF', and is therefore proportional to the accommodation of the eye. A bandwidth of a few tens of Hz is perfectly adequate. The optometer will only give a meaningful signal if the total light flux from the IR sources which falls on the detector is constant. This will be so providing that:

(1) the pupil does not vignette the IR coming into the eye;

(2) the effective pupil diameter does not change as the eye accommodates; and

(3) the degree of accommodation is not so great as to displace or blur the image so that it is larger than the detector.

The detector is sufficiently large that the instrument will correctly measure accommodation of at least ±6 dioptres from the null plane (– 2 D).

10.1.2.5 Sources of error, noise level, and methods of calibration

There are two obvious sources of systematic error which cause infrared optometers to give signals that are about 1.5 D more hyperopic than they should be (Cornsweet and Crane, 1970). These are the chromatic aberration of the eye and retinoscopic artefact, which arise from the fact that the effective plane of reflection is in front of the photoreceptor plane. (Note that the retinoscopic artefact, which is about 0.5 D in humans, can be very large in the small eyes of other species (Glickstein and Millodot, 1970)). The instrumental noise level of the optometer is about 0.1 D rms (root mean square). This is satisfactory, and one suspects that biological factors such as variation in reflectivity of the different patches of retina on which the image falls as the eye moves would make it difficult to interpret much better resolution.

Precise calibration of infrared optometers is not easy. The method we settled on was to make up a realistic model eye whose power could be varied, but without changing the first ('corneal') element. The reason for wishing to leave the corneal element unaltered is that varying its power moves the position of the corneal reflections and so tends to cause artefacts. Alternatively, one can place trial lenses in the output limb of the instrument, but this is a somewhat less direct method. Another approach is to make use of the fact that the ratio between the optometer signal output and the output from the instrument with the electronics set to pupillometer function (but the artificial pupil left 'IN') is purely a matter of geometry and can be calculated. By adjusting the gain to give the appropriate signal output in pupillometer mode, one has, in theory, absolutely calibrated the optometer. This is in fact how we originally used the optometer.

10.1.2.6 Some miscellaneous practical points about the construction of the optometer

The lenses are achromatic doublets for two reasons: to avoid chromatic aberration (or at least minimize it) and also to reduce spherical aberration, because the two halves of the doublet have opposite signs of spherical aberration and to a considerable extent cancel one another out (see Chapter 1 (Fundamental optics) in *Optics guide 5*, available from the Melles Griot Corp., Irvine, California, USA—a very useful source of practical optical information).

Care should be taken about internal reflections in the instrument. The best way of dealing with these would be by antireflection coatings, but these are expensive, and the cheap alternative is to rotate any lens causing a reflection by a modest angle so that the reflection falls outside the optometer pathway. Lens L_2 is rotated for this reason. There is a light trap at point T so that the instrument casing does not reflect or scatter light back into the optical path. This consists of a piece of matt, black sugar paper, folded into a concertina. Black, flocked 'velvet' paper would probably be even better.

The positions of the various elements can be checked optically by assembling the instrument in the appropriate order, and by the sequential use of parallax. For example, one can check that the sources are at the focal plane of the condenser lens by viewing, from some distance, the sources through the condenser lens and checking that there is zero parallax between them and a very distant object. If two objects are at distances $1/D_1$ and $1/D_2$ (metres), then movement of the head by y metres will produce parallax $p = y(D_1 - D_2)$ radians. A parallax of 0.1 deg (approximately 1/600 radian) is easily detected, so if a head movement of $y = 12\,\text{mm}$ (0.012 m) is used a 0.14D mismatch of object positions is easily detected. This is a small error for most purposes.

At least one common material (black, methyl-methacrylate polymer sheet 'Perspex'; 'Plexiglass') has opacity that depends markedly on wavelength: it can be quite transparent in the near-infrared, and needs to be painted with matt black paint. Materials to be used as light traps should be checked for their infrared reflectivity.

10.1.2.7 Optimizing the design

Perhaps the main point to make is whether this is necessary. It may not be: practical instruments only have to be good enough, and the Cumming and Judge optometer design was not optimized. If one should wish to optimize the design, it could be done in the following way. It might seem obvious that one should simply use the highest power IR sources available, provided that they are safe, but this is not necessarily so, because higher power sources may have larger emitter areas, and a smaller source of higher radiance may be better.

The optometer gain, or signal strength per dioptre change in accommodation, depends not only on the factor f_2^2x/f_1 given by eqn 10.4 above, but also on a number of other factors, notably the radiance, R, of the source, the source radius, r, and the focal length f_1 of the condenser lens; because the last three govern the power density, P, in the plane of ST_1:

$$P \propto Rr^2/f_1^2. \tag{10.5}$$

If we include the f_2^2x/f_1 dependence of the optometer signal, S, on the geometry (eqn 10.4), S is given by eqn 10.6, which has been written in terms of the dimensions of the images in the pupil rather than that of the sources. (Because the linear magnification between the sources and their images in the pupil is $r' = rf_2/f_1$ and

$x' = xf_2/f_1$, where r' and x' are the radius and separation of the source images in the pupil, respectively.)

$$S \propto Rx'r'^2; \tag{10.6}$$

where all other factors, notably f_2, are regarded as constants.
Optimizing source-image geometry Because the sources must fit inside the pupil, diameter d:

$$x' + 2r' <= d. \tag{10.7}$$

To find the maximum S compatible with this constraint, note that because S increases monotonically with both x' and r', it is sufficient only to consider the variation of S along the line $x' + 2r' = d$, in which case:

$$S \propto R\,(d - 2r') \tag{10.8}$$

To find the value of r' which gives the maximum S we find the value of r' to make the partial derivative of S with respect to r' zero, and the second partial derivative positive. This is $r'_m = d/3$: in which case $x' = d/3$, i.e. the sources should be overlapped by an amount equal to their radius.

It is always wise when making such calculations to look numerically at how rapidly performance deteriorates with off-optimum values. Consider reducing or increasing the source radius r' by 25%. If r' is reduced to $d/4$ and separation increased to $d/2$ so that the source images are just touching, rather than overlapping, S is reduced by 16%; if r' is increased by 25%, S is reduced by 22%. So the penalty for a modest departure from optimum values is not severe.

Optimizing the source radiance and focal length of the condenser lens Because $r' = rf_2/f_1$, optimizing r' requires an appropriate matching of source radius r and condenser focal length f_1 if a source of half the radius is to be used, the condenser focal length must also be halved. In so far as this matching can be realized, it is better to use the highest radiance source available rather than the highest power source.

In practice, this cannot always be achieved, and so the relative merits of different combinations of sources and condensers would have to be assessed by comparing their values of $R(r/f_1)^3$, i.e. the right-hand side of eqn 10.8 with geometry optimized and expressed in terms of r rather than r'.

Minimizing instrumental noise In as much as the critical noise in the system is in the detector itself rather than the subsequent electronics, and detector noise is proportional to the area of the detector (i.e. to the square of its diameter), reducing the focal length of lens L_4 and the size of the detector will increase the signal-to-noise ratio: although the signal size will be reduced in proportion to the detector diameter, the fall off in noise with the square of the detector diameter will more than compensate for this.

10.1.2.8 Safety

Not a great deal of infrared emerges from the eye, and the sources need to be rather bright. It is therefore prudent to check that the illumination levels are within approved safety limits.

The retinal irradiance with the optometer in use is estimated in the following way. Let P be the instantaneous power output of one of the source diodes and f_1 the focal length of the condenser lens. Then (assuming Lambertian emission) the power density in the parallel beam between the condenser lens and the Maxwellian lens is:

$$P/(\pi f_1^2). \tag{10.9}$$

If f_e is the focal length of the eye and f_2 the focal length of the Maxwellian lens, and we assume the near focal plane of the Maxwellian lens is at the nodal point of the eye, then the areal magnification from the Maxwellian lens to the retina is given by:

$$f_e^2/f_2^2. \tag{10.10}$$

We must also take account of the fact that the beam-splitter reflects only a proportion, r_b, of the incident light to the Maxwellian lens, and that the transmittance, t_e, of the ocular media and absorption of the tissue at the back of the eye, a_e, are not unity.

Putting all these factors together, the retinal power density is given by:

$$I = (P\, f_2^2\, r_b\, t_e\, a_e)/(\pi f_1^2 f_e^2). \tag{10.11}$$

The manufacturer states that the typical total power emission of the diode is 45 mW at the current employed. The other values are: $f_2 = 0.11$ m, $f_1 = 0.08$ m; and $f_e = 0.017$ m, $r_b = 0.30$, $t_e = 0.75$, and $a_e = 0.60$ (the two latter values estimated from Geeraets and Berry (1968). Substituting these values gives:

$$I = 1.3\ \text{mW cm}^{-2}. \tag{10.12}$$

The important point is that this figure is some three orders of magnitude below the threshold values for retinal injury. The threshold retinal irradiance needed to cause injury after 10^3-seconds viewing of an extended 633-nm source is 5 W cm^{-2} and for a 1064-nm source 20 W cm^{-2}. (The peak emission of the sources in the optometer is at 940 nm.) The injury threshold data are taken from Fig. 4. of Sliney and Wolbarsht (1980). The maximum permitted exposure (MPE) levels for indefinite viewing of an extended source (ANSI and British Standards agree) are quoted in terms of the source radiance, on the assumption that the subject views with a 3-mm pupil (see Sliney and Wolbarsht, 1980, Fig 3 and p. 1138). The MPE radiance for 940 nm is $R_{mpe} = 1.87$ W/(cm^2 sr). With the above figures for transmission and absorption this implies an absorbed retinal irradiance of:

$$I_{mpe} = (R_{mpe}\, t_e\, a_{e\pi}\, D^2)/(4f_e^2); \tag{10.13}$$

where D is the pupil diameter (3 mm in this case). This gives $I_{mpe} = 21$ mW cm^{-2}.

The most intense retinal irradiation could well arise from an observer looking directly into the optometer from a nearby point and happening to be focused in or near the plane of the source images—or from looking directly at the sources with the top off the optometer. The point to emphasize here is that the source radiance is high. It is given by R, where:

$$R = P/(\pi A); \tag{10.14}$$

where A is the source area. Since $A = 0.5\,\text{mm}^{-2}$, this gives $R = 2.9\,\text{W}/(\text{cm}^2\,\text{sr})$ which exceeds the recommended limit for indefinite viewing, though not for brief viewing. Nevertheless I recommend that the sources should not be directly viewed, either with the top off the optometer or by looking down the axis of the optometer (other than with one's eye in the correct position).

10.1.3 A retinoscopic optometer

Retinoscopy (the technique the optometrist uses to check a subjective refraction when fitting spectacles) uses the opposite feature of defocus to the Scheiner principle: light from a given point on the retina, which passes through different points of the pupil, ends up (except at the plane conjugate with the retina) at different spatial locations. If we can arrange to have a narrow angle detector observing the pupil, as the retinal image is moved we see a reflex moving across the pupil with a speed and direction that is a measure of the difference between the observation plane and the plane conjugate with the retina. In retinoscopy proper, the narrow angle detector is the observer's eye, viewing through the aperture incorporated in the retinoscope (Fig. 10.5).

In a retinoscopic optometer the narrow angle detector is formed by a stop, lens, and a pair of sensors. This will achieve the same purpose as the observer's eye, if what is measured is the time delay between the activation of one detector and the other. A representative instrument is the one described by Kruger (1979). This optometer has been used not only by Kruger and colleagues but also by Gamlin *et al.* (1989).

As in the Scheiner-principle optometer, light from the source is focused in the plane of the pupil, but in the Kruger optometer (shown in Fig. 10.6) the Maxwellian lens L_1 itself constitutes the field stop that controls retinal image size. An incandescent quartz–halogen source is used, filtered by filter F1 which only passes wavelengths between 800 and 1100 nm. This filter can be replaced by neutral-density filter F2 so that visible light can be used to align the optometer. The chopper, C, is close to the focal plane of lens L_1, and rotation of the chopper therefore creates a moving light/dark edge on the retina. The chopper frequency is 240 Hz.

As in the Scheiner-principle optometer the first lens, L_2, of the output stage is placed so that its near focal plane is in the pupil. A horizontal slit aperture, A, 5 mm high, is placed at the far focal plane of L_2, immediately in front of lens L_3 (focal length 80 mm) which forms an image of the pupil close to its far focal plane. There are two photodetectors centred within this image, one above the other and masked so that their sensitive areas form segments of a circle with the corneal reflex falling in the gap between the segments (see Fig. 10.6).

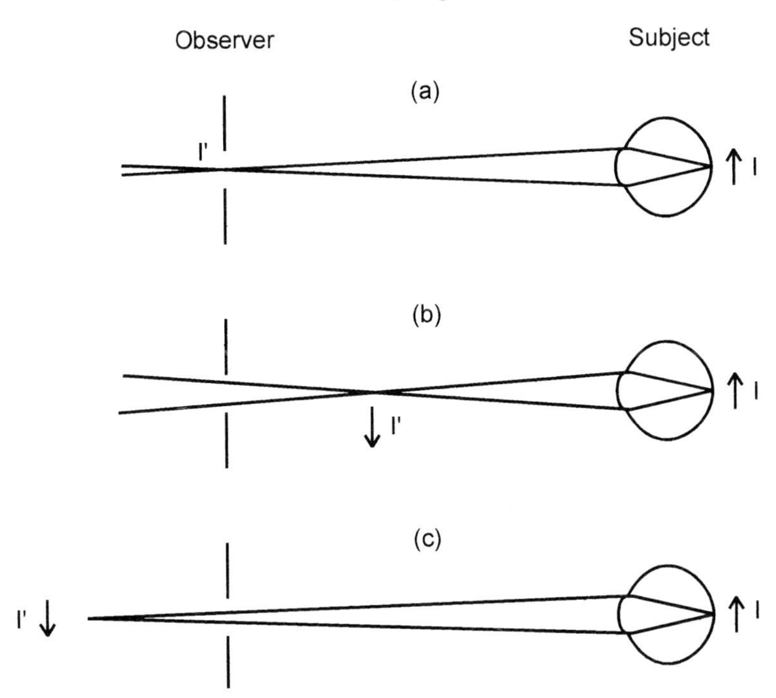

Fig. 10.5 The principle of retinoscopy, after Kruger (1979) with permission. As image, I, moves upwards across the retina, if the eye is focused on the observation plane (a), all parts of the pupil will appear to the observer to be illuminated simultaneously. If the eye is focused closer to the subject (b), the upper pupil is illuminated first; whereas if the eye is focused beyond the observer (c), the lower pupil is illuminated first.

10.1.4 Consideration of the relative merits of Scheiner-principle and retinoscopic optometers

It is possible that the two separate input beams make it somewhat harder to avoid artefacts from corneal reflections in Scheiner optometers than in retinoscopic optometers. More fundamentally, it is possible to show that the chief parameter describing the performance of both kinds of optometer, the 'optical gain' or ratio between the signal the optometer electronics seek to detect and the change in ocular accommodation, is purely a matter of the geometry of the optometer layout; and that with comparable geometry, both schemes should be equally effective. It has already been shown that the gain of the Scheiner-principle optometer is proportional to f_2^2x/f_1 (eqn 10.4), where f_2 and f_1 are the focal lengths of the Maxwellian lens and the collimator lens, respectively, and x is the source separation. I will now show that a similar geometric dependence holds for the retinoscopic optometer, with the focal length of lens L_3 taking the place of that of the collimator lens in the Scheiner design.

In Fig. 10.7 the dashed lines show that the detectors are conjugate to the pupil. The two dashed lines show rays from point A at the upper edge of the pupil, one (ABC) passing through the centre of L_2 and the other (AEF) eccentrically through L_2

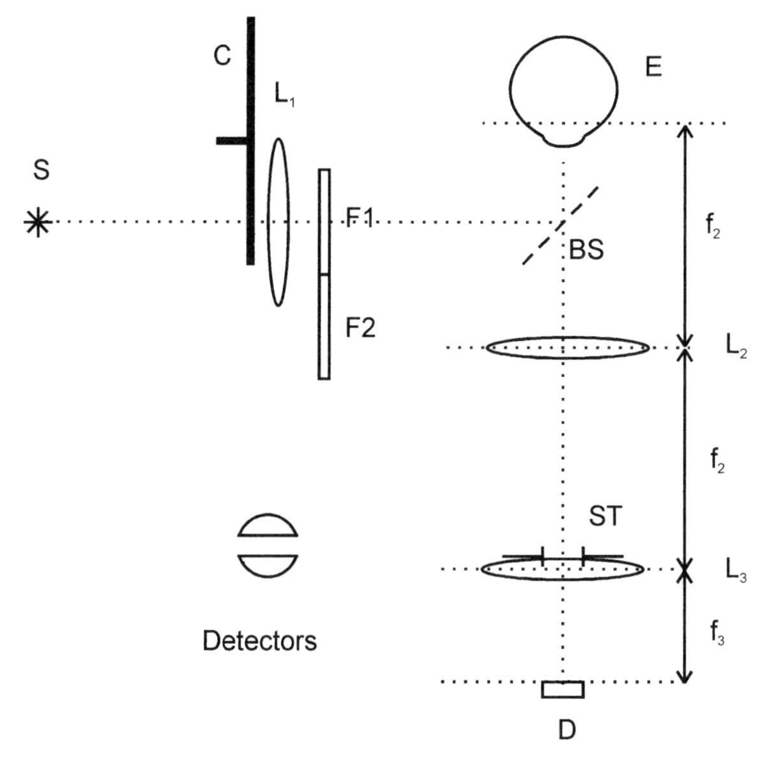

Fig. 10.6 Layout of the Kruger (1979) retinoscopic optometer.

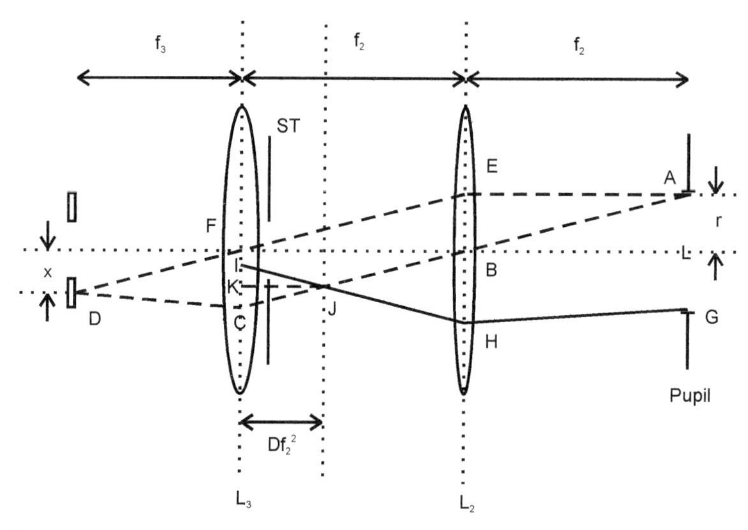

Fig. 10.7 Construction for simplifying the calculation of the optical gain of the retinoscopic optometer.

so that it passes through the centre of L_3. The two rays will be parallel because L_2 is at its focal length from the pupil plane. They will also converge on to point D in the detector plane, because L_3 is at its focal length from the detectors. From similar triangles:

$$r = f_2 x / f_3; \tag{10.15}$$

where r is the eccentricity of A in the pupil and x is the eccentricity of D in the detector plane.

Putting the same diagram to a slightly different use, imagine that the eye is focused on a point distance $1/D$ away, somewhat below the horizontal plane so that ABC is on the upper boundary of the cone of rays re-emitted from the eye, and the solid line GHI shows the lower boundary of the cone of rays re-emitted from the eye. Radiation from the eye focuses at point J, whose distance to the right of lens L_3 is Df_2^2, from the Newtonian lens formula. We can now calculate the half-width, CK or δ of the blur circle in the plane of lens L_3. From similar triangles ABL and CJK:

$$r/f_2 = \delta/(Df_2^2). \tag{10.16}$$

Because the pupil and the detectors are conjugate, all rays from the upper margin A of the pupil, will reach the lower detector, providing they are not occluded by the stop ST, and all unoccluded rays from the lower margin of the pupil will reach the upper detector. Now recall that the elevation of the radiation re-emitted from the eye is changing at a constant rate, because of the action of the chopper on the input beam. At the point in time shown on the figure, only the upper detector is illuminated, but when the beam elevation has increased somewhat the lower detector will also be illuminated. The delay between the signals in the two detectors will be proportional to the width of the blur circle in the plane of stop ST. Although this delay is sensed as a phase shift, to keep the formula dimensionally comparable to that for the Scheiner optometer, the appropriate formula—substituting the value for r from eqn 10.15 in eqn 10.16 and rearranging—is:

$$\delta = Df_2^2 x / f_3. \tag{10.17}$$

This is the same dependence on lens focal lengths as that given by eqn 10.4 for the Scheiner-principle optometer, with lens L_3 taking the place of lens L_1.

A complete analysis would have to include consideration of the power of the sources and the effect of the stops on signal intensity, but this is beyond the scope of this chapter.

10.1.5 The Cornsweet–Crane optometer

The most elegant optometer design ever is that of Crane and Steele (1978) who combined the servo-controlled infrared optometer of Cornsweet and Crane (1970) with the eye tracker of Cornsweet and Crane (1973). The three-dimensional eye tracker (available from Fourward Optical Technologies Inc., San Marcos, Texas) allows one to study the dynamics of accommodation in subjects who are free to move their heads (within limits). Just one of the particularly pleasing features of the Cornsweet and Crane optometer is that it is a 'null' instrument, in which the plane conjugate with the retina is directly determined by a servo that operates to hold the source in that plane. All this makes the 3-D eye tracker a prodigiously complex and therefore expensive machine. It is worth studying its design even if one intends to use a simpler instrument, as some of the features incorporated in it are valuable in the construction of simpler instruments. Some points worth noting are:

- the use of the optical relay;
- the value of using off-the-shelf camera lenses wherever possible;
- the scheme used to attenuate the corneal reflex.

10.2 Pupil

There are at least two general methods for making dynamic measurements of pupil diameter or area. As with dynamic optometry, near-infrared radiation rather than visible light is employed. One method is to focus the input beam in the pupil (using similar optics to those of the dynamic optometers above), so that all the IR radiation incident on the front of the eye passes into the eye. Some is diffusely reflected from the retina and so is re-emitted from the eye; the pupil will then be bright against a dark background. Simply measuring the re-emitted IR flux as a function of time will give a signal that is proportional to pupil area. Because many detectors are sensitive both to IR and visible radiation, to make such devices work properly regardless of room illumination it is necessary to 'chop' the IR illumination, and extract a signal which is the difference in output with the IR sources 'ON' and 'OFF'. To avoid duplicating much of what has been said above, this method is illustrated below (Section 10.2.1) as realized as an additional feature of the Cumming and Judge, Scheiner-principle optometer.

Another method is to diffusely illuminate the front of the eye and detect the boundary between the iris and pupil by means of the differential brightness (iris bright, pupil dark) in a video image of the front of the eye. As an example of this method, the pupillometer of Barbur *et al.* (1987) is discussed in Section 10.2.2. The advantage of methods of this kind is that it is not necessary to control the position of the head and eye so rigidly. Some restriction of head and eye position is probably still necessary to avoid problems with corneal reflections.

The video method has another type of sensitivity to eye movements. Rapid movement of the eye will blur the pupil image unless the duration of illumination is very short.

10.2.1 Pupillometry as an added feature of a Scheiner-principle optometer

The Scheiner-principle optometer described above can be made to work as a pupillometer by opening up the artificial pupil so that the real pupil forms the aperture stop in the output pathway, and suitably altering the way the electronic signals are processed. For operation as an optometer, the two sources are alternately illuminated and the detector circuitry is arranged to amplify that component of the difference between the signal from the left and right half detectors that is alternating in synchrony with the source alternation. For operation as an pupillometer, one source only is illuminated 50% of the time, and that component of the sum of the signals on both halves of the detector, which is due to the source being 'ON', is amplified. The same electronics can serve both purposes with the addition of suitable control logic and analog gates—though the penalty for this economy of components is that the machine cannot function simultaneously as an optometer and a pupillometer.

In principle, it is possible to make a dual function machine, but the electronics need to be more complex. A commercially available device that offers simultaneous measurement of both accommodation and pupil dynamics is the Nidek AA-2000 Accommodometer, 34–14 Maehama Hiroishi–cho, Gamagori, Aichi 443–0038 Japan, which also incorporates a dynamic stimulus whose movement in time is under software control.

10.2.2 A video-image pupillometer

Barbur has described a pupillometer based on a specialized video-image processing board. The Barbur pupillometer largely, but not completely, eliminates the effect of eye movement, by reducing the time during which the pupil is illuminated to 4 ms per video frame: in 4 ms the eye will rotate a degree or so at the fastest point of a modest saccade, and taking 10 mm as the distance between the centre of rotation and the entrance pupil, this will cause blur of one-sixth of a millimetre or so, which may be significant. If one is trying to make 1% accurate measurements of a diameter of around 3 mm it would certainly seem prudent to eliminate data obtained during saccades.

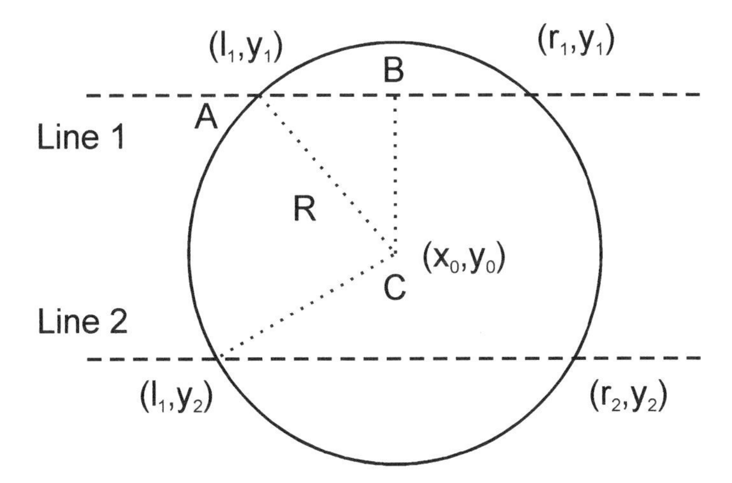

Fig. 10.8 The algorithm for extracting pupil radius used in the pupillometer of Barbur *et al.* (1987).

The positions of the left and right margins of the pupil are found on a selection of horizontal scans across the pupil (Fig. 10.8). By taking pairs of scan lines a known vertical distance apart, it is possible to arrive at estimates of the pupil diameter. By averaging the results of up to 32 such pairs, and rejecting outliers due to image noise, it is possible to arrive at a precise estimate of pupil diameter. Let (l_1, y_1) and (r_1, y_1), and (l_2, y_2) and (r_2, y_2) be the left and right ends of two scans across the pupil. If the pupil image can be assumed to be circular (i.e. subjects with malformations and anomalies of the iris are not used), and its centre is (x_0, y_0), then from symmetry:

$$x_0 = (r_1 + l_1)/2, \tag{10.18}$$

and:

$$x_0 = (r_2 + l_2)/2. \tag{10.19}$$

Applying Pythagoras' theorem to triangle ABC, where B is the midpoint $((r_1 + l_1)/2, y_1)$ of line 1, gives:

$$((r_1 - l_1)/2)^2 + (y_1 - y_0)^2 = R^2. \tag{10.20}$$

And for line 2:

$$((r_2 - l_2)/2)^2 + (y_2 - y_0)^2 = R^2. \tag{10.21}$$

Subtracting these equations gives:

$$y_0 = ((r_1 - l_1)^2/4 + y_1^2 - (r_2 - l_2)^2/4 - y_2^2) \;/\; (2(y_1 - y_2)), \tag{10.22}$$

and hence R.

Barbur stresses the precision of his device, which certainly seems to be excellent, but it is not clear that the precision of the method is inherently superior to that of alternatives. The Barbur device also incorporates software for measuring horizontal and vertical eye movement, which is a valuable additional feature.

Further information about dynamic pupillometry is available in Alexandridis *et al.* (1991).

References

Alexandridis, E., Leendertz, J. A., and Barbur, J. L. (1991). Methods for studying the behaviour of the pupil. *Journal of Psychophysiology*, **5**, 223–39.

Barbur, J. L., Thomson, W. D., and Forsyth, P. M. (1987). A new system for the simultaneous measurement of pupil size and two-dimensional eye movements. *Clinical Vision Sciences*, **2**, 131–42.

Calkins, J. L., Hochheimer, B. F., and D'Anna, S. A. (1980). Potential hazards from specific ophthalmic devices. *Vision Research*, **20**, 1039–53.

Cornsweet, T. N. and Crane, H. D. (1970). Servo-controlled infrared optometer. *Journal of the Optical Society of America*, **60**, 548–54.
Cornsweet, T. N. and Crane, H. D. (1973). Accurate two-dimensional eye tracker using first and fourth Purkinje images. *Journal of the Optical Society of America*, **63**, 921–8.
Crane, H. D. and Steele, C. M. (1978). Accurate three-dimensional eyetracker. *Applied Optics*, **17**, 691–705.
Cumming, B. G. and Judge, S. J. (1986). Disparity-induced and blur-induced convergence eye movement and accommodation in the monkey. *Journal of Neurophysiology*, **55**, 896–914.
Elsner, A. E., Burns, S. A., Weiter, J. J., and Delori, F. C. (1996). Infra-red imaging of sub-retinal structures in the human ocular fundus. *Vision Research*, **36**, 191–205.
Gamlin, P. D. R., Gnadt, J. W., and Mays, L. E. (1989). Lidocaine-induced unilateral inter-nuclear ophthalmoplegia: effects on convergence and conjugate eye movements. *Journal of Neurophysiology*, **62**, 82–95.
Geeraets, W. J. and Berry, E. R. (1968). Ocular spectral characteristics as related to hazards from lasers and other light sources. *American Journal of Ophthalmology*, **66**, 15–20.
Glickstein, M. and Millodot, M. (1970) Retinoscopy and eye size. *Science*, **168**, 605–6.
Henson, D. B. (1983). *Optometric instrumentation*. Butterworth, London.
Kruger, P. B. (1979). Infrared recording retinoscope for monitoring accommodation. *American Journal of Optometry*, **56**, 116–23.
Millodot, M. and Lamont, A. (1974). Refraction of the periphery of the eye. *Journal of the Optical Society of America*, **64**, 110–11.
O'Neill, W. D. and Stark, L. (1968). Triple function ocular monitor. *Journal of the Optical Society of America*, **58**, 570–3.
Pugh, J. R. and Winn, B. (1988). Modification of the Canon Autoref R1 for use as a continuously recording optometer. *Ophthalmic and Physiological Optics*, **8**, 460–4.
Pugh, J. R. and Winn, B. (1989). A procedural guide to the modification of the Canon Autoref R1 for use as a continuously recording optometer. *Ophthalmic and Physiological Optics*, **9**, 451–4.
Sliney, D. H. and Wolbarsht, M. L (1980). Safety standards and measurement techniques for high intensity light sources. *Vision Research*, **20**, 1133–41.
Westheimer, W. (1966) The Maxwellian view. *Vision Research*, **6**, 669–82.

Index

Index